Fundamentals of
Digital Logic
and
Microcontrollers

Fundamentals of
Digital Logic
and
Microcontrollers

Sixth Edition

M. RAFIQUZZAMAN, Ph.D.
Professor
California State Polytechnic University, Pomona
and
President

Rafi Systems, Inc.

Published by John Wiley & Sons, Inc., Hoboken, New Jersey.
Published simultaneously in Canada.

For general information on our other products and services or for technical support, please contact our
Customer Care Department within the United States at (800) 762-2974, outside the U.S. at (317) 572-3993
or fax (317) 572-4002.

Wiley also publishes its books in a variety of electronic formats. Some content that appears in print may
not be available in electronic formats. For more information about Wiley products, visit our web site at
www.wiley.com

Library of Congress Cataloging-in-Publication Data:

Rafiquzzaman, Mohamed.
  [Fundamentals of digital logic and microcomputer design]
  Fundamentals of digital logic and microcontrollers / M. Rafiquzzaman. – Sixth edition.
      pages cm
  Revised edition of: Fundamentals of digital logic and microcomputer design.
  Includes index.
  ISBN 978-1-118-85579-9 (cloth)
  1. Logic circuits. 2. Microcomputers–Design and construction. 3. Electronic digital computers–Circuits.
I. Title.
  TK7888.4.R34 2014
  621.39′5–dc23

                                                                                  2014017642

To my wife, Kusum; son, Tito; daughter-in-law, Trina; and my grandchildren

In memory of my brother Elan

CONTENTS

PREFACE

The fifth edition of the book covered basics of computer engineering and science from digital logic to the design of a complete microcomputer system using Intel 8086 and Motorola 68000. With the growing popularity of microcontrollers, it is now necessary to have a clear understanding of the basic principles of microcontrollers at the undergraduate level. Microcontrollers play an important role in the design of digital systems. They are found in a wide range of applications including office automation systems such as copiers and fax machines, consumer electronics such as microwave ovens, digital instruments, and robotics.

Hence, a typical microcontroller such as Microchip Technology's PIC18F is used to replace the 8086 and 68000 in this edition. Several PIC18F-based simple practical applications using C-language are provided. The sixth edition covers an enhanced version of both combinational and sequential logic design, basics of computer organization and microcontrollers.

Like the fifth edition, emphasis is given on the basic concepts. To cite an example, we clearly point out that computers understand only 0's and 1's. It is, therefore, important that students be familiar with binary numbers. Furthermore, we focus on the fact that computers can normally only add. Hence, all other arithmetic operations such as subtraction are performed via addition. This can be accomplished using two's-complement arithmetic for binary numbers. This topic is, therefore, also included along with a clear explanation of signed and unsigned binary numbers. Basic concepts such as this are illustrated with simple examples throughout this edition.

As in the previous edition, three design levels are covered in this book: device level, logic level, and system level. Device-level design using simple devices such as transistors is included for typical logic gates such as NOT. Logic-level design is the technique in which logic gates are used; design of digital components such as an adder is provided. Finally, system-level applications are covered using a typical microcontroller such as the PIC18F.

Digital systems at the logic level are classified into two types of circuits: combinational and sequential. Combinational circuits have no memory whereas sequential circuits contain memory. Microcontrollers are designed using both combinational and sequential circuits. Therefore, these topics are covered in detail.

This edition of the book contains more details for synthesizing digital logic circuits using a popular hardware description language such as Verilog. An overview of contemporary digital circuit implementation using a popular programmable logic device (PLD) such as field programmable gate array (FPGA) along with the CAD (computer aided design) tools is included.

Several chapters in the previous edition containing digital logic design have been retained and even strengthened in this edition. A few chapters of this new edition are written to present the fundamental concepts of C language programming and interfacing techniques associated with Microchip Technology's PIC18F4321 microcontroller. The PIC18F family continues to be popular. The PIC18F family is an excellent educational tool for acquiring an understanding of both hardware and software aspects of typical microcontrollers.

Several PIC18F-based simple practical applications using C-language are provided. Most of the examples are implemented successfully in the laboratory. In summary, the sixth edition covers an enhanced version of both combinational and sequential logic, basics of computer organization, and microcontrollers.

The following major changes have been provided in this edition:

- Revised Chapters 1 through 5 of the fifth edition to update and strengthen certain topics on both combinational and sequential circuits. Several new examples are included. In addition, certain topics such as timing diagrams, hazards, stability, and Verilog and FPGA are enhanced.

- Chapter 6 of this edition contains design of the CPU, memory, and I/O. Design of both hardwired and microprogrammed CPU is included.

- Chapters 7 through 9 of this edition include microcontroller basics and PIC18F-based applications using C.

In this edition, the book is divided into nine chapters as follow:

Chapter 1 presents an explanation of basic terminologies, fundamental concepts of digital integrated circuits using transistors; a comparison of LSTTL, HC, and HCT IC characteristics, evolution of the microcontroller, and technological forecasts.

Chapter 2 includes various number systems and codes suitable for representing information in typical digital systems.

Chapter 3 covers Boolean algebra along with k-map simplification of Boolean functions. Several examples are included. The basic characteristics of digital logic gates are also presented.

Chapter 4 presents the analysis and design of combinational circuits. Typical combinational circuits such as adder, decoder, encoder, multiplexer, demultiplexer, and ROMs are included. An introduction to PLDs and FPGA is provided. Synthesis of combinational logic design using Verilog is also included.

Chapter 5 covers various types of flip-flops. Analysis and design of sequential circuits such as counters and registers are covered. An overview of topics such as flip-flop set-up time, hold time, and metastability is provided. Finally, synthesis of sequential logic design using Verilog is included.

Chapter 6 introduces basic concepts associated with CPU, memory, and I/O. It also covers the design of CPU including registers, ALU, and the control unit.

Chapter 7 presents basics of microcontroller including organization, memory, I/O, and C programming concepts.

Chapter 8 provides typical C programs which include I/O examples with LEDs and switches. PIC18F-based programmed and interrupt I/O using C are covered in a simplified manner.

Chapter 9 contains simple PIC18F-based applications using C. These include A/D and D/A converters, LCD displays, on-chip timers, DC voltmeter, and motor control using PWM (pulse width modulation).

The book can easily be adopted as a text for "digital logic design" and/or "introduction to microcontrollers" at the undergraduate level in electrical/computer engineering and computer science. The book introduces microcontroller hardware and software to form a bridge between digital logic and an advanced follow-up course in microcontrollers. The book provides an introduction to one of today's popular microcontrollers, the PIC18F family. Although no advanced mathematical background is required, a basic course on DC circuits is required. A first course in C language programming is essential. Since C language programming is prerequisite for the course on microcontrollers, I/O and interfacing using C is included in this book. The audience for this book can also be practicing engineers and scientists in the industry. Practitioners of digital system design in the industry will find more simplified explanations, together with examples and comparison considerations, than are found in manufacturers' manuals.

The author is especially indebted to his colleague, Dr. R. Chandra, of California State Polytechnic University, Pomona for reviewing the manuscript. The author is also grateful to his students Robert Benenyan, Michael Hachache, Cameron Chitsaz, Gagneet Bhatia, Luke Stankiewicz, and others for making constructive suggestions and to CJ Media of California for preparing the final version of the manuscript. Finally, the author is indebted to his deceased parents, who were primarily responsible for his accomplishments.

M. RAFIQUZZAMAN
Pomona, California

1

INTRODUCTION TO DIGITAL SYSTEMS

Digital systems are designed to store, process, and communicate information in digital form. They are found in a wide range of applications, including process control, communication systems, digital instruments, and consumer products. The digital computer, more commonly called the *computer*, is an example of a typical digital system.

A computer manipulates information in digital, or more precisely, binary form. A binary number has only two discrete values — zero or one. Each of these discrete values is represented by the OFF and ON status of an electronic switch called a *transistor*. All computers, therefore, only understand binary numbers. Any decimal number (base 10, with ten digits from 0 to 9) can be represented by a binary number (base 2, with digits 0 and 1).

The basic blocks of a computer are the central processing unit (CPU), the memory, and the input/output (I/O). The CPU of the computer is basically the same as the brain of a human. Computer memory is conceptually similar to human memory. A question asked to a human is analogous to entering a program into the computer using an input device such as the keyboard, and answering the question by the human is similar in concept to outputting the result required by the program to a computer output device such as the printer. The main difference is that human beings can think independently, whereas computers can only answer questions that they are programmed for. Computer hardware refers to components of a computer such as memory, CPU, transistors, nuts and bolts. Programs can perform a specific task such as addition if the computer has an electronic circuit capable of adding two numbers. Programmers cannot change these electronic circuits but can perform tasks on them using instructions.

Computer software, on the other hand, consists of a collection of programs. Programs contain instructions and data for performing a specific task. These programs, written using any programming language such as C, must be translated into binary prior to execution by the computer. This is because the computer only understands binary numbers. Therefore, a translator for converting such a program into binary is necessary. Hence, a translator program called the *compiler* is used for translating programs written in a programming language such as C into binary. These programs in binary form are then stored in the computer memory for execution because computers only understand 1's and 0's. Furthermore, computers can only add. This means that all operations such as subtraction, multiplication, and division are performed by addition.

Due to advances in semiconductor technology, it is possible to fabricate the CPU in a single chip. The result is the *microprocessor*. Both metal oxide semiconductor (MOS) and bipolar technologies were used in the fabrication process. The CPU can be placed on a single chip when MOS technology is used. However, several chips are

Fundamentals of Digital Logic and Microcontrollers, Sixth Edition. M. Rafiquzzaman.
© 2014 John Wiley & Sons, Inc. Published 2014 by John Wiley & Sons, Inc.

required with the bipolar technology. HCMOS (high speed complementary MOS) or BICMOS (combination of bipolar and HCMOS) technology (to be discussed later in this chapter) is normally used these days to fabricate the microprocessor in a single chip. Along with the microprocessor chip, appropriate memory and I/O chips can be used to design a *microcomputer*. The pins on each one of these chips can be connected to the proper lines on the system bus, which consists of address, data, and control lines. In the past, some manufacturers have designed a complete microcomputer on a single chip with limited capabilities. Single-chip microcomputers were used in a wide range of industrial and home applications.

"Microcontrollers" evolved from single-chip microcomputers. The microcontrollers are typically used for dedicated applications such as automotive systems, home appliances, and home entertainment systems. Typical microcontrollers, therefore, include a microcomputer, timers, and A/D (analog to digital) and D/A (digital to analog) converters — all in a single chip. Examples of typical microcontrollers are Intel 8751 (8-bit) / 8096 (16-bit) and Motorola/Freescale HC11 (8-bit) / HC16 (16-bit), and Microchip Technology PIC18F(8-bit)/PIC32(32-bit).

In this chapter, we first define some basic terms associated with the computers. We then describe briefly the evolution of the computers and the microcontrollers. Finally, a typical practical application, and technological forecasts are included.

1.1   Explanation of Terms

Before we go on, it is necessary to understand some basic terms (arranged in alphabetical order).

- *Address* is a pattern of 0's and 1's that represents a specific location in memory or a particular I/O device. An 8-bit microcontroller with 16 address bits can produce 2^{16} unique 16-bit patterns from 0000000000000000 to 1111111111111111, representing 65,536 different address combinations (addresses 0 to 65,535).

- *Addressing mode* is the manner in which the microcontroller determines the operand (data) and destination addresses during execution of an instruction.

- *Arithmetic-logic unit* (ALU) is a digital circuit that performs arithmetic and logic operations on two n-bit digital words. Typical operations performed by an ALU are addition, subtraction, ANDing, ORing, and comparison of two n-bit digital words. The size of the ALU defines the size of the microcontroller. For example, an 8-bit microcontroller contains an 8-bit ALU.

- *Big endian* convention is used to store a 16-bit number such as 16-bit data in two bytes of memory locations as follows: the low memory address stores the high byte while the high memory address stores the low byte. The Motorola/Freescale HC11 8-bit microcontroller follows the big endian format.

- *Bit* is an abbreviation for the term *binary digit*. A binary digit can have only two values, which are represented by the symbols 0 and 1, whereas a decimal digit can have 10 values, represented by the symbols 0 through 9. The bit values are easily implemented in electronic and magnetic media by two-state devices whose states portray either of the binary digits 0 and 1. Examples of such two-state devices are a transistor that is conducting or not conducting, a capacitor that is charged or discharged, and a magnetic material that is magnetized north to south or south to north.

- *Bit size* refers to the number of bits that can be processed simultaneously by the basic arithmetic circuits of a microcontroller. A number of bits taken as a group in this manner is called a *word*. For example, an 8-bit microcontroller can process an 8-bit word. An 8-bit word is referred to as a *byte*, and a 4-bit word is known as a *nibble*.

- *Bus* consists of a number of conductors (wires) grouped to provide a means of communication among different elements in a microcontroller system. The conductors in a bus can be grouped in terms of their functions. A microcontroller normally has an address bus, a data bus, and a control bus. Address bits are sent to memory or to an external device on the *address bus*. Instructions from memory, and data to/from memory or external devices, normally travel on the *data bus*. Control signals such as read/write for the other buses and among system elements are transmitted on the *control bus*. Buses are sometimes *bidirectional*; that is, information can be transmitted in either direction on the bus, but normally in only one direction at a time.

- *Clock* is analogous to human heart beats. The microcontroller requires synchronization among its components, and this is provided by a *clock* or timing circuits.

- The *chip* is an integrated circuit (IC) package containing digital circuits.

- *CPU* (central processing unit) contains several registers (memory elements), an ALU, and a control unit. Note that the control unit translates instructions and performs the desired task. The number of peripheral devices depends on the particular application involved and may even vary within an application.

- *EEPROM* or E^2PROM (electrically erasable programmable ROM) is nonvolatile. EEPROMs can be programmed without removing the chip from the socket. EEPROMs are called read most memories (RMMs) because they have much slower write times than read times. Therefore, these memories are usually suited for applications when mostly reading rather than writing is performed. An example of EEPROM is the 2864 (8K x 8).

- *EPROM* (erasable programmable ROM) is nonvolatile. EPROMs can be programmed and erased. The EPROM chip must be removed from the socket for programming. This memory is erased by exposing the chip to ultraviolet light via a lid or window on the chip. Typical erase times vary between 10 and 30 minutes. The EPROM is programmed by inserting the chip into a socket of the EPROM programmer, and providing proper addresses and voltage pulses at the appropriate pins of the chip. An example of EPROM is the 2764 (8K × 8).

- *Flash memory* is designed using a combination of EPROM and EEPROM technologies. Flash memory is nonvolatile and is invented by Toshiba in mid 1980s. Flash memory can be programmed electrically while embedded on the board. One can change multiple bytes at a time. An example of flash memory is the Intel 28F020 (256K x 8). Flash memory is typically used in cell phones and digital cameras.

- An *FPGA* (field programmable gate array) chip contains an array of digital logic blocks along with input and output blocks which can be connected together via programming using a hardware description language (HDL) such as Verilog or

VHDL. There are two types of components inside an FPGA. These are lookup table (stored in memory), and switch matrices. The concept of FPGA is based on the fact that a combinational circuit can be implemented using memory. In the past, digital logic circuits were built using all hardware (logic gates). It was a time-consuming task to debug the circuits. However, digital circuits implemented using FPGA's are faster to debug since they are programmable. Note that it is much faster to debug software than hardware. Hence, products can be developed using FPGA from conceptual design via prototype to production in a very short time. Therefore, use of FPGA in digital logic is very common these days.

- The term *gate* refers to digital circuits which perform logic operations such as AND, OR, and NOT. In an AND operation, the output of the AND gate is one if all inputs are one; the output is zero if one or more inputs are zero. The OR gate, on the other hand, provides a zero output if all inputs are zero; the output is one if one or more inputs are one. Finally, a NOT gate (also called an *inverter*) has one input and one output. The NOT gate produces one if the input is zero; the output is zero if the input is one.

- *Harvard architecture* is a type of CPU architecture which uses separate instruction and data memory units along with separate buses for instructions and data. This means that these processors can execute instructions and access data simultaneously. Processors designed with this architecture require four buses for program memory and data memory. These are one data bus for instructions, one address bus for addresses of instructions, one data bus for data, and one address bus for addresses of data. The sizes of the address and data buses for instructions may be different from the address and data buses for data. Several microcontrollers including the PIC18F are designed using the Harvard architecture. This is because it is inexpensive to implement these buses inside the chip since both program and data memories are internal to the chip.

- *Instruction set* of a microcontroller is a list of commands that the microcontroller is designed to execute. Typical instructions are ADD, SUBTRACT, and STORE. Individual instructions are coded as unique bit patterns which are recognized and executed by the microcontroller. If a microcontroller has three bits allocated to the representation of instructions, the microcontroller will recognize a maximum of 2^3, or eight, different instructions. The microcontroller will then have a maximum of eight instructions in its instruction set. It is obvious that some instructions will be more suitable than others to a particular application. For example, in a control application, instructions inputting digitized signals to the processor and outputting digital control variables to external circuits are essential. The number of instructions necessary in an application will directly influence the amount of hardware in the chip set and the number and organization of the interconnecting bus lines.

- *Little endian* convention is used to store a 16-bit number such as 16-bit data in two bytes of memory locations as follows: the low memory address stores the low byte while the high memory address stores the high byte. The PIC18F microcontroller follows the little-endian format.

- *Microcomputer* typically consists of a microprocessor (CPU) chip, input and output chips, and memory chips in which programs (instructions and data) are stored.

- *Microcontroller* is implemented on a single chip containing a CPU, memory, and IOP (I/O and peripherals). Note that a typical IOP contains I/O unit of a microcomputer, timers, A/D (analog-to-digital) converter, analog comparators, serial I/O, and other peripheral functions (to be discussed later).

- *Microprocessor* is the CPU of a microcomputer contained in a single chip, and must be interfaced with peripheral support chips in order to function.

- *Pipelining* is a technique that overlaps instruction fetch (instruction read) with execution. This allows a microcontroller's processing operation to be broken down into several steps (dictated by the number of pipeline levels or stages) so that the individual step outputs can be handled by the microcontroller in parallel. Pipelining is often used to fetch the microcontroller's next instruction while executing the current instruction, which speeds up the overall operation of the microcontroller considerably. Microchip technology's PIC18F (8-bit microcontroller) uses a two-stage instruction pipeline in order to speed up instruction execution.

- *Program* contains instructions and data. Two conventions are used to store a 16-bit number such as 16-bit data in two bytes of memory locations. These are called *little endian* and *big endian byte ordering.* In little endian convention, the low memory address stores the low byte while the high memory address stores the high byte. For example, the 16-bit hexadecimal number, 2050 will be stored as two bytes in two 16-bit locations (Hex 5000 and Hex 5001) as follows: Address 5000 will contain 50 while address 5001 will store 20. In big endian convention, on the other hand, the low memory address stores the high byte while the high memory address stores the low byte. For example, the same 16-bit hexadecimal number, 2050 will be stored as two bytes in two 16-bit locations (Hex 5000 and Hex 5001) as follows: Address 5000 will contain 20 while address 5001 will store 50. Motorola / Freescale HC11 (8-bit microcontroller) follows big endian convention. Microchip PIC18F (8-bit microcontroller), on the other hand, follows the little endian format.

- *Random-access memory* (RAM) is a storage medium for groups of bits or words whose contents cannot only be read but can also be altered at specific addresses. A RAM normally provides *volatile storage*, which means that its contents are lost in case power is turned off. There are two types of RAM: static RAM (SRAM) and dynamic RAM (DRAM). *Static RAM* stores data in flip-flops. Therefore, this memory does not need to be refreshed. An example of SRAM is 6116 (2K × 8). *Dynamic RAM*, on the other hand, stores data in capacitors. That is, it can hold data for a few milliseconds. Hence, dynamic RAMs are refreshed typically by using external refresh circuitry. Dynamic RAMs (DRAMs) are used in applications requiring large memory. DRAMs have higher densities than static RAMs (SRAMs). Typical examples of DRAMs are the 4464 (64K × 4), 44256 (256K × 4), and 41000 (1M × 1). DRAMs are inexpensive, occupy less space, and dissipate less power than SRAMs.

- *Read-only memory* (ROM) is a storage medium for the groups of bits called *words*, and its contents cannot normally be altered once programmed. A typical ROM is fabricated on a chip and can store, for example, 2048 eight-bit words, which can be accessed individually by presenting to it one of 2048 addresses. This ROM is referred to as a *2K by 8-bit ROM*. 10110111 is an example of an 8-bit word that might be stored in one location in this memory. A ROM is a *nonvolatile storage* device, which means that its contents are retained in case power is turned off. Because of this characteristic, ROMs are used to store permanent programs (instructions and data).

- *Reduced Instruction Set Computer* (RISC) contains a simple instruction set. In contrast, a *Complex Instruction Set Computer* (CISC) contains a large instruction set. The PIC18F is an RISC-based microcontroller while Motorola/Freescale HC11 is a CISC-based microcontroller.

- *Register* can be considered as volatile storage for a number of bits. These bits may be entered into the register simultaneously (in parallel) or sequentially (serially) from right to left or from left to right, 1-bit at a time. An 8-bit register storing the bits 11110000 is represented as follows:

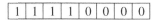

- The *speed power product* (*SPP*) is a measure of performance of a logic gate. It is expressed in picojoules (pJ). SPP is obtained by multiplying the speed (ns) by the power dissipation (mW) of a gate.

- *Transistors* are basically electronic switching devices. There are two types of transistors. These are *bipolar junction transistors* (*BJTs*) and *metal-oxide semiconductor* (*MOS*) transistors. The operation of the BJT depends on the flow of two types of carriers: electrons (*n*-channel) and holes (*p*-channel), whereas the MOS transistor is unipolar and its operation depends on the flow of only one type of carrier, either electrons (*n*-channel) or holes (*p*-channel).

- *von Neumann* (*Princeton*) *architecture* uses a single memory unit and the same bus for accessing both instructions and data. Although CPUs designed using this architecture are slower compared to Harvard architecture since instructions and data cannot be accessed simultaneously because of the single bus, typical microprocessors such as the Pentium use this architecture. This is because memory units such as ROMs, EPROMs, and RAMs are external to the microprocessor. This will require almost half the number of wires on the mother board since address and data pins for only two buses rather than four buses (Harvard architecture) are required. This is the reason Harvard architecture would be very expensive if utilized in designing microprocessors. Note that microcontrollers using Harvard architecture internally will have to use von Neumann architecture externally. Texas Instrument's MSP 430 uses the von Neumann architecture.

1.2   Design Levels

Three design levels can be defined for digital systems: systems level, logic level, and device level.

- *Systems level* is the type of design in which CPU, memory, and I/O chips are interfaced to build a computer.

- *Logic level* is the design technique in which chips containing logic gates such as AND, OR, and NOT are used to design a digital component such as the ALU.

- Finally, *device level* utilizes transistors to design logic gates.

1.3   Combinational and Sequential Circuits

Digital circuits at the logic level can be classified into two types. These are *combinational and sequential.*

   Combinational circuits contain no memory, whereas sequential circuits require memory to remember the present state in order to go to the next state. A binary adder capable of providing the sum upon application of the numbers to be added is an example of a combinational circuit. For example, consider a 4-bit adder. The inputs to this adder will be two 4-bit numbers; the output will be the 4-bit sum. In this case, the adder will generate the 4-bit sum output upon application of the two 4-bit inputs.

   Sequential circuits, on the other hand, require memory. The counter is an example of a sequential circuit. For instance, suppose that the counter is required to count in the sequence 0, 1, 2 and then repeat the sequence. In this case, the counter must have memory to remember the present count in order to go to the next. The counter must remember that it is at count 0 in order to go to the next count, 1. In order to count to 2, the counter must remember that it is counting 1 at the present state. In order to repeat the sequence, the counter must count back to 0 based on the present count, 2, and the process continues. A chip containing sequential circuit such as the counter will have a clock input pin.

   In general, all computers contain both combinational and sequential circuits. However, most computers are regarded as clocked sequential systems. In these computers, almost all activities pertaining to instruction execution are synchronized with clocks.

1.4   Digital Integrated Circuits

The transistor can be considered as an electronic switch. The ON and OFF states of a transistor are used to represent binary digits. Transistors, therefore, play an important role in the design of digital systems. This section describes the basic characteristics of digital devices and logic families. These include diodes, transistors, and a summary of digital logic families. These topics are covered from a very basic point of view. This will allow the readers with some background in digital devices to see how they are utilized in designing digital systems.

1.4.1   Diodes

A diode is an electronic switch. It is a two-terminal device. Figure 1.1 shows the symbolic representation.

   The positive terminal (made with the *p*-type semiconductor material) is called the *anode*; the negative terminal (made with the *n*-type semiconductor material) is called a *cathode.* When a voltage, $V = 0.6$ volt is applied across the anode and the cathode, the switch closes and a current I flows from anode to the cathode.

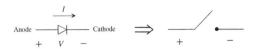

FIGURE 1.1   Symbolic representations of a diode.

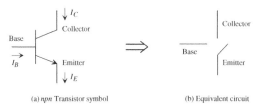

(a) *npn* Transistor symbol   (b) Equivalent circuit

FIGURE 1.2   (a and b) Symbolic representations of an *npn* transistor.

1.4.2   Transistors

A bipolar junction transistor (BJT) or commonly called the *transistor* is also an electronic switch like the diode. Both electrons (*n*-channel) and holes (*p*-channel) are used for carrier flow; hence, the name "bipolar" is used. The BJT is used in transistor logic circuits that have several advantages over diode logic circuits. First of all, the transistor acts as a logic device called an *inverter*. Note that an inverter provides a LOW output for a HIGH input and a HIGH output for a LOW input. Secondly, the transistor is a current amplifier (buffer). Transistors can, therefore, be used to amplify these currents to control external devices such as a light emitting diode (LED) requiring high currents. Finally, transistor logic gates operate faster than diode gates.

There are two types of transistors, namely *npn* and *pnp*. The classification depends on the fabrication process. *npn* transistors are widely used in digital circuits.

Figure 1.2 shows the symbolic representation of an *npn* transistor. The transistor is a three-terminal device. These are base, emitter, and collector. The transistor is a current-controlled switch. This means that an adequate current at the base will close the switch allowing a current to flow from the collector to the emitter.

This current direction is identified on the *npn* transistor symbol in Figure 1.2(a) by a downward arrow on the emitter. Note that a base resistance is normally required to generate the base current.

The transistor has three modes of operation: cutoff, saturation, and active. In digital circuits, a transistor is used as a switch, which is either ON (closed) or OFF (open). When no base current flows, the emitter-collector switch is open and the transistor operates in the cutoff (OFF) mode. On the other hand, when a base current flows such that the voltage across the base and the emitter is at least 0.6 V, the switch closes. If the base current is further increased, there will be a situation in which V_{CE} (voltage across the collector and the emitter) attains a constant value of approximately 0.2 V. This is called the saturation (ON) mode of the transistor. The "active" mode is between the cutoff and saturation modes. In this mode, the base current (I_B) is amplified so that the collector current, $I_C = \beta I_B$, where β is called the gain, and is in the range of 10 to 100 for typical transistors. Note that when the transistor reaches saturation, increasing I_B does not drop V_{CE} below $V_{CE\,(Sat.)}$ of 0.2 V. On the other hand, V_{CE} varies from 0.8 V to 5 V in the active mode. Therefore, the cutoff (OFF) and saturation (ON) modes of the transistor are used in designing digital circuits. The active mode of the transistor in which the transistor acts as a current amplifier (also called *buffer*) is used in digital output circuits.

Operation of the Transistor as an Inverter   Figure 1.3 shows how to use the transistor as an inverter. When $V_{IN} = 0$, the transistor is in cutoff (OFF), and the collector-emitter switch is open. This means that no current flows from $+V_{CC}$ to ground. V_{OUT} is equal to $+V_{CC}$. Thus, V_{OUT} is high.

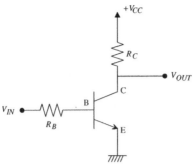

FIGURE 1.3   An inverter.

On the other hand, when V_{IN} is HIGH, the emitter-collector switch is closed. A current flows from $+V_{CC}$ to ground. The transistor operates in saturation, and V_{OUT} = $V_{CE(Sat)}$ = 0.2 V ≈ 0. Thus, V_{OUT} is basically connected to ground.

Therefore, for V_{IN} = LOW, V_{OUT}= HIGH, and for V_{IN} = HIGH, V_{OUT} = LOW. Hence, the *npn* transistor in Figure 1.3 acts as an inverter.

Note that V_{CC} is typically +5 V DC. The input voltage levels are normally in the range of 0 to 0.8 volts for LOW and 2 volts to 5 volts for HIGH. The output voltage levels, on the other hand, are normally 0.2 volts for LOW and 3.6 volts for HIGH.

Light Emitting Diodes (LEDs) and Seven Segment Displays   LEDs are extensively used as outputs in digital systems as status indicators. An LED is typically driven by low voltage and low current. This makes the LED a very attractive device for use with digital systems. Table 1.1 provides the current and voltage requirements of red, yellow, and green LEDs.

Basically, an LED will be ON, generating light, when its cathode is sufficiently negative with respect to its anode. A digital system such as a microcomputer can, therefore, light an LED either by grounding the cathode (if the anode is tied to +5 V) or by applying +5 V to the anode (if the cathode is grounded) through an appropriate resistor value. A typical hardware interface between a microcomputer and an LED is depicted in Figure 1.4. A microcomputer normally outputs 400 µA at a minimum

TABLE 1.1   Current and voltage requirements of LEDs

LEDs	Red	Yellow	Green
Current	10 mA	10 mA	20 mA
Voltage	1.7 V	2.2V	2.4V

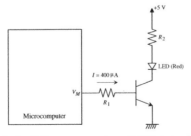

FIGURE 1.4   Microcomputer - LED interface.

voltage, V_M = 2.4 volts for a HIGH. The red LED requires 10 mA at 1.7 volts. A buffer (current amplifier) such as a transistor is required to turn the LED ON. Since the transistor is an inverter, a HIGH input to the transistor will turn the LED ON. We now design the interface; that is, the values of R_1, R_2, and the gain β for the transistor will be determined. Note that the outputs of typical microcontrollers such as the PIC18F are buffered.

A HIGH at the microcomputer output will turn the transistor ON into active mode. This will allow a path of current to flow from the +5 V source through R_2 and the LED to the ground. The appropriate value of R_2 needs to be calculated to satisfy the voltage and current requirements of the LED. Also, suppose that V_{BE} = 0.6 V when the transistor is in active mode. This means that R_1 needs to be calculated with the specified values of V_M = 2.4 V and I = 400 μA. The values of R_1, R_2, and β are calculated as follows:

$$R_1 = \frac{V_M - V_{BE}}{400\,\mu A} = \frac{2.4 - 0.6}{400\,\mu A} = 4.5\ K\Omega$$

Assuming $V_{CE} \cong 0$,

$$R_2 = \frac{5 - 1.7 - V_{CE}}{10\,mA} = \frac{5 - 1.7}{10\,mA} = 330\ \Omega$$

$$\beta = \frac{I_C}{I_B} = \frac{10\,mA}{400\,\mu A} = \frac{10 \times 10^{-3}}{400 \times 10^{-6}} = 25$$

Therefore, the interface design is complete, and a transistor with a minimum β of 25, R_1 = 4.5 KΩ, and R_2 = 330 Ω are required.

An inverting buffer chip such as the 74LS04 can be used in place of a transistor in Figure 1.4. A typical interface of an LED to a microcomputer via an inverter is shown in Figure 1.5. Note that the transistor base resistance is inside the inverter. Therefore, R_1 is not required to be connected to the output of the microcomputer. The symbol —▷— is used to represent an inverter. Inverters will be discussed in more detail later. In Figure 1.5, when the microcomputer outputs a HIGH, the transistor switch inside the inverter closes. A current flows from the +5 V source, through the 330-ohm resistor and the LED, into the ground inside the inverter. The LED is thus turned ON.

Note that if 5V is used to turn the LED ON and 0V to turn it OFF, the LED should be connected as shown in Figure 1.6.

However, if 0 is used to to turn the LED ON and 5V to turn it OFF, the LED should be connected as shown in Figure 1.7.

Note that an LED must not be connected according to the circuit shown in Figure 1.8. This is because the circuit will not provide 1.7V accross the LED and a current of 10 ma through it.

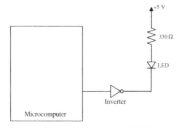

FIGURE 1.5   Microcomputer - LED interface via an inverter.

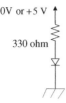

0V or +5 V

330 ohm

FIGURE 1.6   An LED connection to be turned ON by 5V and turned OFF by 0V.

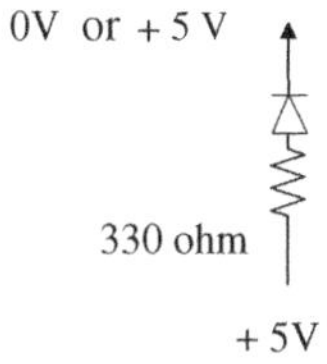

0V or + 5 V

330 ohm

+ 5V

FIGURE 1.7   An LED connection to be turned ON by 5V and turned OFF by 5V.

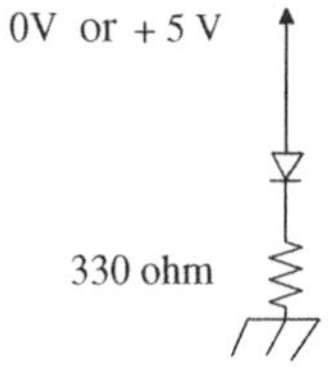

0V or + 5 V

330 ohm

FIGURE 1.8   An invalid LED connection.

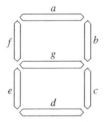

FIGURE 1.9   A seven-segment display.

A seven-segment display can be used to display, for example, decimal numbers from 0 to 9. The name "seven segment" is based on the fact that there are seven LEDs — one in each segment of the display. Figure 1.9 shows a typical seven-segment display.

In Figure 1.9, each segment contains an LED. All decimal numbers from 0 to 9 can be displayed by turning the appropriate segment "ON" or "OFF". For example, a zero can be displayed by turning the LED in segment g "OFF" and turning the other six LEDs in segments a through f "ON." There are two types of seven-segment displays. These are common cathode and common anode. Figure 1.10 shows these display configurations.

In a common cathode arrangement, the microcomputer can send a HIGH to light a segment and a LOW to turn it off. In a common anode configuration, on the other hand, the microcomputer sends a LOW to light a segment and a HIGH to turn it off. In both configurations, R = 330 ohms can be used.

Transistor Transistor Logic (TTL) and Its Variations   The transistor transistor logic (TTL) family of chips evolved from diodes and transistors. This family used to be

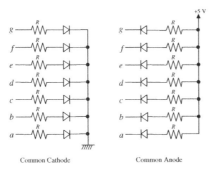

FIGURE 1.10   Seven-segment display configurations.

called *DTL* (*diode transistor logic*). The diodes were then replaced by transistors, and thus the name "TTL" evolved. The power supply voltage (V_{CC}) for TTL is +5 V. The two logic levels are approximately 0 and 3.5 V.

There are several variations of the TTL family. These are based on the saturation mode (saturated logic) and active mode (nonsaturated logic) operations of the transistor. In the saturation mode, the transistor takes some time to come out of the saturation to switch to the cutoff mode. On the other hand, some TTL families define the logic levels in the active mode operation of the transistor and are called *nonsaturated logic*. Since the transistors do not go into saturation, these families do not have any saturation delay time for the switching operation. Therefore, the nonsaturated logic family is faster than saturated logic.

The saturated TTL family includes standard TTL (TTL), high-speed TTL (H-TTL), and low-power TTL (L-TTL). The nonsaturated TTL family includes Schottky TTL (S-TTL), low-power Schottky TTL (LS-TTL), advanced Schottky TTL (AS-TTL), and advanced low-power Schottky TTL (ALS-TTL). The development of LS-TTL made TTL, H-TTL, and L-TTL obsolete. Another technology, called emitter-coupled logic (ECL), utilizes nonsaturated logic. The ECL family provides the highest speed. ECL is used in digital systems requiring ultrahigh speed, such as supercomputers.

The important parameters of the digital logic families are fan-out, power dissipation, propagation delay, and noise margin.

Fan-out is defined as the maximum number of inputs that can be connected to the output of a gate. It is expressed as a number. The output of a gate is normally connected to the inputs of other similar gates. Typical fan-out for TTL is 10. On the other hand, fan-outs for S-TTL, LS-TTL, and ECL, are 10, 20, and 25, respectively.

Power dissipation is the power (milliwatts) required to operate the gate. This power must be supplied by the power supply and is consumed by the gate. Typical power consumed by TTL is 10 mW. On the other hand, S-TTL, LS-TTL, and ECL absorb 22 mW, 2 mW, and 25 mW respectively.

Propagation delay is the time required for a signal to travel from input to output when the binary output changes its value. Typical propagation delay for TTL is 10 nanoseconds (ns). On the other hand, S-TTL, LS-TTL, and ECL have propagation delays of 3 ns, 10 ns, and 2 ns, respectively.

Noise margin is defined as the maximum voltage due to noise that can be added to the input of a digital circuit without causing any undesirable change in the circuit output. Typical noise margin for TTL is 0.4 V. Noise margins for S-TTL, LS-TTL, and ECL are 0.4 V, 0.4 V, and 0.2 V, respectively.

TTL Outputs   There are three types of output configurations for TTL. These are open-collector output, totem-pole output, and tristate (three-state) output.

The open-collector output means that the TTL output is a transistor with nothing connected to the collector. The collector voltage provides the output of the gate. For the open-collector output to work properly, a resistor (called the *pullup resistor*), with a value of typically 1 Kohm, should be connected between the open collector output and a +5 V power supply.

If the outputs of several open-collector gates are tied together with an external resistor (typically 1 Kohm) to a +5 V source, a logical AND function is performed at the connecting point. This is called *wired-AND logic*.

Figure 1.11 shows two open-collector outputs (*A* and *B*) are connected together to a common output point *C* via a 1 K Ω resistor and a +5 V source.

The common-output point *C* is HIGH only when both transistors are in cutoff (OFF) mode, providing *A* = HIGH and *B* = HIGH. If one or both of the two transistors is turned ON, making one (or both open-collector outputs) LOW, this will drive the common output *C* to LOW. Note that a LOW (Ground for example) signal when connected to a HIGH (+5V for example) signal generates a LOW. Thus, *C* is obtained by performing a logical AND operation of the open collector outputs *A* and *B*.

Let us briefly review the totem-pole output circuit shown in Figure 1.12. The circuit operates as follows:

When transistor Q_1 is ON, transistor Q_2 is OFF. When Q_1 is OFF, Q_2 is ON. This is how the totem-pole output is designed. The complete TTL gate connected to the bases of transistors Q_1 and Q_2 is not shown; only the output circuit is shown.

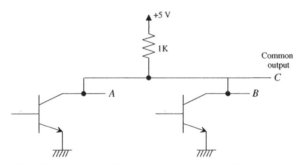

FIGURE 1.11   Two open-collector outputs A and B tied together.

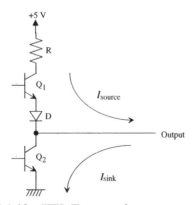

FIGURE 1.12   TTL Totem-pole output.

In the figure, Q_1 is turned ON when the logic gate circuit connected to its base sends a HIGH output. The switches in transistor Q_1 and diode D close while the switch in Q_2 is open. A current flows from the +5 V source through R, Q_1, and D to the output. This current is called I_{source} or output high current, I_{OH}. This is typically represented by a negative sign in front of the current value in the TTL data book, a notation indicating that the chip is losing current. For a low output value of the logic gate, the switches in Q_1 and D are open and the switch in Q_2 closes. A current flows from the output through Q_2 to ground. This current is called I_{sink} or Output Low current, I_{OL}. This is represented by a positive sign in front of the current value in the TTL data book, indicating that current is being added to the chip. Either I_{source} or I_{sink} can be used to drive a typical output device such as an LED. I_{source} (I_{OH}) is normally much smaller than I_{sink} (I_{OL}). I_{source} (I_{OH}) is typically -0.4 mA (or -400 µA) at a minimum voltage of 2.7 V at the output. I_{source} is normally used to drive devices that require high currents. A current amplifier (buffer) such as a transistor or an inverting buffer chip such as 74LS368 needs to be connected at the output if I_{source} is used to drive a device such as an LED requiring high current (10 mA to 20 mA). I_{sink} is normally 8 mA

The totem-pole outputs must not be tied together. When two totem-pole outputs are connected together with the output of one gate HIGH and the output of the second gate LOW, the excessive amount of current drawn can produce enough heat to damage the transistors in the circuit.

Tristate is a special totem-pole output that allows connecting the outputs together like the open-collector outputs. When a totem-pole output TTL gate has this property, it is called a *tristate* (*three state*) *output*. A tristate has three output states:

1. A LOW level state when the lower transistor in the totem-pole is ON and the upper transistor is OFF.
2. A HIGH level when the upper transistor in the totem-pole is ON and the lower transistor is OFF.
3. A third state when both output transistors in the totem-pole are OFF. This third state provides an open circuit or high-impedance state which allows a direct wire connection of many outputs to a common line called the *bus*.

A Typical Switch Input Circuit for TTL   Figure 1.13 shows a switch circuit that can be used as a single bit into the input of a TTL gate. When the DIP switch is open, V_{IN} is HIGH. On the other hand, when the switch is closed, V_{IN} is low. V_{IN} can be used as an input bit to a TTL logic gate for performing laboratory experiments.

1.4.3 MOS Transistors

Metal-Oxide Semiconductor (MOS) transistors occupy less space in the circuit and consume much less power than bipolar junction transistors. Therefore, MOS

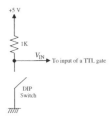

FIGURE 1.13   A typical circuit for connecting an input to a TTL gate.

transistors are used in highly integrated circuits. The MOS transistor is unipolar. This means that one type of carrier flow, either electrons (n-type) or holes (p-type) are used. The MOS transistor works as a voltage-controlled resistance. In digital circuits, a MOS transistor operates as a switch such that its resistance is either very high (OFF) or very low (ON). The MOS transistor is a three-terminal device: gate, source, and drain. There are two types of MOS transistors, namely, nMOS and pMOS. The power supply (V_{CC}) for pMOS is in the range of 17 V to 24 V, while V_{CC} for nMOS is lower than pMOS and can be from 5 V to 12 V. Figure 1.14 shows the symbolic representation of an nMOS transistor. When $V_{GS} = 0$, the resistance between drain and source (R_{DS}) is in the order of megaohms (Transistor OFF state). On the other hand, as V_{GS} is increased, R_{DS} decreases to a few tens of ohms (Transistor ON state). Note that in an MOS transistor, there is no connection between the gate and the other two terminals (source and drain). The nMOS gate voltage (V_{GS}) increases or decreases the current flow from drain to source by changing R_{DS}. Popular 8-bit microprocessors such as the Intel 8085 and the Motorola 6809 were designed using nMOS.

Figure 1.15 depicts the symbol for a pMOS transistor. The operation of the pMOS transistor is very similar to the nMOS transistor except that V_{GS} is typically zero or negative. The resistance from drain to source (R_{DS}) becomes very high (OFF) for $V_{GS} = 0$. On the other hand, R_{DS} decreases to a very low value (ON) if V_{GS} is decreased. pMOS was used in fabricating the first 4-bit microprocessors (Intel 4004/4040) and 8-bit microprocessor (Intel 8008). Basically, in a MOS transistor (nMOS or pMOS), V_{GS} creates an electric field that increases or decreases the current flow between source and drain. From the symbols of the MOS transistors, it can be seen that there is no connection between the gate and the other two terminals (source and drain). This symbolic representation is used in order to indicate that no current flows from the gate to the source, irrespective of the gate voltage.

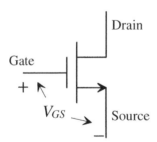

FIGURE 1.14   nMOS transistor symbol.

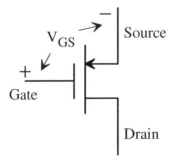

FIGURE 1.15   pMOS transistor symbol.

Operation of the nMOS transistor as an inverter   Figure 1.16 shows an nMOS inverter. When V_{IN} = LOW, the resistance between the drain and the source (R_{DS}) is very high, and no current flows from V_{CC} to the ground. V_{OUT} is therefore high. On the otherhand, when V_{IN} = high, R_{DS} is very low, a current flows from V_{CC} to the source, and V_{OUT} is LOW. Therefore, the circuit acts as an inverter.

Complementary MOS (CMOS)   CMOS dissipates low power and offers high circuit density compared to TTL. CMOS is fabricated by combining nMOS and pMOS transistors together. The nMOS transistor transfers logic 0 well and logic 1 inefficiently. The pMOS transistor, on the other hand, outputs logic 1 efficiently and logic 0 poorly. Therefore, connecting one pMOS and one nMOS transistor in parallel provides a single switch called *transmission gate* that offers efficient output drive capability for CMOS logic gates. The transmission gate is controlled by an input logic level.

    Figure 1.17 shows a typical CMOS inverter. The CMOS inverter is very similar to the TTL totem-pole output circuit. That is, when Q_1 is ON (low resistance), Q_2 is OFF (high resistance), and vice versa. When V_{input} = LOW, Q_1 is ON and Q_2 is OFF. This makes V_{output} HIGH. On the other hand, when V_{input} = HIGH, Q_1 is OFF (high resistance) and Q_2 is ON (low resistance). This provides a low V_{output}. Thus, the circuit works as an inverter.

    Digital circuits using CMOS consume less power than MOS and bipolar transistor circuits. In addition, CMOS provides high circuit density. That is, more circuits can be placed in a chip using CMOS. Finally, CMOS offers high noise immunity. In CMOS, unused inputs should not be left open. Because of the very high input resistance, a floating input may change back and forth between a LOW and a HIGH, creating system problems. All unused CMOS inputs should be tied to V_{CC}, ground, or another high or low signal source appropriate to the device's function. CMOS can operate over a large range of power supply voltages (3 V to 15 V). Two

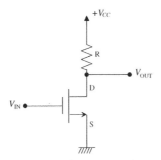

FIGURE 1.16   A typical nMOS inverter.

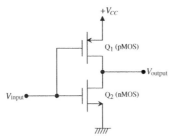

FIGURE 1.17   A CMOS inverter.

CMOS families, namely CD4000 and 54C/74C, were first introduced. CD 4000A is in the declining stage.

There are four members in the CMOS family which are very popular these days: the high-speed CMOS (HC), high-speed CMOS/TTL-input compatible (HCT), advanced CMOS (AC), and advanced CMOS/TTL-input compatible (ACT). The HCT chips have a specifically designed input circuit that is compatible with LS-TTL logic levels (2V for HIGH input and 0.8V for LOW input). LS-TTL outputs can directly drive HCT inputs while HCT outputs can directly drive HC inputs. Therefore, HCT buffers can be placed between LS-TTL and HC chips to make the LS-TTL outputs compatible with the HC inputs.

Several characteristics of 74HC and 74HCT are compared with 74LS-TTL and nMOS technologies in Table 1.2.

Note that in the table, HC and HCT have the same source (I_{OH}) and sink (I_{OL}) currents. This is because in a typical CMOS gate, the ON resistances of the pMOS and nMOS transistors are approximately the same.

The input characteristics of HC and HCT are compared in Table 1.3. The table shows that LS-TTL is not guaranteed to drive an HC input. The LS-TTL output HIGH is grater than or equal to 2.7V while an HC input needs at least 3.15V. Therefore, the HCT input requiring V_{IH} of 2.0V can be driven by the LS-TTL output, providing at least 2.7V; 74HCT244 (unidirectional) and 74HCT245 (bidirectional) buffers can be used.

MOS Outputs   Like TTL, the MOS logic offers three types of outputs. These are push-pull (totem-pole in TTL), open drain (open collector in TTL), and tristate outputs. For example, the 74HC00 contains four independent 2-input NAND gates and includes push-pull output. The 74HC03 also contains four independent 2-input NAND gates, but has open drain outputs. The 74HC03 requires a pull-up resistor for each gate. The 74HC125 contains four independent tri-state buffers in a single chip.

A Typical Switch Input Circuit for MOS Chips   Figure 1.18 shows a switch circuit that can be used as a single bit into the input of a MOS gate. When the DIP switch is open, V_{IN} is HIGH. On the other hand, when the switch is closed, V_{IN} is LOW. V_{IN} can be used as an input bit for performing laboratory experiments. Note that unlike TTL, a 1K resistor is connected between the switch and the input of the MOS gate. This provides for protection against static discharge. This 1-Kohm resistor is not required if the MOS chip contains internal circuitry providing protection against damage to inputs due to static discharge.

TABLE 1.2   Comparison of output characteristics of LS-TTL, nMOS, HC, and HCT

	V_{OH}	I_{OH}	V_{OL}	I_{OL}
LS-TTL	2.7 V	−400 μA	0.5 V	8 mA
nMOS	2.4 V	−400 μA	0.4 V	2 mA
HC	3.7 V	−4 mA	0.4 V	4 mA
HCT	3.7 V	−4 mA	0.4 V	4 mA

TABLE 1.3   Comparison of input characteristics of HC and HCT

	V_{IH}	I_{IH}	V_{IL}	I_{IL}	Fanout
HC	3.15 V	1μA	0.9 V	1μA	10
HCT	2.0 V	1μA	0.8 V	1μA	10

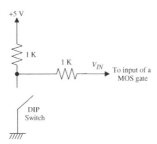

FIGURE 1.18   A typical switch for MOS input.

1.5   Integrated Circuits (ICs)

Device level design utilizes transistors to design circuits called *gates*, such as AND gates and OR gates. One or more gates are fabricated on a single silicon chip by an integrated circuit (IC) manufacturer in an IC package.

An IC chip is packaged typically in a ceramic or plastic package. The commercially available ICs can be classified as small-scale integration (SSI), medium-scale integration (MSI), large-scale integration (LSI), and very large-scale integration (VLSI).

- A single SSI IC contains a maximum of approximately 10 gates. Typical logic functions such as AND, OR, and NOT are implemented in SSI IC chips. The MSI IC, on the other hand, includes from 11 to up to 100 gates in a single chip. The MSI chips normally perform specific functions such as add.

- The LSI IC contains more than 100 to approximately 1000 gates. Digital systems such as 8-bit microprocessors and memory chips are typical examples of LSI ICs.

- The VLSI IC includes more than 1000 gates. More commonly, the VLSI ICs are identified by the number of transistors (containing over 500,000 transistors) rather than the gate count in a single chip. Typical examples of VLSI IC chips include 32-bit microprocessors and one-megabit memories. For example, the Intel Pentium is a VLSI IC containing 3.1 million transistors in a single chip.

An IC chip is usually inserted in a printed-circuit board (PCB) that is connected to other IC chips on the board via pins or electrical terminals. In laboratory experiments or prototype systems, the IC chips are typically placed on breadboards or wire-wrap boards and connected by wires. The breadboards normally have noise problems for frequencies over 4 MHz. Wire-wrap boards are used above 4 MHz. The number of pins in an IC chip varies from ten to several hundred, depending on the package type. Each IC chip must be powered and grounded via its power and ground pins. The VLSI chips such as the Pentium have several power and ground pins. This is done in order to reduce noise by distributing power in the circuitry inside the chip.

The SSI and MSI chips normally use an IC package called *dual in-line package* (DIP). The LSI and VLSI chips, on the other hand, are typically fabricated in surface-mount or pin grid array (PGA) packages. The DIP is widely used because of its low price and ease of installation into the circuit board.

SSI chips are identified as 5400-series (these are for military applications with stringent requirements on voltage and temperature and are expensive) or 7400 series (for commercial applications). Both series have identical pin assignments on chips with the same part numbers, although the first two numeric digits of the part name are different. Typical commercial SSI ICs can be identified as follows:

74S	Schottky TTL
74LS	Low-power Schottky TTL
74AS	Advanced Schottky TTL
74F	Fast TTL (similar to 74AS; manufactured by Fairchild)
74ALS	Advanced low-power Schottky TTL

Note that two digits appended at the end of each of these IC identifications define the type of logic operation performed, the number of pins, and the total number of gates on the chip. For example, 74S00, 74LS00, 74AS00, 74F00, and 74ALS00 perform NAND operation. All of them have 14 pins and contain four independent NAND gates in a single chip.

The gates in the ECL family are identified by the part numbers 10XXX and 100XXX, where XXX indicates three digits. The 100XXX family is faster, requires low power supply, but it consumes more power than the 10XXX. Note that 10XXX and 100XXX are also known as 10K and 100K families.

The commercially available CMOS family is identified in the same manner as the TTL SSI ICs. For example, 74LS00 and 74HC00 (High-speed CMOS) are identical, with 14 pins and containing four independent NAND gates in a single chip. Note that 74HCXX gates have operating speeds similar to 74LS-TTL gates. For example, the 74HC00 contains four independent two-input NAND gates. Each NAND gate has a typical propagation delay of 10 ns and a fanout of 10 LS-TTL.

Unlike TTL inputs, CMOS inputs should never be held floating. The unused input pins must be connected to V_{CC}, ground, or an output. The TTL input contains an internal resistor that makes it HIGH when unused or floating. The CMOS input does not have any such resistor and therefore possesses high resistance. The unused CMOS inputs must be tied to V_{CC}, ground, or other gate outputs. In some CMOS chips, inputs have internal pull-up or pull-down resistors. These inputs, when unused, should be connected to V_{CC} or ground to make the inputs high or low.

The CMOS family has become popular compared to TTL due to better performance. Some major IC manufacturers such as National Semiconductor do not make 7400 series TTL anymore. Although some others, including Fairchild and Texas Instruments still offer the 7400 TTL series, the use of the SSI TTL family (74S, 74LS, 74AS, 74F, and 74ALS) is in the declining stage, and will be obsolete in the future. On the other hand, the use of CMOS-based chips such as 74HC has increased significantly because of their high performance. These chips will dominate the future market.

1.6   CAD (Computer-Aided Design)

Digital logic circuits were used in building the first computers. With the advent of VLSI technology, millions of transistors are contained in the same chip. Hence, it has become a difficult task to design these circuits without using computer-aided design tools.

CAD tools include programs that assist in developing the digital hardware. The CAD tools perform the design process automatically, and come up with an optimized circuit which will satisfy design specifications. The designer is required to provide the precise description of the design in order to obtain the best possible circuit. In order to accomplish this, the designer must have a clear understanding of the theory of digital logic.

CAD tools along with HDL (hardware description language) can be used to design digital logic circuits. FPGAs have become popular in recent years. These logic circuits can be implemented in FPGAs using CAD tools and HDL. These topics are covered in this book in a very simplified manner.

1.7   Evolution of the Microcontroller

Intel Corporation is generally acknowledged as the company that introduced the first microprocessor successfully into the marketplace. Its first microprocessor, the 4004, was introduced in 1971 and evolved from a development effort while making a calculator chip set. The 4004 microprocessor was the central component in the chip set, which was called the MCS-4. The other components in the set were a 4001 ROM, a 4002 RAM, and a 4003 shift register.

Shortly after the 4004 appeared in the commercial marketplace, three other general-purpose microprocessors were introduced: the Rockwell International 4-bit PPS-4, the Intel 8-bit 8008, and the National Semiconductor 16-bit IMP-16. Other companies, such as General Electric, RCA, and Viatron, also made contributions to the development of the microprocessor prior to 1971.

The microprocessors introduced between 1971 and 1972 were the first-generation systems designed using PMOS technology. In 1973, second-generation microprocessors such as the Motorola 6800 and the Intel 8080 (8-bit microprocessors) were introduced. The second-generation microprocessors were designed using NMOS technology. This technology resulted in a significant increase in instruction execution speed over PMOS and higher chip densities. Since then, microprocessors have been fabricated using a variety of technologies and designs. NMOS microprocessors such as the Intel 8085, the Zilog Z80, and the Motorola 6800/6809 were introduced based on second-generation microprocessors. A third generation HMOS microprocessor, introduced in 1978 is typically represented by the Intel 8086 and the Motorola 68000, which are 16-bit microprocessors.

During the 1980's, fourth-generation HCMOS and BICMOS (a combination of bipolar and HCMOS) 32-bit microprocessors evolved. Intel introduced the first commercial 32-bit microprocessor, the problematic Intel 432, which was eventually discontinued. Since 1985, more 32-bit microprocessors have been introduced. These include Motorola's 68020, 68030, 68040, 68060, PowerPC, Intel's 80386, 80486, the Intel Pentium family, Core Duo, and Core2 Duo microprocessors..

The performance offered by the 32-bit microprocessor is more comparable to that of superminicomputers such as Digital Equipment Corporation's VAX11/750 and VAX11/780. Intel and Motorola also introduced RISC microprocessors: the Intel 80960 and Motorola 88100/PowerPC, which had simplified instruction sets. Note that the purpose of RISC microprocessors is to maximize speed by reducing clock cycles per instruction. Almost all computations can be obtained from a simple instruction set. Note that, in order to enhance performance significantly, Intel Pentium Pro and other succeeding members of the Pentium family and Motorola 68060 are designed using a combination of RISC and CISC.

Single-chip microcomputers such as the Intel 8048 evolved during the 80's. Soon afterwards, based on the concept of single-chip microcomputers, Intel introduced the first 8-bit microcontroller---the Intel 8051 which uses Harvard architecture. The 8051 is designed using CISC. The 8051 contains a CPU, memory, I/O, A/D and D/A converters, timer, serial communication interface----- all in a single chip. The microcontrollers became popular during the 80's.

8-bit microcontrollers gained popularity over the last several years. These microcontrollers are small enough for many embedded applications, but also powerful enough to allow a lot of complexity and flexibility in the design process of an embedded system. Several billion 8-bit microcontrollers were sold during the last decade. Several contemporary microcontroller manufacturers use RISC architecture, and thus, provide cost effective approach. In addition, typical 8-bit microcontrollers such as the PIC18F implemented several on-chip enhanced peripheral functions including PWM (Pulse Width Modulation) and flash memories. Note that Motorola/Freescale popular 8-bit microcontroller HC11 does not have on-chip flash memory and PWM functions. PWM function is a very desirable feature for applications such as automotive and motor control. These applications may include driving servo motors. In HC11, timer section is used to generate PWM signals. However, Motorola/ Freescale implemented these features in the HC16 which is a 16-bit microcontroller. Note that the HC11 has been popular because of its rich instruction set.

Like EEPROM, flash memory can be programmed and erased electrically. Flash memory is very popular these days compared to EEPROM. Note that EEPROM can be erased one byte at a time while flash memory can be erased only in blocks.

Table 1.4 provides a comparison of the basic features of some of the typical microcontrollers. Microchip has introduced several different versions of the PIC18F microcontroller over the years. All members of the PIC18F family basically contain the same instruction set. However, certain features such as memory sizes, number of I/O ports, A/D channels and PWM modules may vary from one version to another. In this book, a specific PIC18F chip such as the PIC18F4321 will be considered later in detail.

1.8 Typical Microcontroller Applications

Some of the typical microcontroller applications include the following:

- Automotive
- to operate devices such as a microwave oven, a radiator fan in a car, or servo motors used to move the handles on a foosball table.

TABLE 1.4   Comparison of basic features of typical microcontrollers

	PIC18F	MSP 430	HC11	AVR
Manufacturer	Microchip Technology	Texas Instruments	Motorola / Freescale	Atmel
Introduced	2000; the first PIC in 1989.	Late 1990s	1985	1996
Size	8-bit	16-bit	8-bit	8-bit
Architecture	Harvard	von Neumann	von Neumann	Harvard
Design approach	RISC	RISC	CISC	RISC
On-chip flash memory	Yes	Yes	No	Yes. First to offer on-chip flash.
On-chip PWM(Pulse Width Modulation)	Yes	Yes	No	Yes
CPU Clock	40-MHz (Maximum)	1-MHz (Maximum)	4-MHz (Maximum)	20-MHz (Maximum)
Total Instructions	75	27	144	123
Total Addressing modes	6	7	6	5

- Barcode readers
- Hotel card key writers
- Robotics

In the following, a microcontroller-based temperature control systems is first described. Since microcontrollers are widely used as "embedded controllers" in embedded applications, the basic concepts associated with embedded controllers are then considered.

1.8.1   A Simple Microcontroller Application

To put microcontrollers into perspective, it is important to explore a simple application.

For example, consider the microcontroller-based dedicated controller shown in Figure 1.19. Suppose that it is necessary to maintain the temperature of a furnace to a desired level to maintain the quality of a product. Assume that the designer has decided to control this temperature by adjusting the fuel. This can be accomplished using a typical microcontroller such as the PIC18F along with the interfacing components as follows. Temperature is an analog (continuous) signal. It can be measured by a temperature-sensing (measuring) device such as a thermocouple. The thermocouple provides the measurement in millivolts (mV) equivalent to the temperature.

Since microcontrollers only understand binary numbers (0's and 1's), each analog mV signal must be converted into a binary number using the microcontroller's on-chip analog-to-digital (A/D) converter. Note that the PIC18F contains on-chip A/D converter. The PIC18F does not include on-chip digital-to-analog (D/A) converter. However, the D/A converter chip can be interfaced to the PIC18F externally.

First, the millivolt signal is amplified by a mV/V amplifier to make the signal compatible for A/D conversion. A microcontroller such as the PIC18F can be programmed to solve an equation with the furnace temperature as an input. This equation compares the temperature measured with the temperature desired which can be entered into the microcontroller using the keyboard. The output of this equation will provide the appropriate opening and closing of the fuel valve to maintain the appropriate temperature. Since this output is computed by the microcontroller, it is a binary number. This binary output must be converted into an analog current or voltage signal.

The D/A (digital-to-analog) converter chip inputs this binary number and converts it into an analog current (I). This signal is then input into the current/

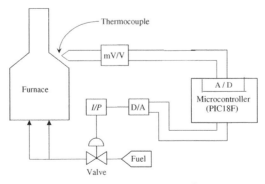

FIGURE 1.19   Furnace temperature control.

pneumatic (*I/P*) transducer for opening or closing the fuel input valve by air pressure to adjust the fuel to the furnace. The furnace temperature desired can thus be achieved. Note that a transducer converts one form of energy (electrical current in this case) to another form (air pressure in this example).

1.8.2 Embedded Controllers

Embedded microcontroller systems, also called *embedded controllers*, are designed to manage specific tasks. Once programmed, the embedded controllers can manage the functions of a wide variety of electronic products. In embedded applications, the microcontrollers are embedded in the host system, their presence and operation are basically hidden from the host system.

Typical embedded control applications include office automation products such as copiers, laser products, fax machines, and consumer electronics such as VCRs and microwave ovens. Applications such as printers typically utilize a microcontroller. The RISC microcontrollers are ideal for these types of applications. Note that the Personal Computer interfaced to the printer is the host.

RISC microcontrollers such as the PIC18F are well suited for applications including robotics, controls, instrumentation, and consumer electronics. The key features of the RISC microcontrollers that make them ideal for these applications are their relatively low level of integration in the chip, and instruction pipeline architecture. These characteristics result in low power consumption, fast instruction execution, and fast recognition of interrupts.

Although microcontrollers including PIC18F are considered ideal for many embedded applications, sometimes they might not be able to perform certain tasks. For example, applications such as laser printers require a high performance microprocessor with on-chip floating-point hardware. The PowerPC RISC microprocessor with on-chip floating-point hardware is ideal for these types of applications. Note that the personal computer interfaced to the laser printer is the host. The PIC18F will not be suitable for such an application since it does not provide floating-point instructions.

2

NUMBER SYSTEMS AND CODES

This chapter describes the basics of number systems, codes, and error detection/correction. Note that adequate coverage of *Number Systems and Codes* is included in this chapter. This will provide sufficient background to understand the concepts described in the chapters that follow.

2.1 Number Systems

A computer, like all digital systems, utilizes two states to represent information. These two states are given the symbols 0, and 1. It is important to remember that these 0's and 1's are symbols for the two states and have no inherent numerical meanings of their own. These two digits are called *binary digits* (bits) and can be used to represent numbers of any magnitude. Digital systems including the microcontroller carry out all arithmetic and logic operations internally using binary numbers. Because binary numbers are long, a more compact form using some other number system is preferable to represent them. The computer user finds it convenient to work with this compact form. Hence, it is important to understand the various number systems used with computers. These are described in the following sections.

2.1.1 General Number Representation

In general, N can be written as a string of digits whose integer parts$(d_{p-1}, d_{p-2}, \ldots, d_1, d_0)$ and fractional parts $(d_{-1} \cdots d_{-q})$ are separated by the radix or decimal point ($\bullet$). In this format, the number N is represented as

$$N = d_{p-1} d_{p-2} \cdots d_1 d_0 \bullet d_{-1} \cdots d_{-q} \tag{2.1}$$

If a number has no fractional portion, then the number is called an integer number or an integer. Conversely, if the number has no integer portion, then the number is called a *fractional number or a fraction*. The number, N can also be represented in the following form:

$$N = d_{p-1} \times b^{p-1} + d_{p-2} \times b^{p-2} + \cdots + d_0 \times b^0 + d_{-1} \times b^{-1} + \cdots + d_{-q} \times b^{-q} \tag{2.2}$$

where b is the base or radix of the number system, the d's are the digits of the number system, p is the number of integer digits, and q is the number of fractional digits.

Decimal Number System   The decimal number system has a base or radix of 10 and has 10 allowable digits, 0 through 9. As an example, consider the number 125.532_{10}.

$$125.532_{10} = 1 \times 10^2 + 2 \times 10^1 + 5 \times 10^0 + 5 \times 10^{-1} + 3 \times 10^{-2} + 2 \times 10^{-3}$$

Comparing the right-hand side of Equation 2.2 with equation 2.1 yields $b = 10$, $p = 3$, $q = 3$, $d_2 = 1$, $d_1 = 2$, $d_0 = 5$, $d_1 = 5$, $d_2 = 3$, and $d_3 = 2$.

Binary Number System   The binary number system has a base or radix of 2 and has two allowable digits, 0 and 1. From Equation 2.1, a 4-bit binary number 1110_2 can be interpreted as an integer decimal as follows:

$$1110_2 = 1 \times 2^3 + 1 \times 2^2 + 1 \times 2^1 + 0 \times 2^0 = 14_{10}$$

This conversion from binary to integer decimal can be obtained by inspecting the binary number as follows:

Note that bits 0, 1, 2, and 3 have corresponding weighting values of 1, 2, 4, and 8. Because a binary number only contains 1's and 0's, adding the weighting values of only the bits of the binary number containing 1's will provide its decimal value. The decimal value of 1110_2 is 14_{10} (2 + 4 + 8) because bits 1, 2, and 3 have binary digit 1, whereas bit 0 contains 0.

Therefore, the decimal (integer) value of any binary number can be readily obtained by just adding the weighting values for the bit positions containing 1's. Furthermore, the value of the least significant bit (bit 0) determines whether the number is odd or even. For example, if the least significant bit is 1, the number is odd; otherwise, the number is even.

Next, consider converting a mixed binary number (containing both integer and fractional parts), 101.01_2 into decimal using equation 2.1 as follows:

$$101.01_2 = 1 \times 2^2 + 0 \times 2^1 + 1 \times 2^0 + 0 \times 2^{-1} + 1 \times 2^{-2} \qquad (2.3)$$

The decimal or base 10 value of 101.01_2 is found from the right-hand side of Equation 2.3 as $4 + 0 + 1 + 0 + \frac{1}{4} = 5.25_{10}$.

One can determine whether a binary number is odd or even by inspecting the least significant bit (bit 0) of the number; the number is even if bit 0 is zero while the number is odd if bit 0 is one. For example, in order to find whether a binary number is odd or even, check the least significant bit as follows:

The number is odd, since the least significant bit is 1.

Decimal value = $32 + 8 + 2 + 1 = 43_{10}$, which is odd.

TABLE 2.1   Octal digits along with the coresponding binary values

Octal digits	Binary values
0	000
1	001
2	010
3	011
4	100
5	101
6	110
7	111

Similarly, in order to find whether the following binary number is odd or even, check the least significant bit as follows:

$$128 \quad 64 \quad 32 \quad 16 \quad 8 \quad 4 \quad 2 \quad 1 \longleftarrow \text{Weighting}$$
$$1 \quad 0 \quad 1 \quad 0 \quad 0 \quad 0 \quad 1 \searrow 0_2$$

The number is even, since the least significant bit is 0.

Decimal value $= 128 + 32 + 2 = 162_{10}$, which is even.

Octal Number System   The radix or base of the octal number system is 8. There are eight digits, 0 through 7, allowed in this number system. Table 2.1 shows the octal digits along with their corresponding binary values.

Consider the octal number 25.32_8, which can be interpreted as:

$$2 \times 8^1 + 5 \times 8^0 + 3 \times 8^{-1} + 2 \times 8^{-2}$$

The decimal value of this number is found by completing the summation as follows:

$$16 + 5 + 3 \times 1/8 + 2 \times 1/64 = 16 + 5 + 0.375 + 0.03125 = 21.4062$$

Hexadecimal Number System   The hexadecimal or base-16 number system has 16 individual digits. Each of these digits, as in all number systems, must be represented by a single unique symbol. The digits in the hexadecimal number system are 0 through 9 and the letters A through F. Letters were chosen to represent the hexadecimal digits greater than 9 because a single symbol is required for each digit. Table 2.2 lists the 16 digits of the hexadecimal number system and their corresponding binary and decimal values.

TABLE 2.2   Hexadecimal number system and their corresponding binary and decimal values.

Decimal	Hexadecimal	Binary
0	0	0000
1	1	0001
2	2	0010
3	3	0011
4	4	0100
5	5	0101
6	6	0110
7	7	0111
8	8	1000
9	9	1001
10	A	1010
11	B	1011
12	C	1100
13	D	1101
14	E	1110
15	F	1111

2.1.2   Converting Numbers from One Base to Another

Binary-to-Decimal Conversion and Vice Versa

Consider converting 1100.01_2 to its decimal equivalent. As before,

$$1100.01_2 = 1 \times 2^3 + 1 \times 2^2 + 0 \times 2^1 + 0 \times 2^0 + 0 \times 2^{-1} + 1 \times 2^{-2}$$

$$= 8 + 4 + 0 + 0 + 0 + .25$$

$$= 12.25_{10}$$

Next, convert the decimal number 12.25_{10} back to its binary equivalent.

Note that continuous division by 2, keeping track of the remainders, provides a simple method of converting a decimal number to its binary equivalent. Hence, to convert decimal 12_{10} to its binary equivalent 1100_2, proceed as follows:

quotient	+	remainder
$\dfrac{12}{2} = 6$	+	0
$\dfrac{6}{2} = 3$	+	0
$\dfrac{3}{2} = 1$	+	1
$\dfrac{1}{2} = 0$	+	1

$$1\ 1\ 0\ 0\,_2$$

Now convert the fractional part 0.25_{10} to its binary equivalent. Note that continuous multiplication by 2 and keeping track of the integer parts will accomplish this as follows:

$$
\begin{array}{cc}
0.25 & 0.50 \\
\times 2 & \times 2 \\
\hline
\textcircled{0}50 & \textcircled{1}00 \\
\downarrow & \downarrow \\
0 & 1
\end{array}
$$

Thus $0.25_{10} = 0.01_2$. Therefore $12.25_{10} = 1100.01_2$.

Unfortunately, binary-to-decimal fractional conversions are not always exact. Suppose that it is desired to convert 0.3615 into its binary equivalent:

$$
\begin{array}{ccccc}
0.3615 & 0.7230 & 0.4460 & 0.8920 & 0.7840 \\
\times 2 & \times 2 & \times 2 & \times 2 & \times 2 \\
\hline
\textcircled{0}7230 & \textcircled{1}4460 & \textcircled{0}8920 & \textcircled{1}7840 & \textcircled{1}5680 \\
\downarrow & \downarrow & \downarrow & \downarrow & \downarrow \\
0 & 1 & 0 & 1 & 1
\end{array}
$$

The answer is $0.01011..._2$. As a check, let us convert back:

$$0.01011_2 = 0 \times 2^{-1} + 1 \times 2^{-2} + 0 \times 2^{-3} + 1 \times 2^{-4} + 1 \times 2^{-5}$$

$$= 0 + 0.25 + 0 + 0.0625 + 0.03125$$

$$= 0.34375$$

The difference is 0.3615 - 0.34375 = 0.01775. This difference is caused by the neglected remainder 0.5680. The neglected remainder (0.5680) multiplied by the smallest computed term (0.03125) gives the total error:

$$0.5680 \times 0.03125 = 0.01775$$

Note that the same procedure applies for converting a decimal number to other number systems such as octal or hexadecimal; Continuous division (or multiplication) by the appropriate base (8 or 16) and keeping track of remainders (or interger parts) converts a number from decimal to the selected number systems.

Binary-to-Octal Conversion and Vice Versa   One can convert a number from binary to octal representation easily by taking the binary digits in groups of 3 bits to an octal digit.

The octal digit is obtained by considering each group of 3 bits as a separate binary number capable of representing the octal digits 0 through 7. The radix point remains in its original position. The following example illustrates the procedure.

Suppose that it is desired to convert 1001.11_2 into octal form. First take the groups of 3 bits starting at the decimal point. Where there are not enough leading or trailing bits to complete the triplet, 0's are appended. Now each group of 3 bits is converted to its corresponding octal digit.

$$\underset{1}{\underbrace{001}} \; \underset{1}{\underbrace{001}} \; . \; \underset{6}{\underbrace{110}}_2 \; = 11.6_8$$

The conversion back to binary from octal is simply the reverse of the binary-to-octal process. For example, conversion from 11.6_8 to binary is accomplished by expanding each octal digit to its equivalent binary values as shown:

$$\underset{001}{\underbrace{1}} \; \underset{001}{\underbrace{1}} \; . \; \underset{110}{\underbrace{6}}$$

Octal-to-Decimal Conversion and Vice Versa   Consider converting the octal number 230_8 into its decimal equivalent and vice versa. This can be accomplished as follows:

$$230_8 = 2 \times 8^2 + 3 \times 8^1 + 0 \times 8^0$$

$$= 128 + 24 + 0 = 152_{10}$$

Now to convert 152_{10} back to 230_8 :

	quotient	+	remainder
$\dfrac{152}{8} =$	19	+	0
$\dfrac{19}{8} =$	2	+	3
$\dfrac{2}{8} =$	0	+	2
			2 3 0

Thus, $152_{10} = 230_8$.

Binary-to-Hexadecimal Conversion and Vice Versa The conversions between hexadecimal and binary numbers are done in exactly the same manner as the conversions between octal and binary, except that groups of 4 are used. The following examples illustrate this:

$$1011011_2 = \underset{5}{\underline{0101}} \quad \underset{B}{\underline{1011}} = 5B_{16}$$

Note that the binary integer number is grouped in 4-bit units, starting from the least significant bit. Zeros are added with the most significant 4 bits if necessary. As with octal numbers, for fractional numbers this grouping into 4 bits is started from the radix point. Now consider converting $2AB_{16}$ into its binary equivalent as follows:

$$2AB_{16} = \begin{matrix} 2 & A & B \\ \downarrow & \downarrow & \downarrow \\ 0010 & 1010 & 1011 \end{matrix}$$

$$= 001010101011_2$$

Hexadecimal-to-Decimal Conversion and Vice Versa Consider converting the hexadecimal number $23A_{16}$ into its decimal equivalent and vice versa. This can be accomplished as follows:

$$23A_{16} = 2 \times 16^2 + 3 \times 16^1 + 10 \times 16^0$$

$$= 512 + 48 + 10 = 570_{10}$$

Note that in the equation, the value 10 is substituted for A. Now to convert 570_{10} back to $23A_{16}$:

	quotient	+	remainder
$\dfrac{570}{16} =$	35	+	A
$\dfrac{35}{16} =$	2	+	3
$\dfrac{2}{16} =$	0	+	2

$$2 \quad 3 \quad A$$

Thus, $570_{10} = 23A_{16}$.

2.2 Unsigned and Signed Binary Numbers

An unsigned binary number has no arithmetic sign. Unsigned binary numbers are therefore always positive. Typical examples are your age or a memory address which are always positive numbers. An 8-bit unsigned binary integer represents all numbers from 00_{16} through FF_{16} (0_{10} through 255_{10}).

Signed binary numbers, on the other hand, include both positive and negative numbers. The techniques used to represent the signed integers are:

* Sign-magnitude approach
* Ones complement approach
* Twos complement approach

TABLE 2.3   4-bit integers represented in Sign-magnitude Form

Four-bit integers	Interpretation using sign-magnitude
0000	+0
0001	+1
0010	+2
0011	+3
0100	+4
0101	+5
0110	+6
0111	+7
1000	−0
1001	−1
1010	−2
1011	−3
1100	−4
1101	−5
1110	−6
1111	−7

Because the sign of a number can be either positive or negative, only one bit, referred to as the *sign bit*, is needed to represent the sign. The widely used sign convention is that if the sign bit is zero, the number is positive; otherwise it is negative. (The rationale behind this convention is that the quantity $(-1)^s$ is positive when s = 0 and is negative when s = 1). Also, in all three approaches, the most significant bit of the number is considered as the sign bit.

In sign-magnitude representation, the most significant bit of the given n-bit binary number holds the sign, and the remaining $n - 1$ bits directly give the magnitude of the negative number. For example, the sign-magnitude representation of +7 is 0111 and that of -4 is 1100. Table 2.3 represents 4-bit integers and their meanings in sign-magnitude form.

In Table 2.3, the sign-magnitude approach represents a signed number in a natural manner. With 4 bits we can only represent numbers in the range $-7 \leq x \leq +7$. In general, if there are *n* bits, then we can cover all numbers in the range $\pm(2^{n-1} - 1)$. Note that with $n - 1$ bits, any value from 0 to $2^{n-1} - 1$ can be represented. However, this approach leads to a confusion because there are two representations for the number zero (0000 means +0; 1000 means -0).

In complement approach, positive numbers have the same representation as they do in the sign-magnitude representation. However, in this technique negative numbers are represented in a different manner. Before we proceed, let us define the term *complement* of a number. The complement of a number *A*, written as $\bar{A}$ (or A'), is obtained by taking bit-by-bit complement of *A*. In other words, each 0 in *A* is replaced with 1 and vice versa. For example, the complement of the number 0100_2 is 1011_2 and that of 1111_2 is 0000_2. In the ones complement approach, a negative number, -*x*, is the complement of its positive representation. For example let us find the ones complement representation of 0100_2 ($+4_{10}$). The complement of 0100 is 1011, and this denotes the negative number -4_{10}. Table 2.4 summarizes 4-bit integers and their interpretations using ones complement numbers.

TABLE 2.4   Four-bit integers represented in Ones Complement Form

Four-bit integers	interpretation using ones complement
0000	+0
0001	+1
0010	+2
0011	+3
0100	+4
0101	+5
0110	+6
0111	+7
1000	−7
1001	−6
1010	−5
1011	−4
1100	−3
1101	−2
1110	−1
1111	−0

From Table 2.4, the ones complement approach does not handle negative numbers naturally. In other words, if the number is negative (when the sign bit is 1), its magnitude is not obvious from its ones complement. To determine its magnitude, one needs to take its ones complement. For example, consider the number 110110. The most significant bit indicates that this is a negative number. Because the number is negative, its magnitude cannot be obtained by directly looking at 110110. Instead, one needs to take the ones complement of 110110 to obtain 001001. The value of 001001 as a sign-magnitude number is +9. On the other hand, 110110 represents -9 in ones complement form. Like the sign-magnitude representation, the ones complement approach does not increase the range of numbers covered by a fixed number of bit patterns. For example, 4 bits cover the range -7 to +7. The same range is obtained with sign-magnitude representation. Note that the confusion of two distinct representations for zero exists in the ones complement approach. Now, let us discuss the two's complement approach.

In this method, positive integers are represented in the same manner as they are in the sign-magnitude method. In other words, if the sign bit is zero, the number is positive and its magnitude can be directly obtained by looking at the remaining $n - 1$ bits. However, a negative number -x can be represented in twos complement form as follows:

Represent +x in sign magnitude form and call this result y

- Take the ones complement of y to get $\bar{y}$ (or y')
- $\bar{y} + 1$ is the twos complement representation of -x.

Table 2.5 lists 4-bit integers along with their twos complement forms. From Table 2.5, it can be concluded that:

- Twos complement form does not provide two representations for zero.
- Twos complement form covers up to -8 in the negative side, and this is more than can be achieved with the other two methods. In general, with n bits, and using twos complement approach, one can cover all the numbers in the range $-(2^{n-1})$ to $+(2^{n-1} - 1)$.

TABLE 2.5   Four-Bit integers represented in Twos complement Form

Four-bit Integers	Interpretation using two's complement integers
0000	0
0001	+1
0010	+2
0011	+3
0100	+4
0101	+5
0110	+6
0111	+7
1000	−8
1001	−7
1010	−6
1011	−5
1100	−4
1101	−3
1110	−2
1111	−1

Example 2.1

Represent the following decimal numbers in twos complement form. Use 7 bits to represent the numbers.
(a)   +39
(b)   -43

Solution

(a) Because the number +39 is positive, its twos complement representation is the same as its sign-magnitude representation as shown here:

$$y = \underbrace{0}_{+}\ \overset{2^5}{1}\ \overset{2^4}{0}\ \overset{2^3}{0}\ \overset{2^2}{1}\ \overset{2^1}{1}\ \overset{2^0}{1}$$
$$\underbrace{}_{39}$$

(b) In this case, the given number −43 is negative. The twos complement form of the number can be obtained as follows:

   1.   Step 1:   Represent +43 in sign magnitude form

$$y = \underbrace{0}_{+}\ \overset{2^5}{1}\ \overset{2^4}{0}\ \overset{2^3}{1}\ \overset{2^2}{0}\ \overset{2^1}{1}\ \overset{2^0}{1}$$
$$\underbrace{}_{43}$$

   2.   Step 2:   Take the ones complement of y:
$$\bar{y} = 1\ 0\ 1\ 0\ 1\ 0\ 0$$

   3.   Step 3:   Add one to $\bar{y}$ to get the final answer.

$$\begin{array}{r} 1010100 \\ +\ \underline{\qquad 1} \\ 1010101 \end{array}$$

It should be pointed out that 11111111_2 is $+255_{10}$ when interpreted as an unsigned number. On the other hand, 11111111_2 is -1_{10} when interpreted as a signed number. Note that typical 16-bit microprocessors such as Motorola 68000 have separate unsigned and signed multiplication and division instructions as follows: MULU (multiply two unsigned numbers), MULS (multiply two signed numbers), DIVU (divide two unsigned numbers), and DIVS (divide two signed numbers). It is important for the programmer to clearly understand how to use these instructions.

For example, suppose that it is desired to compute $(X^2)/255$. Now, if X is a signed 8-bit number, the programmer should use MULS instruction to compute X * X which is always unsigned (square of a number is always positive), and then use DIVU to compute $(X^2)/255$ (16-bit by 8-bit unsigned divide) since 255_{10} is positive. But, if the programmer uses DIVS, then both X * X and $255_{10}(FF_{16})$ will be interpreted as signed numbers. FF_{16} will be interpreted as -1, and the result will be wrong. On the other hand, if X is an unsigned number, the programmer needs to use MULU and DIVU to compute $(X^2)/255$.

2.3 Codes

Codes are used extensively with computers to define alphanumeric characters and other information. Some of the codes used with computers are described in the following sections.

2.3.1 Binary-Coded-Decimal Code (8421 Code)

The 10 decimal digits 0 through 9 can be represented by their corresponding 4-bit binary numbers. The digits coded in this fashion are called *binary-coded-decimal* (BCD) digits in 8421 code, or BCD digits. Two unpacked BCD bytes are usually packed into a byte to form "packed BCD." For example, two unpacked BCD bytes 02_{16} and 05_{16} can be combined as a packed BCD byte 25_{16}. The concept of packed and unpacked BCD numbers are explained later in this section. Table 2.6 provides the bit encodings of the 10 decimal numbers.

The six possible remaining 4-bit codes as shown in Table 2.6 are not used and represent invalid BCD codes if they occur.

TABLE 2.6   BCD bit encodings of the 10 decimal numbers

Decimal numbers	BCD bit encoding
0	0000
1	0001
2	0010
3	0011
4	0100
5	0101
6	0110
7	0111
8	1000
9	1001

Consider obtaining the binary and BCD equivalents of the decimal number 35 as follows:

$$
\begin{array}{cc}
\underbrace{3} & \underbrace{5} \\
\downarrow & \downarrow \\
0011 & 0101
\end{array}
$$

Note that decimal number 35 is represented as 00100011 in binary. In contrast, decimal number 35 is represented as 00110101 in BCD. It should be pointed out that it is very convenient to display binary outputs of digital systems in BCD. For example, digital systems such as an adder perform all operations in binary, and provide addition result in binary. It is very inconvenient for the users to interpret the binary result. However, converting the binary result into BCD, and then displaying the result in BCD on two seven-segment displays (in this case) can easily be interpreted by the user.

2.3.2 Alphanumeric Codes

A computer must be capable of handling nonnumeric information if it is to be very useful. In other words, a computer must be able to recognize codes that represent numbers, letters, and special characters. These codes are classified as alphanumeric or character codes. A complete and adequate set of necessary characters includes these:

- 26 lowercase letters
- 26 uppercase letters
- 10 numeric digits (0-9)
- About 25 special characters, which include + / # ×, and so on.

This totals 87 characters. To represent 87 characters with some type of binary code would require at least 7 bits. With 7 bits there are $2^7 = 128$ possible binary numbers; 87 of these combinations of 0 and 1 bits serve as the code groups representing the 87 different characters.

The 8-bit byte has been universally accepted as the data unit for representing character codes. The two common alphanumeric codes are known as the *American Standard Code for Information Interchange (ASCII)* and the *Extended Binary-Coded Decimal Interchange Code (EBCDIC)*. ASCII is typically used with computers including microcontrollers. IBM uses EBCDIC code. Although EBCDIC is an obsolete code, it will be used here for illustrative purposes. Eight bits are used to represent characters, although 7 bits suffice, because the eighth bit is frequently used to test for errors and is referred to as a *parity bit*. It can be set to 1 or 0, so that the number of 1 bits in the byte is always odd or even.

Table 2.7 shows a list of ASCII and EBCDIC codes. Some EBCDIC codes do not have corresponding ASCII codes. Note that decimal digits 0 through 9 are represented by 30_{16} through 39_{16} in ASCII. On the other hand, these decimal digits are represented by $F0_{16}$ though $F9_{16}$ in EBCDIC.

A computer program is usually written for code conversion when input/output devices of different codes are connected to the computer. For example, suppose it is desired to enter a number 5 into a computer via an ASCII keyboard and print this data on an EBCDIC printer. The ASCII keyboard will generate 35_{16} when the number 5 is pushed. The ASCII code 35_{16} for the decimal digit 5 enters into the computer and resides in the computer's memory. To print the digit 5 on the EBCDIC printer, a program must be written that will convert the ASCII code 35_{16} for 5 in its EBCDIC code $F5_{16}$. The output of this program is $F5_{16}$. This will be input to the EBCDIC

TABLE 2.7   ASCII and EBCDIC codes in Hex

Character	ASCII	EBCDIC	Character	ASCII	EBCDIC	Character	ASCII	EBCDIC	Character	ASCII	EBCDIC
@	40			60		blank	20	40	NUL	00	
A	41	C1	a	61	81	!	21	5A	SOH	01	
B	42	C2	b	62	82	"	22	7F	STX	02	
C	43	C3	c	63	83	#	23	7B	ETX	03	
D	44	C4	d	64	84	$	24	5B	EOT	04	37
E	45	C5	e	65	85	%	25	6C	ENQ	05	
F	46	C6	f	66	86	&	26	50	ACK	06	
G	47	C7	g	67	87	'	27	7D	BEL	07	
H	48	C8	h	68	88	(	28	4D	BS	08	16
I	49	C9	i	69	89	)	29	5D	HT	09	05
J	4A	D1	j	6A	91	*	2A	5C	LF	0A	25
K	4B	D2	k	6B	92	+	2B	4E	VT	0B	
L	4C	D3	l	6C	93	,	2C	6B	FF	0C	
M	4D	D4	m	6D	94	-	2D	60	CR	0D	15
N	4E	D5	n	6E	95	.	2E	4B	SO	0E	
O	4F	D6	o	6F	96	/	2F	61	SI	0F	
P	50	D7	p	70	97	0	30	F0	DLE	10	
Q	51	D8	q	71	98	1	31	F1	DC1	11	
R	52	D9	r	72	99	2	32	F2	DC2	12	
S	53	E2	s	73	A2	3	33	F3	DC3	13	
T	54	E3	t	74	A3	4	34	F4	DC4	14	
U	55	E4	u	75	A4	5	35	F5	NAK	15	
V	56	E5	v	76	A5	6	36	F6	SYN	16	
W	57	E6	w	77	A6	7	37	F7	ETB	17	
X	58	E7	x	78	A7	8	38	F8	CAN	18	
Y	59	E8	y	79	A8	9	39	F9	EM	19	
Z	5A	E9	z	7A	A9	:	3A		SUB	1A	
[	5B		{	7B		;	3B	5E	ESC	1B	
\	5C		|	7C	4F	<	3C	4C	FS	1C	
]	5D		}	7D		=	3D	7E	GS	1D	
^	5E		~	7E		>	3E	6E	RS	1E	
_	5F	6D	DEL	7F	07	?	3F	6F	US	1F	

printer. Because the printer only understands EBCDIC codes, it inputs the EBCDIC code $F5_{16}$ and prints the digit 5.

Let us now discuss packed and unpacked BCD codes in more detail. For example, in order to enter data 24 in decimal into a computer, the two keys (2 and 4) will be pushed on the ASCII keyboard. This will generate 32 and 34 (32 and 34 are ASCII codes in hexadecimal for 2 and 4 respectively) inside the computer. A program can be written to convert these ASCII codes into unpacked BCD 02 and 04, and then convert to packed BCD 24 or to binary inside the computer to perform the desired operation.

2.3.3   Excess-3 Code

The excess-3 representation of a decimal digit d can be obtained by adding 3 to its value. All decimal digits and their excess-3 representations are listed in Table 2.8. The excess-3 code is an unweighted code because its value is obtained by adding three to the corresponding binary value. The excess-3 code is self-complementing. For example, decimal digit 0 in excess-3 (0011) is ones complement of 9 in excess three (1100). Similarly, decimal digit 1 is ones complement of 8, and so on. This is why some older computers used excess three code. Conversion between excess-3 and decimal numbers is illustrated below:

Decimal number   1   9   8   3

Excess-3 Representation   0100   1100   1011   0110

TABLE 2.8   Excess-3 representation of decimal digits

Decimal digits	Excess-3 representation
0	0011
1	0100
2	0101
3	0110
4	0111
5	1000
6	1001
7	1010
8	1011
9	1100

2.3.4   Gray Code

Sometimes codes can also be constructed using a property called *reflected symmetry*. One such code is known as *Gray code*. The Gray code is used in Karnaugh maps for simplifying combinational logic design. This topic is covered in Chapter 4. Before we proceed, we briefly explain the concept of reflected symmetry. Consider the two bits 0 and 1, and stack these two bits. Assume that there is a plane mirror in front of this stack and produce the reflected image of the stack as shown in the following:

$$
\begin{array}{c}
0 \\
1 \\
\text{mirror} \leftarrow \overline{} \\
1 \\
0
\end{array}
$$

Appending a zero to all elements of the stack above the plane mirror and append a one to all elements of the stack that lies below the mirror will provide the following result:

$$
\begin{array}{rl}
\text{Appended zeros} \left\{ \begin{array}{cc} 0 & 0 \\ 0 & 1 \end{array} \right. \\
\hline
\text{Appended ones} \left\{ \begin{array}{cc} 1 & 1 \\ 1 & 0 \end{array} \right.
\end{array}
$$

Now, removal of the plane mirror will result in a stack of 2-bit Gray Code as follows:

$$
\begin{array}{cc}
0 & 0 \\
0 & 1 \\
1 & 1 \\
1 & 0
\end{array}
$$

Here, any two adjacent bit patterns differ only in one bit. For example, the patterns 11 and 10 differ only in the least significant bit. Two-bit Gray code along with correspoonding decimal and binary equivalents are provided in Figure 2.1.

Decimal	Binary	Gray code
0	00	00
1	01	01
2	10	11
3	11	10

FIGURE 2.1   Two-bit Gray code along with corresponding decimal and binary equivalents.

Repeating the reflection operation on the stack of two-bit binary patterns, a three-bit Gray code can be obtained. Two adjacent binary numbers differ in only one bit. The result is shown in Figure 2.2.

Applying the reflection process to the 3-bit Gray code, 4-bit Gray Code can be obtained. This is shown in Figure 2.3.

The Gray code is useful in instrumentation systems to digitally represent the position of a mechanical shaft. In these applications, one bit change between characters is required. For example, suppose a shaft is divided into eight segments and each shaft is assigned a number. If binary numbers are used, an error may occur while changing segment 7 (0111_2) to segment 8 (1000_2). In this case, all 4 bits need to be changed. If the sensor representing the most significant bit takes longer to change, the result will be 0000_2, representing segment 0. This can be avoided by using Gray code, in which only one bit changes when going from one number to the next.

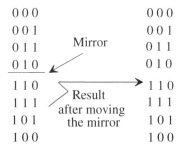

Decimal	Binary	Gray code
0	000	000
1	001	001
2	010	011
3	011	010
4	100	110
5	101	111
6	110	101
7	111	100

FIGURE 2.2   Three-bit Gray code along with corresponding decimal and binary equivalents.

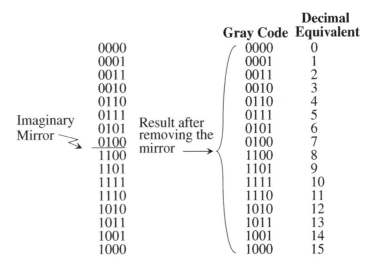

	Gray Code	Decimal Equivalent
0000	0000	0
0001	0001	1
0011	0011	2
0010	0010	3
0110	0110	4
0111	0111	5
0101	0101	6
0100	0100	7
1100	1100	8
1101	1101	9
1111	1111	10
1110	1110	11
1010	1010	12
1011	1011	13
1001	1001	14
1000	1000	15

Imaginary Mirror →   Result after removing the mirror →

Decimal	Binary	Gray code
0	0000	0000
1	0001	0001
2	0010	0011
3	0011	0010
4	0100	0110
5	0101	0111
6	0110	0101
7	0111	0100
8	1000	1100
9	1001	1101
10	1010	1111
11	1011	1110
12	1100	1010
13	1101	1011
14	1110	1001
15	1111	1000

FIGURE 2.3   Four-bit Gray code along with corresponding decimal and binary equivalents.

2.3.5   Unicode

Basically, Computers work with numbers. Note that letters and other characters are stored in computers as numbers; a number is assigned to each one of them.

Before invention of unicode, there were numerous encoding systems for assigning these numbers. It is not possible for a single encoding system to cover all the languages in the world. For example, a single encoding system was not able to assign all the letters, punctuation, and common technical symbols. Typical encoding systems can conflict with each other. For example, two different characters can be assigned with the same number in two different encoding systems. Also, different numbers can be assigned to the same character in two different encodings. These types of assignments of numbers can create problems for certain computers such as servers which need to support several different encodings. Hence, when data is transferred between different encodings or platforms, that data may be corrupted.

Unicode avoids this by assigning a unique number to each character regardless of the platform, the program, or the language. More information on Unicode can be obtained from the web site at www.unicode.org.

Unicode is a computing industry standard for the consistent encoding, representation and handling of text expressed in most of the world's writing systems. Developed in conjunction with the Universal Character Set standard and published in book form as *The Unicode Standard*, the latest version of Unicode contains a repertoire of more than 110,000 characters covering 100 scripts. The standard consists of a set of code charts for visual reference, an encoding method and set of standard character encodings, a set of reference data computer files, and a number of related items, such as character properties, rules for normalization, decomposition collation, rendering, and bidirectional display order (for the correct display of text containing both right-to-left scripts, such as Arabic and Hebrew, and left-right scripts). As of September 2013, the most recent version is *Unicode 6.3*. The standard is maintained by the Unicode Consortium.

Unicodes's success at unifying character sets has led to its widespread and predominant use in the internationalization and localization of computer software. The standard has been implemented in many recent technologies, including modern operating systems, XML, the Java programming language, and the Microsoft .NET Framework.

Unicode can be implemented by different character encodings. The most commonly used encodings are UTF-8, UTF-16 and the now-obsolete UCS-2. UTF-8 uses one byte for any ASCII characters, all of which have the same code values in both UTF-8 and ASCII encoding, and up to four bytes for other characters. UCS-2 uses a 16-bit code unit (two 8-bit bytes) for each character but cannot encode every character in the current Unicode standard. UTF-16 extends UCS-2, using one 16-bit unit for the characters that were representable in UCS-2 and two 16-bit units (4 x 8 bit) to handle each of the additional characters.

2.4 Fixed-Point and Floating-Point Representations

A number representation assuming a fixed location of the radix point is called *fixed-point representation*. The range of numbers that can be represented in fixed-point notation is severely limited. The following numbers are examples of fixed-point numbers:

$$0110.1100_2, \ 51.12_{10}, \ DE.2A_{16}$$

In typical scientific computations, the range of numbers is very large. Floating-point representation is used to handle such ranges. A floating-point number is represented as $N \times r^p$, where N is the mantissa or significand, r is the base or radix of the number system, and p is the exponent or power to which r is raised.

Some examples of numbers in floating-point notation and their fixed-point decimal equivalents are:

fixed-point numbers	floating-point representation
0.0167_{10}	0.167×10^{-1}
1101.101_2	0.1101101×2^4
$BE.2A9_{16}$	$0.BE2A9 \times 16^2$

In converting from fixed-point to floating-point number representation, we normalize the resulting mantissas, that is, the digits of the fixed-point numbers are shifted so that the highest-order nonzero digit appears to the right of the decimal point, and consequently a 0 always appears to the left of the decimal point. This convention is normally adopted in floating-point number representation. Because all numbers will be assumed to be in normalized form, the binary point is not required to be represented in the computers.

Typical 32-bit microprocessors such as the Intel 80486/Pentium and the Motorola 68040 and PowerPC contain on-chip floating-point hardware. This means that these microprocessors can be programmed using instructions to perform operations such as addition, subtraction, multiplication, and division using floating-point numbers.

2.5 Arithmetic Operations

As mentioned before, computers can only add. Therefore, all other arithmetic operations are typically accomplished via addition. All numbers inside the computer are in binary form. These numbers are usually treated internally as integers, and any fractional arithmetic must be implemented by the programmer in the program. The arithmetic and logic unit (ALU) in the computer's CPU performs typical arithmetic and logic operations. The ALUs perform function such as addition, subtraction, magnitude comparison, ANDing, and ORing of two binary or packed BCD numbers. The procedures involved in executing these functions are discussed now to provide an understanding of the basic arithmetic operations performed in a typical microcontroller. The logic operations are covered in Chapter 3.

2.5.1 Binary Arithmetic

Addition The addition of two binary numbers is carried out in the same way as the addition of decimal numbers. However, only four possible combinations can occur when adding two binary digits (bits):

augend	+	addend	=	carry	sum	decimal value
0	+	0	=	0	0	0
0	+	1	=	0	1	1
1	+	0	=	0	1	1
1	+	1	=	1	0	2

The following are some examples of binary addition. The corresponding decimal additions are also included.

$$
\begin{array}{ll}
010 & (2) \\
+\ 011 & (3) \\
\hline
101 & (5)
\end{array}
\qquad
\begin{array}{ll}
111 \leftarrow \text{carry} & \\
101.11 & (5.75) \\
+\ 011.10 & (3.50) \\
\hline
1\ 001.01 & (9.25)
\end{array}
$$

final carry

Addition is the most important arithmetic operation in microprocessors because the operations of subtraction, multiplication, and division as they are performed in most modern digital computers use only addition as their basic operation.

The addition of two unsigned numbers is performed in the same way as illustrated above. Also, the addition of two numbers in the sign-magnitude form is performed in the same manner as ordinary arithmetic. For example, if both numbers have the same signs, the two numbers are added and the common sign is assigned to the result. On the other hand, if the numbers have opposite signs, the number with smaller magnitude is subtracted from the number with larger magnitude and the result is assigned with the sign of the number with larger magnitude. For example, (-14) + (+18) = + (18 - 14) = +4. This is performed by subtracting the smaller magnitude 14 from the higher magnitude 18 and the sign of the larger magnitude 18 (+ in this case) is assigned to the result. The same rules apply to binary numbers in sign-magnitude form.

Subtraction   As mentioned before, computers can usually only add binary digits; they cannot directly subtract. Therefore, the operation of subtraction in microcontrollers is performed using addition with complement arithmetic. In general, the b's complement of an m-digit number, M is defined as $b^m - M$ for $M \neq 0$ and 0 for $M = 0$. Note that for base 10, $b = 10$ and 10^m is a decimal number with a 1 followed by m 0's. For example, 10^4 is 10000; 1 followed by four 0's. On the other hand, $b = 2$ for binary and 2^m indicates 1 followed by m 0's. For example, 2^3 means 1000 in binary.

The (b -1)'s complement of an m-digit number, M is defined as $(b^m - 1) - M$. Therefore, the b's complement of an m-digit number, M can be obtained by adding 1 to its (b - 1)'s complement. Next, let us illustrate the concept of complement arithmetic by means of some examples. Consider a 4-digit decimal number, 5786. In this case, $b = 10$ for base 10 and $m = 4$ since there are four digits.

$$10\text{'s complement of } 5786 = 10^4 - 5786 = 10000 - 5786 = 4214$$

Now, let us obtain 10's complement of 5786 using (10-1)'s or 9's complement arithmetic as follows:

$$9\text{'s complement of } 5786 = (10^4 - 1) - 5786 = 9999 - 5786 = 4213$$

Hence, 10's complement of 5786 = 9's complement of 5786 + 1 = 4213 + 1 = 4214.

Next, let us determine the 2's complement of a three-bit binary number, 010 via its one's complement.

In this case, $b = 2$ for binary and $m = 3$ since there are three bits in the number.

The 2's complement of 010 (+2) can be obtained using its 1's complement as follows:

$$1\text{'s complement of } 010 = (2^3 - 1) - 010 = 111 - 010 = 101$$

$$2\text{'s complement of } 101 = 101 + 1 = 110$$

Now, let us find the decimal equivalent of 110_2. Since the most significant bit of 110 is 1, the two's complement of 110 must be taken. Hence, $110_2 = -2_{10}$. Therefore, two's complement of a number negates the number being complemented. This will be explained later in this section.

From the above procedure for obtaining the 1's complement of 010, it can be concluded that the 1's complement of a binary number can be achieved by subtracting each bit of the binary number from 1. This means that when subtracting a bit (0 or 1) from 1, one can have either $1 - 0 = 1$ or $1 - 1 = 0$; that is, the 1's complement of 0 is 1 and the 1's complement of 1 is 0. In general, the 1's complement of a binary number can be obtained by changing 0's to 1's and 1's to 0's.

The procedure for performing X-Y (both X and Y are in base 2) using 1's complement can be performed as follows:

1. Add the minuend X to the 1's complement of the subtrahend Y.

2. Check the result in step 1 for a carry. If there is a carry, add 1 to the least significant bit to obtain the result. If there is no carry, take the 1's complement of the number obtained in step 1 and place a negative sign in front of the result.

For example, consider two 6-bit numbers (arbitrarily chosen), $X = 010011_2 = 19_{10}$ and $Y = 110001_2 = 49_{10}$. X-Y= 19 - 49 = -30 in decimal. The operation X-Y using 1's complement can be performed as follows:

$$X = 010011$$

Add 1's complement of Y = 001110

100001

Since there is no carry, Result = - (1's Complement of 100001) = $-011110_2 = -30_{10}$.

Next consider, $X = 101100_2 = 44$ and $Y = 011000_2 = 24_{10}$. In decimal, X-Y = 44 -24 = 20.

Using 1's complement, X-Y can be obtained as follows: X = 101100

Add 1's Complement of Y = 100111

Carry →1 010011

Since there is a carry, Result = 010011 + 1 = $+010100_2 = + 20_{10}$.

Next, let us describe the procedure of subtracting decimal numbers using addition. This process requires the use of the 10's complement form. The 10's complement of a number can be obtained by subtracting the number from 10.

Consider the decimal subtraction 7 - 4 = 3. The 10's complement of 4 is 10 - 4 = 6. The decimal subtraction can be performed using the 10's complement addition as follows:

$$
\begin{array}{rr}
\text{minuend} & 7 \\
\text{10's complement of subtrahend} & + 6 \\
\hline
& 13
\end{array}
$$

Ignore final carry of 1 to obtain

the subtraction result of 3.

When a larger number is subtracted from a smaller number, there is no carry to be discarded. Consider the decimal subtraction 47 = 3. The 10's complement of 7 is 10-7 = 3. Therefore,

$$
\begin{array}{rr}
\text{minuend} & 4 \\
\text{10's complement of subtrahend} & + 3 \\
\hline
& 7
\end{array}
$$

no final carry

When there is no final carry, the final answer is the negative of the 10's complement of 7. Therefore, the correct result of subtraction is - (10-7) = -3.

The same procedures can be applied for performing binary subtraction. In the case of binary subtraction, the twos complement of the subtrahend is used.

As mentioned before, the twos complement of a binary number is obtained by replacing each 0 with a 1 and each 1 with a 0 and adding 1 to the resulting number. The first step generates a ones complement or simply the complement of a binary number.

For example, the ones complement of 10010101 is 01101010. Note that the ones complement of a binary number can be obtained by using inverters; eight inverters are required for generating ones complement of an 8-bit number.

The twos complement of a binary number is formed by adding 1 to the ones complement of the number. For example, the twos complement of 10010101 is found as follows:

$$
\begin{array}{rr}
\text{binary number} & 10010101 \\
\text{1's complement} & 01101010 \\
\text{add 1} & +\ 1 \\
\hline
\text{2's complement} & 01101011
\end{array}
$$

Now, using the twos complement, binary subtraction can be carried out.

Consider the following subtraction using the normal (pencil and paper) procedure:

$$
\begin{array}{rrl}
\text{minuend} & 0101 & (5) \\
\text{subtrahend} & -0011 & (-3) \\
\hline
\text{result} & 0010_2 & = 2_{10}
\end{array}
$$

Using the twos complement subtraction,

$$
\begin{array}{rr}
\text{minuend} & 0101 \\
\text{2's complement of subtrahend} & 1101 \\
\hline
& 0010
\end{array}
$$

Final Carry = 1

The final answer is 0010 (decimal 2) by discarding the final carry as shown with decimal subtraction using 10's complement. One can determine the sign of the result by inspecting the most significant bit (sign bit) of the result. Since the sign bit of the result, 0010 is 0, the rsult in decimal is +2.

Consider another example provided below. Using pencil and paper method:

$$
\begin{array}{rrl}
\text{minuend} & 0101 & (5) \\
\text{subtrahend} & -\ 0110 & (-6) \\
\hline
\text{result} & -\ 0001 & (-1)
\end{array}
$$

Using the twos complement,

$$
\begin{array}{rr}
\text{minuend} & 0101 \\
\text{Add 2's complement of subtrahend} & 1010 \\
\hline
\text{result} & 1111
\end{array}
$$

Since the sign bit of 1111 is 1, the final answer is -(twos complement of 1111) = -1_{10}.

Computers including microcontrollers use two's complement of a binary number to represent its negative value, and obtain the sign of the result from the most significant bit. Hence, microcontrollers typically handle signed numbers by using the most significant bit of a number as the sign bit. If this bit is zero, the number is positive; if this bit is one, the number is negative. This convention will be used in this book.

For example, the number $+22_{10}$ can be represented using 8 bits as:

$$+22_{10} = \underbrace{0}00101110_2$$

$\uparrow$
sign bit
(positive)

Hence, twos complement of $+22_{10} = -22_{10} = \underbrace{1}1101010_2$

$\uparrow$
sign bit
(negative)

Next, the concept of overflow will be explained. An error (indicated by overflow in a microcontroller) may occur while performing twos complement arithmetic. The overflow arises from the representation of the sign flag by the most significant bit of a binary number in signed binary operation. The computer automatically sets an overflow bit to 1 if the result of an arithmetic operation is too big for the computer's maximum word size; otherwise it is reset to 0. To clearly understand the concept of overflow, consider the following examples for 8-bit numbers. Let C_7 be the carry out of the most significant bit (sign bit) and C_6 be the carry out of the previous (bit 6) data bit (seventh bit). We will show by means of numerical examples that as long as C_7 and C_6 are the same, the result is always correct. If, however, C_7 and C_6 are different, the result is incorrect and sets the overflow bit to 1. Now consider the following cases.

Case 1: C_7 and C_6 are the same.

$$
\begin{array}{ll}
0\,0\,0\,0\,0\,1\,1\,0 & 06_{16} \\
0\,0\,0\,1\,0\,1\,0\,0 & +14_{16} \\ \hline
0\ \ 0\,0\,0\,1\,1\,0\,1\,0 & 1A_{16}
\end{array}
$$

$C_7 = 0$

$C_6 = 0$

$$
\begin{array}{ll}
0\,1\,1\,0\,1\,0\,0\,0 & 68_{16} \\
1\,1\,1\,1\,1\,0\,1\,0 & -06_{16} \\ \hline
1\ \ 0\,1\,1\,0\,0\,0\,1\,0 & 62_{16}
\end{array}
$$

$C_7 = 1$

$C_6 = 1$

Therefore when C_7 and C_6 are either both 0 or both 1, a correct answer is obtained.

Case 2: C_7 and C_6 are different.

$$
\begin{array}{ll}
0\,1\,0\,1\,1\,0\,0\,1 & 59_{16} \\
0\,1\,0\,0\,0\,1\,0\,1 & +45_{16} \\ \hline
0\ \ 1\,0\,0\,1\,1\,1\,1\,0 & -62_{16}\ ?
\end{array}
$$

$C_7 = 0$

$C_6 = 1$

$C_6 = 1$ and $C_7 = 0$ give an incorrect answer because the result shows that the addition of two positive numbers is negative.

$$
\begin{array}{c}
1\,0\,1\,1\,0\,1\,1\,0 \qquad -4A_{16} \\
\underline{1\,0\,0\,0\,0\,0\,0\,1} \qquad \underline{-7F_{16}} \\
1\quad 0\,0\,1\,1\,0\,1\,1\,1 \qquad +37_{16}\ ?
\end{array}
$$

$C_7 = 1$ $C_6 = 0$

$C_6 = 0$ and $C_7 = 1$ provide an incorrect answer because the result indicates that the addition of two negative numbers is positive. Hence, the overflow bit will be set to zero if the carries C_7 and C_6 are the same, that is, if both C_7 and C_6 are either 0 or 1. On the other hand, the overflow flag will be set to 1 if the carries C_7 and C_6 are different. The answer is incorrect when the overflow bit is set to 1. Thus, Overflow $= C_7 \oplus C_6$. In general, Overflow $= C_f \oplus C_p$.

Note that the symbol $\oplus$ represents exclusive-OR logic operation. Exclusive-OR means that when two inputs are the same (both one or both zero), the output is zero. On the other hand, if two inputs are different, the output is one. The overflow can be considered as the output while C_p (Previous Carry) and C_f (Final Carry) are the two inputs. The exclusive-OR operation is covered in Chapter 3.

While performing signed arithmetic using pencil and paper, one must consider the overflow bit to ensure that the result is correct. An overflow of one after a signed operation indicates that the result is too large to be accommodated in the number of bits assigned. One must increase the number of bits for the correct result.

Example 2.2

Perform the following signed operations and comment on the results. Assume twos complement numbers.

(a) $A = 1010_2$, $B = 0111_2$. Find $A - B$.
(b) Perform $(-3_{10}) - (-2_{10})$ using twos complement and 4 bits.

Solution

(a) The most significant bit of A is 1, so A is a negative number whereas B is a positive number.

$$
\begin{array}{r}
A = \quad 1\,0\,1\,0 \qquad (-6_{10}) \\
\text{Add 2's complement of } B = +\ 1\,0\,0\,1 \qquad -(+7_{10}) \\
\hline
0\,0\,1\,1 = 3_{10} \qquad -13_{10}
\end{array}
$$

$C_3 = 1$ $C_2 = 0$

Because C_3 and C_2 are different, there is an overflow and the result is incorrect. Four bits are too small to hold the correct answer. If we increase the number of bits for A and B to 5, the correct result can be obtained as follows:

$$A = -6 = 11010_2$$
$$B = +7 = 00111_2$$

$$
\begin{array}{rl}
A = & 1\,1\,0\,1\,0_2 \\
\text{Add 2's complement of B} = + & 1\,1\,0\,0\,1_2 \\
\hline
& 1\,0\,0\,1\,1_2
\end{array}
$$

$C_4 = 1 \quad \longleftarrow \quad C_3 = 1$

The result is correct because C_4 and C_3 are the same. The most significant bit of the result is 1. This means that the result is negative. Therefore, to express the result in base-10, one must take the twos complement and convert the binary number to decimal and place a negative sign in front of it. Thus, twos complement of $10011_2 = -01101 = -13_{10}$.

(b)

$$
\begin{array}{rll}
& -3_{10} = & 1\,1\,0\,1\,_2 & (-3_{10}) \\
\text{Add 2's complement of } -2_{10} = & +0\,0\,1\,0\,_2 & -(-2_{10}) \\
\hline
& 1\,1\,1\,1 & -1_{10}
\end{array}
$$

$C_3 = 0 \quad \longleftarrow \quad C_2 = 0$

C_2 and C_3 are the same, so the result is correct. The most significant bit of the result is 1. This means that the result is negative. To find the result in decimal, one must take twos complement of the result and place a negative sign in front of it.
Twos complement of $1111_2 = -1_{10}$

Unsigned Multiplication Multiplication of two binary numbers can be carried out in the same way as is done with the decimal numbers using pencil and paper. Consider the following example:

$$
\begin{array}{lll}
\text{Multiplicand} & \longrightarrow 0110 & (6_{10}) \\
\text{Multiplier} & \longrightarrow 0101 \times & (5_{10}) \\
\hline
& 0110 & \\
& 0000 & \\
& 0110 & \text{partial products} \\
& 0000 & \\
\hline
& 0011110 & (30_{10}) \\
\end{array}
$$

Final product

Several multiplication algorithms are available. Multiplication of two unsigned numbers can be accomplished via repeated addition. For example, to multiply 4_{10} by 3_{10}, the number 4_{10} can be added twice to itself to obtain the result, 12_{10}.

Signed Multiplication Signed multiplication can be performed using various algorithms. A simple algorithm follows. Assume that M (multiplicand) and Q (multiplier) are in two's-complement form. Also, assume that M_n and Q_n are the most significant bits (sign bits) of the multiplicand (M) and the multiplier (Q) respectively. The sign bit of the product is determined as $M_n \oplus Q_n$, where M_n and Q_n are the most significant bits (sign bits) of the multiplicand (M) and the multiplier (Q), respectively. To perform signed multiplication, proceed as follows:

1. If $M_n = 1$, compute the twos complement of M; else, keep M unchanged.
2. If $Q_n = 1$, compute the twos complement of Q; else, keep Q unchanged.
3. Multiply the $n - 1$ bits of the multiplier and the multiplicand using unsigned multiplication.
4. The sign of the result, $S_n = M_n \oplus Q_n$.
5. If $S_n = 1$, compute the two's-complement of the result obtained in step 3.

Next, consider a numerical example. Assume that M and Q are two's-complement numbers. Suppose that $M = 1100_2$ and $Q = 0111_2$. Because $M_n = 1$, take the two's-complement of $M = 0100_2$; because $Q_n = 0$, do not change Q. Multiply 0111_2 and 0100_2 using the unsigned multiplication method discussed before. The product is 00011100_2. The sign of the product $S_n = M_n \oplus Q_n = 1 \oplus 0 = 1$. Hence, take the two's-complement of the product 00011100_2 to obtain 11100100_2, which is the final answer: -28_{10}.

Unsigned Division Binary division is carried out in the same way as the division of decimal numbers. As an example, consider the following division:

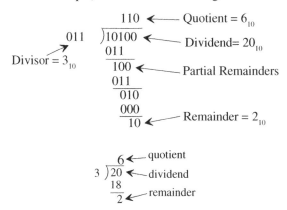

Division between unsigned numbers can be accomplished via repeated subtraction. For example, consider dividing 7_{10} by 3_{10} as follows:

Dividend	Divisor	Subtraction result	Counter
7_{10}	3_{10}	$7 - 3 = 4$	1
		$4 - 3 = 1$	$1 + 1 = 2$

Quotient = Counter value = 2
Remainder = subtraction result = 1

Here, one is added to a counter whenever the subtraction result is greater than the divisor. The result is obtained as soon as the subtraction result is smaller than the divisor.

Signed Division   Signed division can be performed using various algorithms. A simple algorithm follows. Assume that DV (Dividend) and DR (Divisor) are in two's-complement form. For the first case, perform unsigned division using repeated subtraction of the magnitudes without the sign bits. The sign bit of the quotient is determined as $DV_n \oplus DR_n$, where DV_n and DR_n are the most significant bits (sign bits) of the dividend (DV) and the divisor (DR) respectively. To perform signed division, proceed as follows:

1.   If $DV_n = 1$, compute the twos complement of DV, else keep DV unchanged.
2.   If $DR_n = 1$, compute the twos complement of DR, else keep DR unchanged.
3.   Divide the n - 1 bits of the dividend by the divisor using unsigned division algorithm (repeated subtraction).
4.   The sign of the Quotient, $Q_n = DV_n \oplus DR_n$. The sign of the remainder is the same as the sign of the dividend unless the remainder is zero. The following numerical examples illustrate this:
      The general equation for division can be used for signed division. Note that the general equation for division is *dividend = quotient * divisor + remainder*. For example, consider dividend = -9, divisor = 2. Three possible solutions are shown below:
      (a) $-9 = -4 * 2 - 1$, Quotient = -4, Remainder = -1.
      (b) $-9 = -5 * 2 + 1$, Quotient = -5, Remainder = $+1$.
      (c) $-9 = -6 * 2 + 3$, Quotient = -6, Remainder = $+3$.
      However, the correct answer is shown in (a), in which, the Quotient = -4 and the remainder = -1. Hence, for signed division, the sign of the remainder is the same as the sign of the dividend, unless the remainder is zero.
5.   If $Q_n = 1$, compute the two's-complement of the quotient obtained in step 3, else keep the quotient unchanged.

The above algorithm will be verified using numerical examples provided in the following:.

i)   Signed division with zero remainder

Assume 4-bit numbers

Dividend = $+6 = 0110_2$ Divisor = -2 = Twos complement of $2 = 1110_2$

Since, the sign bit of Dividend is 0, do not change dividend. Because the sign bit of divisor is 1, take 2's complement of 1110 which is 0010. Now, divide 0110 by 0010 using repeated subtraction as follows:

DIVIDEND	DIVISOR	SUBTRACTION RESULT USING 2'S COMPLEMENT	COUNTER
0110	0010	0110-0010=0100	0001
		0100-0010=0010	0010
		0010-0010=0000	0011

Result of unsigned division: Quotient = Counter value = 0011_2,
               Remainder = Subtraction result = 0000_2
Result of signed division 6 (0110) divided by -2 (1110):
Sign of the quotient = (Sign of dividend)$\oplus$(Sign of divisor) = $0 \oplus 1 = 1$
Hence, Quotient = 2's complement of $0011_2 = 1101 = -3_{10}$, Remainder = 0000_2

ii)   Signed division with nonzero remainder

Assume 4-bit numbers

Dividend = -5 = Twos complement of $0101_2 = 1011_2$ Divisor = -2 = Twos complement of $2 = 1110_2$

Since, the sign bit of Dividend is 1, take 2's complement of 1011 which is 0101. Because the sign bit of divisor is 1, take 2's complement of 1110 which is 0010. Now, divide 0101 by 0010 using repeated subtraction as follows:

DIVIDEND	DIVISOR	SUBTRACTION RESULT USING 2'S COMPLEMENT	COUNTER
0101	0010	0101-0010=0011	0001
		0011-0010=0001	0010

Result of unsigned division: Quotient = Counter value = 0010_2,
                              Remainder = Subtraction result = 0001_2
Result of signed division -5 (1011) divided by -2 (1110):
Sign of the quotient = (Sign of dividend)$\oplus$(Sign of divisor) = $1 \oplus 1 = 0$. Hence, do not take twos complement of Quotient.
Quotient $= 0010 = +2_{10}$ Remainder has the same sign as the dividend which is negative (bit 3 = 1). Hence, Remainder = 2's complement of $0001_2 = 1111_2 = -1_{10}$.
          Next, using decimal numbers, it will be verified why the sign of the remainder is the same as the sign of the dividend unless the remainder is 0. This can be verified by the following numerical examples using decimal numbers:

Case 1: when the remainder is 0

i) Assume both dividend and divisor are positive
Dividend = +6 Divisor = +2
Result: Quotient = +3 Remainder = 0

ii) Assume dividend is negative and divisor is positive.
Dividend = -6 Divisor = +2
Result: Quotient = -3 Remainder = 0

iii) Assume dividend is positive and divisor is negative.
Dividend = +6 Divisor = -2
Result: Quotient = -3 Remainder = 0

iv) Assume both dividend and divisor are negative.
Dividend = -6 Divisor = -2
Result: Quotient = +3 Remainder = 0

Case 2: when the remainder is nonzero

Since, Dividend = Quotient x Divisor + Remainder
Hence, Remainder = Dividend - Quotient x Divisor.

i) Assume both dividend and divisor are positive.

Dividend = +5 Divisor = +2

Result: Quotient = +2, Remainder can be obtained from the equation,

Remainder = Dividend - Quotient x Divisor. Hence, Remainder = +5 - (+2 x +2) = +1.
ii) Assume dividend is negative and divisor is positive.

Dividend = -5 Divisor = +2

Result: Quotient = -2 Remainder can be obtained from the equation, Remainder = Dividend - Quotient x Divisor. Hence, Remainder = -5 - (-2 x +2) = - 1.

iii) Assume dividend is positive and divisor is negative.

Dividend = +5 Divisor = -2

Result: Quotient = -2 Remainder can be obtained from the equation, Remainder = Dividend - Quotient x Divisor. Hence, Remainder = +5 - (-2 x --2) = + 1.

iv) Assume both dividend and divisor are negative.

Dividend = -5 Divisor = -2

Result: Quotient = +2 Remainder can be obtained from the equation, Remainder = Dividend - Quotient x Divisor. Hence, Remainder = -5 - (+2 x -2) = - 1.

From above, the sign of the remainder is the same as the sign of the dividend unless the remainder is zero.

2.5.2 BCD Arithmetic

Many computers have instructions to perform arithmetic operations using packed BCD numbers. Next, we consider some examples of packed BCD addition and subtraction.

BCD Addition The two cases that may occur while adding two packed BCD numbers are considered next. Consider adding packed BCD numbers 25 and 33:

25	0010	0101
+33	0011	0011
58	0101	1000

In this example, none of the sums of the pairs of decimal digits exceeded 9; therefore, no decimal carries were produced. For these reasons, the BCD addition process is straightforward and is actually the same as binary addition.

Now, consider the addition of 8 and 4 in BCD

8	0000	1000	
+4	0000	0100	
12	0000	1100	← invalid code group for BCD

The sum 1100 does not exist in BCD code. It is one of the six forbidden or invalid 4-bit code groups. This has occurred because the sum of two digits exceeds 9. Whenever this occurs, the sum has to be corrected by the addition of 6 (0110) to skip over the six invalid code groups.

For example,

$$
\begin{array}{cccc}
8 & 0000 & 1000 & \\
+4 & 0000 & 0100 & \\
\hline
12 & 0000 & 1100 & \text{invalid sum} \\
& +0000 & 0110 & \text{add 6 for correction} \\
\hline
& \underbrace{0001}_{1} & \underbrace{0010}_{2} & \text{BCD for 12} \\
\end{array}
$$

As another example, add packed BCD numbers 56 and 81:

$$
\begin{array}{cccc}
56 & 0101 & 0110 & \text{BCD for 56} \\
+81 & 1000 & 0001 & \text{BCD for 81} \\
\hline
137 & 1101 & 0111 & \text{invalid sum in 2nd digit} \\
& +0110 & & \text{add 6 for correction} \\
\hline
\underbrace{0001}_{1} & \underbrace{0011}_{3} & \underbrace{0111}_{7} & \leftarrow \text{correct answer 137} \\
\end{array}
$$

Therefore, it can be concluded that addition of two BCD digits is correct if the binary sum is less than or equal to 1001_2 (9 in decimal). A binary sum greater than 1001_2 results into an invalid BCD sum; adding 0110_2 to an invalid BCD sum provides the correct sum with an output carry of 1. Furthermore, addition of two BCD digits (each digit having a maximum value of 9) along with carry will provide a sum not exceeding 19 in decimal (10011_2). Hence, a correction is necessary for the following:

i)   If the binary sum is greater than or equal to decimal 16 (This will generate a carry of one)

ii)   If the binary sum is 1010_2 through 1111_2.

For example, consider adding packed BCD numbers 97 and 39:

$$
\begin{array}{cccc}
& 111 & \leftarrow \text{Intermediate Carries} & \\
97 & 1001 & 0111 & \text{BCD for 97} \\
+39 & 0011 & 1001 & \text{BCD for 39} \\
\hline
136 & 1101 & 0000 & \text{invalid sum} \\
& +0110 & +0110 & \text{add 6 for correction} \\
\hline
\underbrace{0001}_{1} & \underbrace{0011}_{3} & \underbrace{0110}_{6} & \leftarrow \text{correct answer 136} \\
\end{array}
$$

BCD Subtraction   Subtraction of packed BCD numbers can be accomplished in a number of different ways. One method is to add the 10's complement of the subtrahend to the minuend using packed BCD addition rules, as described earlier.

One way of finding the 10's complement of a d-digit packed BCD number N is to take the twos complement of each digit individually, producing a number N_1. Then, ignoring any carries, add the d-digit factor M to N_1, where the least significant digit of M is 1010 and all remaining digits of M are 1001.

As an example, consider subtracting 26_{10} from 84_{10} using BCD subtraction. This can be accomplished as follows:

26_{10}	$\underbrace{0010}_{2}$	$\underbrace{0110}_{6}$

Now, the 10's complement of 26_{10} can be found according to the rules by individually determining the twos complement of 2 and 6, adding the 10's complement factor, and discarding any carries. The twos complement of 2 is 1110, and the twos complement of 6 is 1010.

Therefore,

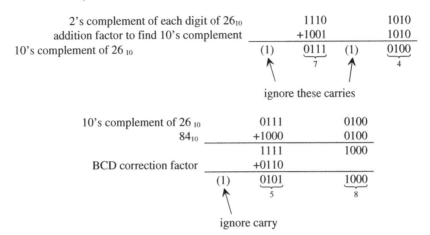

Therefore, the final answer is 58_{10}.

2.6 Error Correction and Detection

In digital systems, it is possible that the transmitted information is not received correctly. Note that a computer is a digital system in which information transfer can take place in many ways. For example, data may be moved from a CPU register to another device or vice versa. When the transmitted data is not received correctly at the receiving end, an error occurs. One possible cause for such errors is noise problems during transmission. To avoid these problems, error detection and correction may be necessary. In a digital system, an error occurs when a 0 is changed to a 1 and vice versa. Correction of this error means replacement of a 1 with 0 and vice versa. The reliability of digital data depends on the methods employed for error detection and correction.

The simplest way to detect the presence of an error is by adding a single bit, called the *parity* bit, to the message bits and then transmitting the message along with the parity bit. The parity bit is usually computed in two ways: even parity and odd parity. In the even parity method, the parity bit is added in such a way that after its inclusion, the number of 1's in the message together with the parity bit is an even number. On the other hand, in an odd parity scheme, the parity bit is added in such a way that the number of 1's in the message and the parity bit is an odd number. For example, suppose a message to be transmitted is 0110. If even parity is used by the transmitting computer, the transmitted data along with the parity will be 00110. On the other hand, if odd parity is used, the data to be transmitted will be 10110. The parity computation can be implemented in hardware by using exclusive-OR gates (to be discussed in Chapter 3). Usually for a given message, the parity bit is generated using either an even or odd parity scheme by the transmitting computer. The message is then transmitted along with the parity bit. At the receiving end, the parity is checked by the receiving computer. If there is a discrepancy, the data received will obviously be incorrect. For example, suppose that the message bits are 1101. The even parity bit for this message is 1. The transmitted data will be

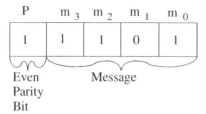

Suppose that an error occurs in the least significant bit; that is m_0 is changed from 1 to 0 during transmission. The received data will be

1	1	1	0	0

The receiving computer performs parity check on this data by counting the number of ones and finds it to be an odd number, three. Therefore, an error is detected.

With a single parity bit, an error due to a single bit change can be detected. Errors due to 2-bit changes during transmission will go undetected. In such situations, multiple parity bits are used. One such technique is the "Hamming code," which uses 3 parity bits for a 4-bit message. The extended Hamming code with additional Parity bits is popular in computer memory sytems where it is known as SECDED (*Single Error Correction, Double Error Detection*). The SECDED code uses 8-bit data plus five parity bits.

Questions and Problems

2.1 Convert the following unsigned binary numbers into their decimal equivalents:
(a) 01110101_2
(b) 1101.101_2
(c) 1000.111_2

2.2 Convert the following numbers into binary:
(a) 152_{10}
(b) 343_{10}

2.3 Convert the following numbers into octal:
(a) 1843_{10}
(b) 1766_{10}

2.4 Convert the following numbers into hexadecimal
(a) 1987_{10}
(b) 3072_{10}

2.5 Convert the following binary numbers into octal and hexadecimal numbers:
(a) 1101011100101
(b) 11000011100110000011

2.6 Using 8 bits, represent the integers -48 and 52 in
(a) sign magnitude form
(b) ones complement form
(c) twos complement form

2.7 Identify the following unsigned binary numbers as odd or even without converting them to decimal: 11001100_2; 00100100_2; 01111001_2.

2.8 Convert 532.372_{10} into its binary equivalent.

2.9 Convert the following hex numbers to binary: $15FD_{16}$; $26EA_{16}$.

2.10 Provide the BCD bit encodings for the following decimal numbers:
(a) 11264
(b) 8192

2.11 Represent the following numbers in excess-3:
(a) 678
(b) 32874
(c) 61440

2.12 What is the excess-3 equivalent of octal 1543?

2.13   Represent the following binary numbers in BCD:
        (a)   0001 1001 0101 0001
        (b)   0110 0001 0100 0100 0000

2.14   Express the following binary numbers into excess-3:
        (a)   0101 1001 0111
        (b)   0110 1001 0000

2.15   Perform the following unsigned binary addition. Include the answer in decimal.

$$1\ 0\ 1\ 1 \bullet 0\ 1$$
$$+\ 0\ 1\ 1\ 0 \bullet 0\ 1\ 1$$

2.16   Perform the indicated arithmetic operations in binary. Assume that the numbers are in decimal and represented using 8 bits. Express results in decimal. Use twos complement approach for carrying out all subtractions.

        (a)   14   (c)   32
                 +17   −14

        (b)   34   (d)   34
                 +28   −42

2.17   Using twos complement, perform the following subtraction: $3AFA_{16} - 2F1E_{16}$. Include answer in hex.

2.18   Using 9's and 10's complement arithmetic, perform the following arithmetic operations:
        (a) $254_{10} - 132_{10}$
        (b) $783_{10} - 807_{10}$

2.19   Perform the following arithmetic operations in binary using 6 bits. Assume all numbers are signed decimal. Use twos complement arithmetic. Indicate if there is any overflow.

        (a)   14   (b)   7   (c)   27
                 + 8   +(−7)   +(−19)

        (d)   (−24)   (e)   19   (f)   (−17)
                +(−19)   −(−12)   −(−16)

2.20   Perform the following unsigned multiplication in binary using a minimum number of bits required for each decimal number using pencil and paper method:

$$12 \times 52$$

2.21   Perform the following unsigned division in binary using a minimum number of bits required for each decimal number:

$$3 \overline{\smash{\big)}\ 14}$$

2.22 Obtain the bit encodings of the following decimal numbers and then perform the indicated arithmetic operations using BCD:

      (a)   54   (b)   782   (c)   82
              +48   +219   −58

2.23 Obtain the bit encodings of the following decimal numbers and then perform addition using BCD:
999 + 999

2.24 Find the odd parity bit for the following binary message to be transmitted: 10110000.

2.25 Repeat Problem 2.20 using repeated addition.

2.26 Repeat Problem 2.21 using repeated subtraction.

2.27 Assume 2 two's-complement signed numbers, $M = 11111111_2$ and $Q = 1111100_2$. Perform signed multiplication using the algorithm described in Section 2.5.1.

2.28 Using the signed division algorithm described in section 2.5.1, find the quotient and remainder of (-25)/3.

2.29 If a transmitting computer sends an 8-bit binary message 11000111 using an even parity bit. Write the 9-bit data with the parity bit in the most significant bit. If the receiving computer receives the 9-bit data as 110000111, is the 8-bit message received correctly? Comment.

2.30 Perform the following binary addition operation; assume that the numbers are in two's complement form. Express result in decimal. Determine sign and overflow flags.

$$(0000\ 1111) + (\ 1111\ 1100)$$

3

BOOLEAN ALGEBRA AND DIGITAL LOGIC GATES

This chapter describes fundamentals of logic operations, Boolean algebra, minimization techniques, and implementation of basic digital circuits.

Digital circuits contain hardware elements called *gates* that perform logic operations on binary numbers. Devices such as transistors can be used to perform the logic operations. Boolean algebra is a mathematical system that provides the basis for these logic operations. George Boole, an English mathematician, introduced this theory of digital logic. The term *Boolean variable* is used to mean the two-valued binary digit 0 or 1.

3.1 Basic Logic Operations

Boolean algebra uses three basic logic operations namely, NOT, OR, and AND. These operations are described next.

3.1.1 NOT Operation

The NOT operation inverts or provides the ones complement of a binary digit. This operation takes a single input and generates one output. The NOT operation of a binary digit provides the following result:

$$\text{NOT } 0 = 1$$
$$\text{NOT } 1 = 0$$

Therefore, NOT of a Boolean variable A, written as $\overline{A}$ (or A') is 1 if and only if A is 0. Similarly, A is 0 if and only if A is 1. This definition may also be specified in the form of a truth table:

Input	Output
A	$\overline{A}$
0	1
1	0

The truth table contains the inputs and outputs of a digital logic circuit. The truth table is used to verify the correct operation of the digital circuit. Note that $\overline{A}$ will be used in this book to represent NOT A.

The symbolic representation of an electronic circuit that implements a NOT operation is shown in Figure 3.1. The NOT gate is also referred to as an *inverter* because

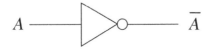

FIGURE 3.1   Symbol for a NOT gate.

Fundamentals of Digital Logic and Microcontrollers, Sixth Edition. M. Rafiquzzaman.
© 2014 John Wiley & Sons, Inc. Published 2014 by John Wiley & Sons, Inc.

it inverts the voltage levels. As discussed in Chapter 1, a transistor acts as an inverter. A 0-volt at the input generates a 5-volt output; a 5-volt input provides a 0-volt output.

The 74HC04 (or 74LS04) is a commercially available 14-pin inverter chip. It contains six independent inverters in the same chip as shown in Figure 3.2.

Computers normally include a NOT instruction to perform the ones complement of a binary number on a bit-by-bit basis. An 8-bit computer can perform NOT operation on an 8-bit binary number. For example, the computer can execute a NOT instruction on an 8-bit binary number 01101111 to provide the result 10010000. The computer utilizes an internal electronic circuit consisting of eight inverters to invert the 8-bit data in parallel.

3.1.2   OR operation

The OR operation for two variables A and B generates a result of 1 if A or B, or both, are 1. However, if both A and B are zero, then the result is 0.

A plus sign + (logical sum) or ∨ symbol is normally used to represent OR. The four possible combinations of ORing two binary digits are

$$0 + 0 = 0$$
$$0 + 1 = 1$$
$$1 + 0 = 1$$
$$1 + 1 = 1$$

The truth table for the OR operation is provided below:

Inputs		
A	*B*	*Output = A + B*
0	0	0
0	1	1
1	0	1
1	1	1

Figure 3.3 shows the symbolic representation of an OR gate.

Logic gates using diodes provide good examples to understand how semiconductor devices are utilized in logic operations. Figure 3.4 shows a two-input-diode OR gate.

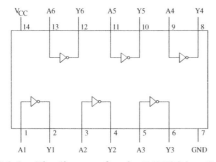

FIGURE 3.2   Pin diagram for the 74HC04 or 74LS04.

FIGURE 3.3   Symbol for an OR gate.

Although diodes are obsolete, and transistors are used in designing logic gates, diodes are utilized here merely for illustrative purposes. Note that the diode (see Chapter 1) is a switch, and it closes when there is a voltage drop of 0.6 V between the anode and the cathode. Suppose that a voltage range 0 to 2 V is considered as logic 0 and a voltage of 3 to 5 V is logic 1. If both A and B are at logic 0 (say 1.5 V) with a voltage drop across the diodes of 0.6 V to close the diode switches, a current flows from the inputs through R to ground, output C will be at 1.5 V - 0.6 V = 0.9 V (logic 0). On the other hand, if one or both inputs are at logic 1 (say 4.5 V) the output C will be at 4.5 - 0.6 V = 3.9 V (logic 1). Therefore, the circuit acts as an OR gate.

The 74HC32 (or 74LS32) is a commercially available quad 2-input 14-pin OR gate chip. This chip contains four 2-input/1-output independent OR gates as shown in Figure 3.5.

To understand the logic OR operation, consider Figure 3.6. V is a voltage source, A and B are switches, and L is an electrical lamp. L will be turned ON if either switch A or B or both are closed; otherwise, the lamp will be OFF. Hence, $L = A + B$. Computers normally contain an OR instruction to perform the OR operation between two binary numbers. For example, the computer can execute an OR instruction to OR $3A_{16}$ with 21_{16} on a bit-by-bit (bit-wise) basis:

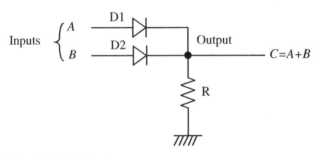

FIGURE 3.4   Diode OR gate.

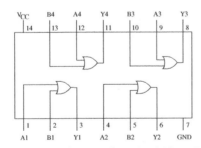

FIGURE 3.5   Pin diagram for 74HC32 or 74LS32.

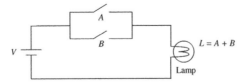

FIGURE 3.6   An example of the OR operation.

$$3A_{16} = 0011\ \ 1010$$
$$21_{16} = 0010\ \ 0001$$
$$\overline{\phantom{3A_{16} = }0011\ \ 1011}$$
$$\underbrace{}_{3}\ \ \underbrace{}_{B}{}_{16}$$

The computer typically utilizes eight two-input OR gates to accomplish this.

3.1.3   AND operation

The AND operation for two variables A and B generates a result of 1 if both A and B are 1. However, if either A or B, or both, are zero, then the result is 0.

The dot · and ^ symbol are both used to represent the AND operation.

The AND operation between two binary digits is provided below:

$$0 \cdot 0 = 0$$
$$0 \cdot 1 = 0$$
$$1 \cdot 0 = 0$$
$$1 \cdot 1 = 1$$

The truth table for the AND operation is provided below:

Inputs		
A	*B*	*Output = A · B = AB*
0	0	0
0	1	0
1	0	0
1	1	1

Figure 3.7 shows the symbolic representation of an AND gate. Figure 3.8 shows a two-input diode AND gate.

As we did for the OR gate, let us assume that the range 0 to +2 V represents logic 0 and the range 3 to 5 V is logic 1. Now, if A and B are both HIGH (say 3.3 V) and the anode of both diodes at 3.9 V, the switches in D1 and D2 close. A current flows from +5 V through resistor R to +3.3 V input to ground. The output C will be HIGH (3.9 V). On the other hand, if a low voltage (say 0.5 V) is applied at A and a

FIGURE 3.7   AND gate symbol.

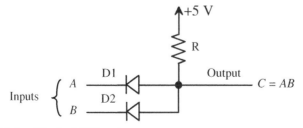

FIGURE 3.8   Diode AND gate.

high voltage (3.3V) is applied at B. The value of *R* is selected in such a way that 1.1 V appears at the anode side of D1; at the same time 3.9 V appears at the anode side of D2. The switches in both diodes will close because each has a voltage drop of 0.6 V between the anode and cathode. A current flows from the +5 V input through R and the diodes to ground. Output *C* will be low (1.1 V) because the output will be lower of the two voltages. Thus, it can be shown that when either one or both inputs are low, the output is low, so the circuit works as an AND gate. As mentioned before, diode logic gates are easier to understand, but they are not normally used these days.

The 74HC08 (or 74LS08) is a commercially available quad 2-input 14-pin AND gate chip. This chip contains four 2-input/1-output independent AND gates as shown in Figure 3.9. To illustrate the logic AND operation, consider Figure 3.10. The lamp *L* will be on when both switches *A* and *B* are closed; otherwise, the lamp *L* will be turned OFF. Hence, $L = A \cdot B$

Computers normally have an instruction to perform the AND operation between two binary numbers. For example, the computer can execute an AND instruction to perform ANDing 31_{16} with $A1_{16}$ on a bit-by-bit basis (bit-wise) as follows:

$$31_{16} = 0011\ 0001$$
$$A1_{16} = 1010\ 0001$$
$$\overline{\phantom{A1_{16} = 1010\ 0001}}$$
$$0010\ 0001$$
$$\underbrace{}_{2}\ \underbrace{}_{1}{}_{16}$$

The computer utilizes eight two-input AND gates to accomplish this.

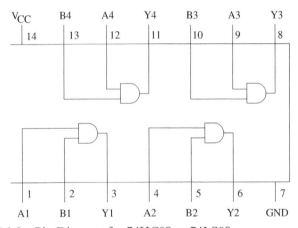

FIGURE 3.9   Pin Diagram for 74HC08 or 74LS08.

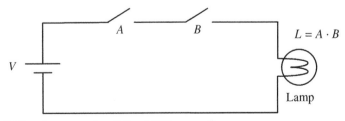

FIGURE 3.10   An example of the AND operation.

3.2   Other Logic Operations

The four other important logic operations are NOR, NAND, Exclusive-OR (XOR) and Exclusive-NOR (XNOR).

3.2.1   NOR operation

Inverted OR is called *NOR*. This means that the NOR output is produced by inverting the output of an OR operation. Figure 3.11 shows a NOR gate along with its truth table. Figure 3.12 shows the symbolic representation of a NOR gate. In the figure, the small circle at the output of the NOR gate is called the *inversion bubble*. The 74HC02 (or 74LS02) is a commercially available quad 2-input 14-pin NOR gate chip. This chip contains four 2-input/1-output independent NOR gates as shown in Figure 3.13.

3.2.2   NAND operation

Invered AND is called *NAND*. This means that the NAND output is generated by inverting the output of an AND operation. Figure 3.14 shows a NAND gate and its truth table. Figure 3.15 shows the symbolic representation of a NAND gate.

The 74HC00 (or 74LS00) is a commercially available quad 2-input/1-output 14-pin NAND gate chip. This chip contains four 2-input/1-output independent NAND gates as shown in Figure 3.16.

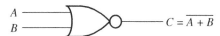

NOR gate Truth Table		
A	B	$C = \overline{A + B}$
0	0	1
0	1	0
1	0	0
1	1	0

FIGURE 3.11   A NOR gate with its truth table.

FIGURE 3.12   NOR gate symbol.

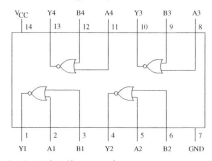

FIGURE 3.13   Pin diagram for 74HC02 or 74LS02.

NAND gate Truth Table		
A	B	$C = \overline{AB}$
0	0	1
0	1	1
1	0	1
1	1	0

FIGURE 3.14   A NAND gate and its truth table.

3.2.3 Exclusive-OR operation (XOR)

The Exclusive-OR operation (XOR) generates an output of 1 if the inputs are different and 0 if the inputs are the same. The $\oplus$ or $\forall$ symbol is used to represent the XOR operation. The XOR operation between two binary digits is

$$0 \oplus 0 = 0$$
$$0 \oplus 1 = 1$$
$$1 \oplus 0 = 1$$
$$1 \oplus 1 = 0$$

Most computers have an instruction to perform the bit-by-bit (bit-wise) XOR operation. Consider XORing $3A_{16}$ with 21_{16}.

$$3A_{16} = 0011\ 1010$$
$$21_{16} = 0010\ 0001$$
$$\underbrace{0001}_{1}\ \underbrace{1011}_{B}\ {}_{16}$$

It is interesting to note that XORing any number with another number of the same length but with all 1's will generate the ones complement of the original number. For example, consider XORing 31_{16} with FF_{16}:

$$31_{16}\qquad 0011\ 0001$$
$$\text{1's complement of } 31_{16}\qquad \underbrace{1100}_{C}\ \underbrace{1110}_{E}\ {}_{16}$$

$$31_{16} \oplus FF_{16}\qquad 0011\ 0001$$
$$\qquad\qquad\qquad 1111\ 1111$$
$$\qquad\qquad \underbrace{1100}_{C}\ \underbrace{1110}_{E}\ {}_{16}$$

The truth table for Exclusive-OR operation is provided below:

Inputs		Output
A	B	$C = A \oplus B$
0	0	0
0	1	1
1	0	1
1	1	0

$$C = \overline{AB}$$

FIGURE 3.15   NAND gate symbol.

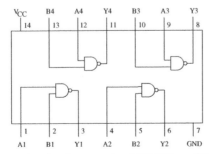

FIGURE 3.16   Pin diagram for 74HC00 or 74LS00.

The equation for output (C) in terms of the inputs (A,B) can be obtained by inspecting the truth table. From the truth table, $C = 1$ for

$$\overline{A} \ (A = 0) \text{ AND } B \ (B = 1)$$

OR

$$A \ (A = 1) \text{ AND } \overline{B} \ (B = 0).$$

Therefore,

$$C = A \oplus B = \overline{A}B + A\overline{B}$$

Figure 3.17 shows the implementation of an XOR gate using AND, OR, and INVERTER gates. Figure 3.18 shows the symbolic representation of the Exclusive-OR gate.

The 74HC86 (or 74LS86) is a commercially available quad 2-input 14-pin Exclusive-OR gate chip. This chip contains four 2-input/1-output independent exclusive-OR gates as shown in Figure 3.19.

3.2.4 Exclusive-NOR Operation (XNOR)

Inverted Exclusive-OR is called *Exclusive-NOR*. This means that the XNOR output is generated by inverting the output of an XOR operation. Figure 3.20 shows its symbolic

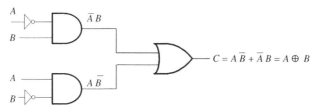

FIGURE 3.17   Implementation of the Exclusive-OR gate.

FIGURE 3.18   XOR symbol.

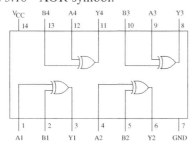

FIGURE 3.19   Pin diagram for 74HC86 or 74LS86.

	XNOR gate Truth Table	
A	B	C
0	0	1
0	1	0
1	0	0
1	1	1

FIGURE 3.20   Exclusive-NOR symbol along with its truth table.

representation along with the truth table. The XNOR operation is represented by the symbol $\odot$. Therefore, $C = \overline{A \oplus B} = A \odot B$. The XNOR operation is also called *equivalence*. From the truth table, output C is 1 if both A and B are 0's or if both A and B are 1's; otherwise, C is 0. That is, $C = 1$, for $\overline{A}$ ($A = 0$) AND $\overline{B}$ ($B = 0$) OR A ($A = 1$) AND B ($B = 1$). Hence, $C = A \odot B = \overline{A}\ \overline{B} + AB$.

The 74HC266 (or 74LS266) is a quad 2-input/1-output 14-pin Exclusive-NOR gate chip. This chip contains four 2-input/1-output independent Exclusive-NOR gates shown in Figure 3.21.

Note that the symbol C is chosen arbitrarily in all the above logic operations to represent the output of each logic gate. Also, note that all logic gates (except NOT) can have at least two inputs with only one output. The NOT gate, on the other hand, has one input and one output.

3.3 IEEE Symbols for Logic Gates

The Institute of Electrical and Electronics Engineers (IEEE) recommends rectangular shape symbols for logic gates: The original logic symbols have been utilized for years and will be retained in the rest of this book. IEEE symbols for gates are listed below:

Gate	Common Symbol	IEEE Symbol
AND	$f = AB$	$f = AB$
OR	$f = A + B$	$f = A + B$
NOT	$f = \overline{A}$	$f = \overline{A}$
NAND	$f = \overline{AB}$	$f = \overline{AB}$
NOR	$f = \overline{A + B}$	$f = \overline{A + B}$
Exclusive-OR	$f = A \oplus B$	$f = A \oplus B$
Exclusive-NOR	$f = \overline{A \oplus B}$	$f = \overline{A \oplus B}$

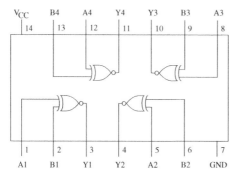

FIGURE 3.21   Pin Diagram for 74HC266 or 74LS266.

3.4   Positive and Negative Logic

The inputs and outputs of logic gates are represented by either logic 0 or logic 1. There are two ways of assigning voltage levels to the logic levels, positive logic and negative logic. The positive logic convention assigns a HIGH (*H*) voltage for logic 1 and LOW (*L*) voltage for logic 0. On the other hand, in the negative logic convention, a logic 1 = LOW (*L*) voltage and logic 0 = HIGH (*H*) voltage.

The IC data sheets typically define these levels in terms of voltage levels rather than logic levels. The designer decides on whether to use positive or negative logic. As an example, consider a gate with the following truth table:

A	*B*	*f*
L	*L*	*H*
L	*H*	*H*
H	*L*	*H*
H	*H*	*L*

Using positive logic, ($H = 1$ and $L = 0$) the following table is obtained:

A	*B*	*f*
0	0	1
0	1	1
1	0	1
1	1	0

This is the truth table for a NAND gate. However, negative logic, ($H = 0$ and $L = 1$) provides the following table:

A	*B*	*f*
1	1	0
1	0	0
0	1	0
0	0	1

This is the truth table for a NOR gate. Note that converting from positive to negative logic and vice versa for logic gates basically provides the *dual* (discussed later in this chapter) of a function. This means that changing 0's to 1's and 1's to 0's for both inputs and outputs of a logic gate, the logic gate is converted from a NOR gate to a NAND gate as shown in the example.

Note that positive logic and active high logic are equivalent (HIGH = 1, LOW = 0). On the other hand, negative logic and active low logic are equivalent

(HIGH = 0, LOW = 1). A signal is "active high" if it performs the required function when HIGH (H = 1). An "active low" signal, on the other hand, performs the required function when LOW (L = 0). A signal is said to be asserted when it is active. A signal is disasserted when it is not at its active level.

Active levels may be associated with inputs and outputs of logic gates. For example, an AND gate performs a logical AND operation on two active HIGH inputs and provides an active HIGH output. This also means that if both the inputs of the AND gate are asserted, the output is asserted.

3.5 Boolean Algebra

Boolean algebra provides basis for logic operations using binary variables. Alphabetic characters are used to represent the binary variables. A binary variable can have either true or complement value. For example, the binary variable A can be either A and/or $\overline{A}$ in a Boolean function.

A Boolean function is an operation expressing logical operations between binary variables. The Boolean function can have a value of 0 or 1. As an example of a Boolean function, consider the following:

$$f(A,B,C) = \overline{A}\,\overline{B} + C$$

Here, the Boolean function f is 1 if both $\overline{A}$ and $\overline{B}$ are 1 or C is 1; otherwise, f is 0. Note that $\overline{A}$ means that if $A = 1$, then $\overline{A} = 0$. Thus, when $B = 1$, then $\overline{B} = 0$. It can, therefore, be concluded that f is 1 when $A = 0$ and $B = 0$ or $C = 1$.

A truth table can be used to represent a Boolean function. As mentioned before, the truth table lists inputs and outputs. Furthermore, the truth table provides the value of the Boolean function (f) as 1 or 0 for each combination of the input binary variables (A,B,C). Table 3.1 provides the truth table for the Boolean function $f(A,B,C) = \overline{A}\,\overline{B} + C$. In the truth table, if $A = 1$, $B = 1$, and $C = 0$, $f = 0.0 + 0 = 0$. Note that Table 3.1 contains three input variables (A, B, C) and one output variable (f). As mentioned in sections 3.2.3 and 3.2.4, output, f can be obtained directly from the truth table as $f = \overline{A}\,\overline{B}\,\overline{C} + \overline{A}\,\overline{B}\,C + \overline{A}\,BC + A\,\overline{B}\,C + ABC$. However, the function, f can be simplified as $f = \overline{A}\,\overline{B} + C$ using the Boolean identities to be discussed in Section 3.5.1. This simplified form of output,

TABLE 3.1   Truth Table for $f = \overline{A}\,\overline{B} + C$

A	B	C	f
0	0	0	1
0	0	1	1
0	1	0	0
0	1	1	1
1	0	0	0
1	0	1	1
1	1	0	0
1	1	1	1

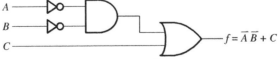

FIGURE 3.22   Logic diagram for $f = \overline{A}\,\overline{B} + C$.

f can be implemented using fewer gates resulting in lower implementation cost. Note that the original equation for f obtained from the truth table and the simplified form for f are equivalent since they have the same truth table of Table 3.1.

Next, the Boolean function, f can be represented in the form of a logic diagram or schematic. Figure 3.22 shows the logic diagram or schematic for $f = \overline{A}\,\overline{B} + C$. The Boolean expression , $f = \overline{A}\,\overline{B} + C$ contains two terms, $\overline{A}\,\overline{B}$ and C, which are inputs to logic gates. Each term may include a single or multiple variables, called *literals* that may or may not be complemented. For example , $f = \overline{A}\,\overline{B} + C$ contains three literals, $\overline{A}$, $\overline{B}$, and C. Note that a variable and its complement are both called literals. For two variables, the literals are A, B, $\overline{A}$, and $\overline{B}$.

3.5.1   Boolean Identities

Here is a list of Boolean identities that are useful in simplifying Boolean expressions:

1.	a) $A + 0 = A$	b) $A \cdot 1 = A$	
2.	a) $A + 1 = 1$	b) $A \cdot 0 = 0$	
3.	a) $A + A = A$	b) $A \cdot A = A$	
4.	a) $A + \overline{A} = 1$	b) $A \cdot \overline{A} = 0$	
5.	a) $(\overline{\overline{A}}) = A$		
6.	Commutative Law:		
	a) $A + B = B + A$	b) $A \cdot B = B \cdot A$	
7.	Associative Law:		
	a) $A + (B + C) = (A + B) + C$	b) $A \cdot (B \cdot C) = (A \cdot B) \cdot C$	
8.	Distributive Law:		
	a) $A \cdot (B + C) = A \cdot B + A \cdot C$	b) $A + B \cdot C = (A + B) \cdot (A + C)$	
9.	DeMorgan's Theorem:		
	a) $\overline{A + B + C \cdots} = \overline{A} \cdot \overline{B} \cdot \overline{C} \cdots$	b) $\overline{A \cdot B \cdot C \cdots} = \overline{A} + \overline{B} + \overline{C} + \cdots$	

In the list, each identity identified by b) on the right is the dual of the corresponding identify a) on the left. Note that the dual of a Boolean expression is obtained by changing 1's to 0's and 0's to 1's if they appear in the equation, and AND to OR and OR to AND on both sides of the equal sign.

For example, consider identity 4. Relation 4a is the dual of relation 4b because the AND in the expression is replaced by an OR and then, 0 by 1.

The Duality Principle of Boolean algebra states that a Boolean expression is unchanged if the dual of both sides of the equal sign is taken. Consider, for example, the Boolean function, $f = \overline{B} + \overline{A}\,\overline{B}$

$$f = \overline{B} \bullet (1 + \overline{A})$$
$$= \overline{B}$$

The dual of f,
$$f_D = \overline{B} \bullet (\overline{A} + \overline{B})$$
$$f_D = \overline{B} \bullet \overline{A} + \overline{B} \bullet \overline{B} = \overline{B}\overline{A} + \overline{B}$$
$$= \overline{B}(\overline{A} + 1) = \overline{B}$$

Hence, $f = f_D$. In order to verify some of the identities, consider the following examples:

i)   Identity 2a) $A + 1 = 1$. For $A = 0$, $A + 1 = 0 + 1 = 1$. For $A = 1$, $A + 1 = 1 + 1 = 1$
ii) Identity 4b) $A \cdot \overline{A} = 0$. If $A = 1$, then $\overline{A} = 0$. Hence, $A \cdot \overline{A} = 1 . 0 = 0$

iii) Identity 8b) $A + B \cdot C = (A + B) \cdot (A + C)$ is very useful in manipulating Boolean expressions. This identity can be verified by means of a truth table as follows:

A	B	C	$B \cdot C$	$A+B$	$A+C$	$A+B \cdot C$	$(A+B) \cdot (A+C)$
0	0	0	0	0	0	0	0
0	0	1	0	0	1	0	0
0	1	0	0	1	0	0	0
0	1	1	1	1	1	1	1
1	0	0	0	1	1	1	1
1	0	1	0	1	1	1	1
1	1	0	0	1	1	1	1
1	1	1	1	1	1	1	1

iv) Identities 9a) and 9b) (DeMorgan's Theorem) are useful in determining one's complement of a Boolean expression. DeMorgan's theorem for two variables (A,B) can be verified by means of a truth table as follows:

A	B	$\bar{A}$	$\bar{B}$	$\bar{A} \cdot \bar{B}$	$A+B$	$\overline{A+B}$	$A \cdot B$	$\overline{A \cdot B}$	$\bar{A}+\bar{B}$
0	0	1	1	1	0	1	0	1	1
0	1	1	0	0	1	0	0	1	1
1	0	0	1	0	1	0	0	1	1
1	1	0	0	0	1	0	1	0	0

De Morgan's Theorem can be expressed in a general form for n variables as follows:

$$\overline{A+B+C+D+...} = \bar{A} \cdot \bar{B} \cdot \bar{C} \cdot \bar{D} \cdot ...$$

$$\overline{A \cdot B \cdot C \cdot D \cdot ...} = \bar{A}+\bar{B}+\bar{C}+\bar{D}+...$$

The logic gates except for the inverter can have more than two inputs if the logic operation performed by the gate is commutative and associative (Identities 6a and 7a). For example, the OR operation has these two properties as follows: $A + B = B + A$ (Commutative) and $(A+B) + C = A+ (B + C) = A + B + C$ (Associative). This can be verified using a truth table. Thus, the OR gate can have more than two inputs. Also, this means that the OR gate inputs can be interchanged. Figure 3.23(a) shows implementation of $f = A + B +C$ using two two-input OR gates. However, implementation of the single three-input OR gate of Figure 3.23(b) requires fewer transistors and less time delay. In contrast, implementation of the three-input OR gate with two two-input OR gates (Figure 3.23(a)) requires more transistors and more time delays. Hence, designers typically implement multiple-input OR gates using a single OR gate (Figure 3.23(b)). The same rule applies to multiple-input AND gates. Similarly, using the identities 6b and 7b, it can be shown that the AND gate can also have more than two inputs. Note that the NOR and NAND operations, on the other hand, are commutative, but not associative. Therefore, it is not possible to have NOR and NAND gates with more than two inputs. However, NOR

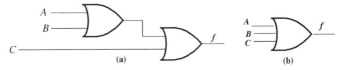

FIGURE 3.23 Obtaining a three input OR gate (a) using two two-input OR gates (b) using a Three-input OR gate.

and NAND gates with more than two inputs can be obtained by using inverted OR and inverted AND respectively. The Exclusive-OR and Exclusive-NOR operations are both commutative and associative. Thus, these gates can have more than two inputs. However, Exclusive-OR and Exclusive-NOR gates with more than two inputs are uncommon from the hardware point of view.

3.5.2 Simplification Using Boolean Identities

Although there are no defined set of rules for minimizing a Boolean expression, appropriate identities can be used to accomplish this. Consider the Boolean function

$$f = ABCD + \overline{A}BCD + \overline{BC}$$

This equation can be implemented using logic gates as shown in Figure 3.24(a). The expression can be simplified by using identities as follows:

$$f = BCD(A + \overline{A}) + \overline{BC}$$

$$= BCD \cdot 1 + \overline{BC} \qquad\qquad \text{By identity 4a)}$$

$$= BCD + \overline{BC} \qquad\qquad \text{By identity 1b)}$$

Assume $BC = E$, then $\overline{BC} = \overline{E}$

Hence, $f = ED + \overline{E}$

$$= (E + \overline{E})(\overline{E} + D) \qquad\qquad \text{By identity 8b)}$$

$$= \overline{E} + D \qquad\qquad \text{By identity 4a)}$$

Substituting $\overline{E} = \overline{BC}$, $f = \overline{BC} + D$

The simplified form is implemented using logic gates in Figure 3.24(b). The logic diagram in Figure 3.24(b) requires only one NAND gate and an OR gate. This implementation is inexpensive compared to the circuit of Figure 3.24(a). Both logic circuits perform the same function. The following truth table can be used to show that the outputs produced by both circuits are equivalent:

A	B	C	D	$f = ABCD + \overline{A}BCD + \overline{BC}$	$f = \overline{BC} + D$
0	0	0	0	1	1
0	0	0	1	1	1
0	0	1	0	1	1
0	0	1	1	1	1
0	1	0	0	1	1
0	1	0	1	1	1
0	1	1	0	0	0
0	1	1	1	1	1
1	0	0	0	1	1
1	0	0	1	1	1
1	0	1	0	1	1
1	0	1	1	1	1
1	1	0	0	1	1
1	1	0	1	1	1
1	1	1	0	0	0
1	1	1	1	1	1

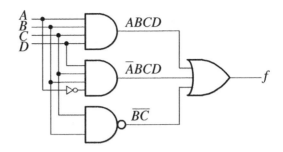

(a) Implementation of $f = ABCD + \bar{A}BCD + \overline{BC}$

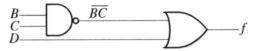

(b) implementation of the simplified function $f = \overline{BC} + D$

FIGURE 3.24   Implementation of Boolean function using logic gates.

The following are some more examples for simplifying Boolean expressions using identities:

i) $f = \bar{x} + \bar{y} + \overline{xy} + \overline{xy}z = \overline{xy} + \overline{xy} + \overline{xy}z = \overline{xy} + \overline{xy}z = \overline{xy}\ (1 + z) = \overline{xy}$

ii) $f = \overline{abcd} + \bar{a}cd + \bar{b}cd + (1 \oplus ab)\ cd = \overline{abcd} + cd\ (\bar{a} + \bar{b}) + \overline{abcd} = \overline{abcd} + \overline{abcd} + \overline{abcd} = \overline{abcd}$

iii) $F = XY + X\bar{Z} + XZ = X\ (Y + \bar{Z} + Z) = X\ (Y + 1) = X \cdot 1 = X$

iv) $F = A\ \overline{BC} + AB + \bar{A}\ \bar{C} = A\ (\bar{B} + \bar{C}\ ) + AB + \bar{A}\ \bar{C} = A\ \bar{B} + A\ \bar{C} + AB + \bar{A}\ \bar{C}$

   $= A\ (B + \bar{B}\ ) + \bar{C}(A + \bar{A}) = A + \bar{C}$

v) $f = x + \overline{x}y + x + \bar{y} = x + \overline{x}y + \bar{y} = (x + \bar{x})(x + y) + \bar{y} = x + y + \bar{y} = x + 1 = 1$

vi) $f = A\ (B \oplus 1)\ (\bar{A} + B) = A\ \bar{B}\ (\bar{A} + B) = A\ \bar{B}\ \bar{A} + A\ \bar{B}B = 0$

vii) $F = B\ (A + B) + A\ \bar{B} + \bar{B} = AB + BB + A\ \bar{B} + \bar{B} = AB + B + A\ \bar{B} + \bar{B} = 1 + AB + A\ \bar{B} = 1$

viii) $f = (x + y + z)\ (\bar{x}\ y + \bar{y}\ z) = \overline{x}yx + \overline{x}yy + \overline{x}yz + \overline{y}zx + \overline{y}z\ y + \overline{y}z\ z$

   $= \overline{x}y + \overline{x}yz + \overline{y}zx + \overline{y}z = \overline{x}y(1 + z) + \overline{y}z(x + 1) = \overline{x}y + \overline{y}z$

ix) $f = xy + xy\bar{z} + \bar{x}\ \bar{y} = xy\ (1 + \bar{z}) + \bar{x}\ \bar{y} = xy + \bar{x}\ \bar{y} = \overline{x \oplus y}$

x) $F = \bar{A}BC + ABC + B\bar{C} = BC(\bar{A} + A) + B\bar{C} = BC + B\bar{C} = B\ (C + \bar{C}) = B \cdot 1 = B$

xi) Show that $f = \overline{(a + \bar{b})(\bar{a} + b)}$ can be implemented using a single logic gate.

Solution: Using DeMorgan's theorem, $f = \overline{(a + \bar{b})(\bar{a} + b)}$

   $= \overline{(a + \bar{b})} + \overline{(\bar{a} + b)} = (\bar{a}.\bar{\bar{b}}) + (\bar{\bar{a}}.\bar{b}) = \bar{a}b + a\bar{b} = a \oplus b$

xii) Show that $f = \overline{(A + B)(E + F)}$ can be implemented using two AND and one OR gates.

Solution: $f = \overline{(AB)(\overline{EF})} = AB + EF$ using DeMorgan's theorem.

xiii) Express $f=(X+\overline{X}Z)(X + Z)$ using a single two-input gate.
Solution: $f=(X+\overline{X})(X + Z)(X + Z)$ using distributive law. Also, $(X + Z)(X + Z)$
$= (X + Z)$ Hence, $f= X+Z$

xiv) Express f given $\overline{f}=(\overline{A} + \overline{B} + \overline{C}) + \overline{ABC}$ using only one three input gate.

Solution: Using DeMorgan's theorem,

$$f=\overline{\overline{f}} = \overline{(\overline{A}+\overline{B}+\overline{C})+\overline{ABC}}$$
$$= (\overline{\overline{ABC}})+(\overline{\overline{ABC}})=ABC$$

xv) Simplify each of the following Boolean expressions as much as possible using identities:

(a) $f(x,y,z) = \overline{y} z + x \overline{y}z + \overline{x} \overline{y}z$

Solution: $f(x,y,z) = \overline{y} z (1 + x + \overline{x}) = \overline{y}z$

(b) $F(A,B) = (A + \overline{A}B)(A + B) (\overline{A})$

Solution: $F(A,B) = (A + \overline{A}) (A + B) (A + B) (\overline{A}) = (A+B)(\overline{A})=A\overline{A} +\overline{A} B =\overline{A} B$

(c) $f(x,y,z) = x \overline{z} + z + \overline{y}z$

Solution: $f(x,y,z) = x \overline{z} + z (1 +\overline{y}) = x \overline{z} + z = z + x \overline{z} = (z + x) (z +\overline{z}) = x + z$

(d) $F(A,B) = (A + \overline{B})(A + B) (\overline{AB})$

Solution: $F(A,B) = (AA + A\overline{B} + AB + B \overline{B}) (\overline{A} +\overline{B}) = (A + A (\overline{B} + B) + 0)(\overline{A} +\overline{B})$
$= A (\overline{A} +\overline{B}) = A\overline{A} + A\overline{B} =A\overline{B}$

(e) $F (A,B,C) = \overline{B} C + \overline{A} \overline{B}C + B \overline{C}$

Solution: $F(A,B,C) = \overline{B} C (1+\overline{A}) + B \overline{C} = \overline{B} C + B \overline{C} = B \oplus C$

(f) $F (A,B,C) = ((A+ \overline{B}) +(A+ \overline{B}) C) (\overline{A} \overline{B} \overline{C})$

Solution:
Let $D = A+ \overline{B}$. Hence $F (A,B,C) = (D + DC) (\overline{A} \overline{B} \overline{C}) = D (1+C) (\overline{A} \overline{B} \overline{C}) = D (\overline{A} \overline{B} \overline{C})$
Therefore, $F (A,B,C) = (A+ \overline{B}) (\overline{A} \overline{B} \overline{C}) = A (\overline{A} \overline{B} \overline{C}) +\overline{B} (\overline{A} \overline{B} \overline{C})= 0 +(\overline{A} \overline{B} \overline{C}) = \overline{A} \overline{B} \overline{C}$

3.5.3 Consensus Theorem

The Consensus Theorem is expressed as $AB + \overline{A}C + BC = AB + \overline{A}C$

The theorem states that the AND term BC can be eliminated from the expression if one of the literals such as B is ANDed with the true value of another literal (A) and the other term C is ANDed with its complement . This theorem can sometimes be applied to simplify Boolean equations. The Consensus Theorem can be proved as follows:

$$AB+\overline{A}C+BC = AB+\overline{A}C+BC\left(A +\overline{A}\right)$$

$$= AB+\overline{A}C+ABC+\overline{A}BC$$

$$= AB+ABC+\overline{A}C+\overline{A}BC$$

$$= AB(1+C)+\overline{A}C(1+B)$$

$$= AB+\overline{A}C$$

The dual of the Consensus Theorem can be expressed as
$$(A + B)(\overline{A} + C)(B + C) = (A + B)(\overline{A} + C)$$

To illustrate how a Boolean expression can be manipulated by applying the Consensus Theorem, consider the following:

$$f = (B + \overline{D})(\overline{B} + C)$$

$$= B\overline{B} + BC + \overline{B}\,\overline{D} + C\overline{D}$$

$$= BC + \overline{B}\,\overline{D} + C\overline{D}, \text{ since } B\overline{B} = 0$$

Because C is ANDed with B, and $\overline{D}$ is ANDed with its complement $\overline{B}$, by using the Consensus Theorem, $C\overline{D}$ can be eliminated. Thus, $f = BC + \overline{B}\,\overline{D}$.

The Consensus Theorem can be used in logic circuits for avoiding undesirable behavior. To illustrate this, consider the logic circuits in Figure 3.25. In Figure 3.25(a), the output is one i) if B and C are 1 and $A = 0$ or ii) if B and C are 1 and $A = 1$.

Suppose that in Figure 3.25(a), $B = 1$, $C = 1$, and $A = 0$. Assume that the propagation delay time of each gate is 10 ns (nanoseconds). The circuit output f will be 1 after 30 ns (3 gate delays). Now, if input A changes from 0 to 1, the outputs of NOT gate 1 and AND gate 2 will be 0 and 1 respectively after 10 ns. This will make output $f = 1$ after 20 ns. The output of AND gate 3 will be low after 20 ns, which will not affect the output of f.

Now, assume that B and C stay at 1 while A changes from 1 to 0. The outputs of NOT gate 1 and AND gate 2 will be 1 and 0 respectively after 10 ns. Because the output of AND gate 3 is 0 from the previous case, this will change output of OR gate 4 to 0 for a brief period of time. After 10 ns, the output of AND gate 3 changes to 1, making the output of f HIGH (desired value). Note that, for $B = 1$, $C = 1$, and $A = 0$, the output f should have stayed at 1 from the equation . However, f changed to zero for a short period of time. This change is called a *glitch* or *hazard* and occurs from the gate delays in a circuit. Glitches can cause circuit malfunction and should be eliminated. Application of the Consensus Theorem gets rid of the glitch. By adding the redundant term BC, the modified logic circuit for f is obtained. Figure 3.25(b) shows the logic circuit. Now, consider the case in which the glitch occurs in Figure 3.25(a) when B and C stay at 1 while A changes from 1 to 0. For the circuit in Figure 3.25(b) the glitch will disappear, because $BC = 1$ throughout any changes in values of A and $\overline{A}$. Thus, minimization of logic gates might not always be desirable; rather, a circuit without any hazards would be the main objective of the designer.

There are two types of hazards: static and dynamic. Static hazard occurs when a signal should remain at one value, but instead it oscillates a few times before settling back to its original value. Dynamic hazard occurs, when a signal should make a clean transition to a new logic value, but instead it oscillates between the two logic values before making the transition to its final value. Both types of hazards occur because of *races* in the various paths of a circuit. A *race* is a situation in which signals traveling through two or more paths compete with each other to affect a common signal. It is, therefore, possible for the final signal value to be determined by the winner of the race. One way to eliminate races is by applying the consensus Theorem as illustrated in the preceding example.

3.5.4 Complement of a Boolean Function

The complement of a function f can be obtained algebraically by applying DeMorgan's Theorem. It follows from this theorem that the complement of a function can also be derived by taking the dual of the function and complementing each literal.

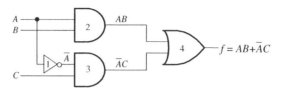

(a) Logic circuit for $f = AB + \overline{A}C$

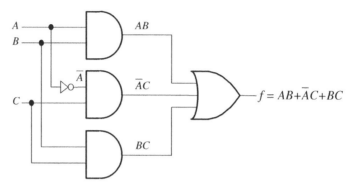

(b) Logic circuit for $f = AB + \overline{A}C + BC$

FIGURE 3.25   Logic circuit for the Consensus Theorem.

Example 3.1

Find the complement of the function $f = \overline{C}(AB + \overline{A}\overline{B}D + \overline{A}B\overline{D})$
    i)   Using DeMorgan's Theorem
    ii)   By taking the dual and complementing each literal

Solution

        Using DeMorgan's Theorem as many times as required, the complement of the function can be obtained:

$$\overline{f} = \overline{\overline{C}\left(AB + \overline{A}\,\overline{B}D + \overline{A}B\overline{D}\right)}$$

$$= \overline{\overline{C}} + \overline{\left(AB + \overline{A}\,\overline{B}D + \overline{A}B\overline{D}\right)}$$

$$= C + \left(\overline{AB} \cdot \overline{\overline{A}\,\overline{B}D} \cdot \overline{\overline{A}B\overline{D}}\right)$$

$$= C + (\overline{A} + \overline{B})(A + B + \overline{D})(A + \overline{B} + D)$$

By taking the dual and complementing each literal, we have:

    The dual of f: $\overline{C} + (A + B)(\overline{A} + \overline{B} + D)(\overline{A} + B + \overline{D})$

    Complementing each literal: $C + (\overline{A} + \overline{B})(A + B + \overline{D})(A + \overline{B} + D) = \overline{f}$

3.6   Standard Representations

The standard representations of a Boolean function typically contain either logical product (AND) terms called *minterms* or logical sum (OR) terms called *maxterms*.

Expressing a Boolean function using literals may lead to errors. However, it is very convenient to express a Boolean function in terms of minterms or maxterms. Furthermore, these standard representations make the minimization procedures easier. The standard representations are also called *Canonical forms.*

There are four terms associated with two variables, A and B. These terms are: $\overline{A}\ \overline{B}$, $\overline{A}B$, $A\overline{B}$, and AB. Each of these four terms is called a *minterm*. Hence, a minterm is a product (AND) term of all variables in which each variable can be either complemented or uncomplemented. Similarly, there are eight minterms for three variables, A, B, and C. These minterms are $\overline{A}\ \overline{B}\ \overline{C}$, $\overline{A}\ \overline{B}\ C$, $\overline{A}\ B\ \overline{C}$, $\overline{A}\ B\ C$, $A\ \overline{B}\ \overline{C}$, $A\ \overline{B}\ C$, $A\ B\ \overline{C}$, and ABC. These product terms represent numeric values from 0 through 7. In general, there are 2^n minterms for n variables.

A minterm is represented by the symbol m_j, where the subscript j is the decimal equivalent of the binary number of the minterm. For example, the decimal equivalents (j) of the binary numbers represented by the four minterms of two variables, A and B, are 0 ($\overline{A}\ \overline{B}$), 1($\overline{A}\ B$), 2($A\ \overline{B}$), and 3 ($AB$). Therefore, the symbolic representations of the four minterms of two variables are m_0, m_1, m_2, and m_3 as follows:

A	B	Minterm	Symbol
0	0	$\overline{A}\ \overline{B}$	m_0
0	1	$\overline{A}B$	m_1
1	0	$A\overline{B}$	m_2
1	1	AB	m_3

In general, the 2^n minterms of n variables are: $m_0, m_1, m_2, \ldots, m_p$ where $p = 2^n-1$. As an example for $n = 2$, the minterms are m_0, m_1, m_2, and m_3.

It has been shown that a Boolean function can be defined by a truth table. A Boolean function can also be exressed in terms of minterms. For example, consider the following truth table:

A	B	$f(A,B)$
0	0	1
0	1	0
1	0	1
1	1	1

One can determine the function f by logically summing (ORing) the product terms for which f is 1. Therefore,

$$f(A,B) = \overline{A}\ \overline{B} + A\ \overline{B} + AB$$

This is called the *Sum-of-Products* (SOP) expression. A logic diagram of a sum-of-products expression contains several AND gates followed by a single OR gate. In terms of minterms, $f(A,B)$ can be represented as:

$$f(A,B) = \Sigma m(0, 2, 3)$$

The symbol Σ denotes the logical sum (OR) of the minterms.

A maxterm, on the other hand, is obtained from the logical sum of all the variables after complementing each variable. The four maxterms of two variables are $A + B$, $A + \overline{B}$, $\overline{A} + B$, and $\overline{A} + \overline{B}$. Each maxterm is represented by the symbol M_j, where subscript j is the decimal equivalent of the binary number of the maxterm. The symbolic representations of the four maxterms of two variables are M_0, M_1, M_2, and M_3 as follows:

A	B	Maxterm	Symbol
0	0	$A + B$	M_0
0	1	$A + \bar{B}$	M_1
1	0	$\bar{A} + B$	M_2
1	1	$\bar{A} + \bar{B}$	M_3

In the preceding figure, consider maxterm M_2 as an example. Since $A = 1$ and $B = 0$, the maxterm M_2 is found as $A + B$ by taking the logical sum of complement of A (since $A = 1$) and true value of B (since $B = 0$). In general, there are n maxterms (M_0, M_1, ... , M_{n-1}) for p variables, where $n = 2^p$.

The relationship between minterm and maxterm can be established by using DeMorgan's Theorem. Consider, for example, minterm m_1 and maxterm M_1 for two variables:

$$m_1 = \bar{A}B,\ M_1 = A + \bar{B}$$

Taking the complement of m_1,

$$m_1 = \overline{\bar{A}B}$$

$$= \bar{\bar{A}} + \bar{B} \text{ by DeMorgan's Theorem}$$

$$= A + \bar{B}$$

$$= M_1$$

Therefore, $m_1 = \overline{M_1}$ or $\overline{m_1} = M_1$. This implies that $m_j = \overline{M_j}$ or $\overline{m_j} = M_j$. That is, a minterm is the complement of its corresponding maxterm and vice versa.

In order to represent a Boolean function in terms of maxterms, consider the following truth table:

A	B	$f(A,B)$	$\bar{f}(A,B)$
0	0	1	0
0	1	0	1
1	0	0	1
1	1	0	1

Taking the logical sum of minterms of $\bar{f}(A,B)$,

$$\bar{f}(A,B) = \bar{A}B + A\bar{B} + AB$$

$$= m_1 + m_2 + m_3$$

$$= \sum m(1,2,3)$$

By taking complement of $\bar{f}(A,B)$,

$$f(A,B) = \bar{\bar{f}}(A,B) = \overline{m_1 + m_2 + m_3} = \overline{m_1} \cdot \overline{m_2} \cdot \overline{m_3}$$

$$= M_1 \cdot M_2 \cdot M_3 \text{ (since } M_j = \overline{m_j})$$

$$= (A + \bar{B})(\bar{A} + B)(\bar{A} + \bar{B})$$

This is called the *product-of-sums* (POS) expression. The logic diagram of a *product-of-sums* expression contains several OR gates followed by a single AND gate. Hence, $f(A,B) = \Pi M(1, 2, 3)$ where the symbol Π represents the logical product (AND) of maxterms M_1, M_2, and M_3 in this case. Note that one can express a Boolean function in terms of maxterms by inspecting a truth table and then logically ANDing the maxterms for which the Boolean function has a value of 0.

A Boolean function that is not expressed in terms of sums of minterms or product of maxterms can be represented by a truth table. The function can then be expressed in terms of minterms or maxterms. For example, consider $f(A,B,C) = A + B\overline{C}$. The function $f(A,B,C)$ is not in a sum of minterms or product of maxterms form, since each term does not include all three literals of A, B, and C. The truth table for $f(A,B,C)$ can be determined as follows:

A	B	C	$f = A + B\overline{C}$
0	0	0	0
0	0	1	0
0	1	0	1
0	1	1	0
1	0	0	1
1	0	1	1
1	1	0	1
1	1	1	1

From the truth table, the sum of minterm form ($f = 1$) is:

$$f = \Sigma m(2, 4, 5, 6, 7) = \overline{A}B\overline{C} + A\overline{B}\,\overline{C} + A\overline{B}C + AB\overline{C} + ABC$$

From the truth table, the product of maxterm form ($f = 0$) is:

$$f = \Pi M(0, 1, 3) = (A + B + C)(A + B + \overline{C})(A + \overline{B} + \overline{C})$$

Note that $f(A,B,C) = \Sigma m(0, 1, 3)$, is obtained by the logical sum of minterms for $f(A,B,C) = 0$. Also, note that a function containing all minterms is 1. This means that in the above truth table, if $f(A,B,C) = 1$ for all eight combinations of A, B, and C, then $f(A,B,C) = \Sigma m(0, 1, 2, 3, 4, 5, 6, 7) = 1$.

As mentioned before, the logic diagram of a sum of minterm form contains several AND gates and a single OR gate. This is illustrated by the logic diagram for $f = \Sigma m(2, 4, 5, 6, 7) = \overline{A}B\overline{C} + A\overline{B}\,\overline{C} + A\overline{B}C + AB\overline{C} + ABC$ as shown in figure 3.26(a). Similarly, the logic diagram of a product of maxterm expression form contains several OR gates and a single AND gate. This is illustrated by the logic diagram for $f = \Pi M(0, 1, 3) = (A + B + C)(A + B + \overline{C})(A + \overline{B} + \overline{C})$ as shown in figure 3.26(b).

Note that it is assumed in Figures 3.25(a) and 3.25(b) that both true and complemented inputs are available.

Finally, consider converting $f(A,B,C) = (A + B + C)(A + B + \overline{C})(A + \overline{B} + \overline{C})$ into maxterm form. First, complement each literal of the right side of the equation and obtain the follwing:

$$f(A,B,C) = (\overline{A} + \overline{B} + \overline{C})(\overline{A} + \overline{B} + C)(\overline{A} + B + C)$$

Next, convert the right side of the above equation into decimal to obtain the maxterm form as follows:

$$f(A,B,C) = \Pi M(0, 1, 3) \text{ which is equivalent to } \overline{f}(A,B,C) = \Sigma m(0,1,3)$$

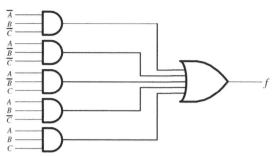

FIGURE 3.26 (a)   Logic diagram of a sum of minterms.

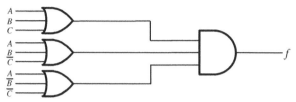

FIGURE 3.26 (b)   Logic diagram of a product of maxterms.

Example 3.2

Using the following truth table, express the Boolean function f in terms of sum-of-products (minterms) and product-of-sums (maxterms):

A	B	C	f
0	0	0	0
0	0	1	1
0	1	0	1
0	1	1	1
1	0	0	0
1	0	1	0
1	1	0	1
1	1	1	0

Solution

From the truth table, $f = 1$ for minterms m_1, m_2, m_3, and m_6. Therefore, the Boolean function f can be expressed by taking the logical sum (OR) of these minterms as follows:

$$f = \Sigma m(1, 2, 3, 6) = \overline{A}\,\overline{B}C + \overline{A}B\overline{C} + \overline{A}BC + AB\overline{C}$$

Now, let us express f in terms of maxterms. By inspecting the truth table, $f = 0$ for maxterms M_0, M_4, M_5, and M_7. Therefore, the function f can be obtained by logically ANDing these maxterms as follows:

$$f = \Pi M(0, 4, 5, 7) = (A + B + C)(\overline{A} + B + C)(\overline{A} + B + \overline{C})(\overline{A} + \overline{B} + \overline{C})$$

3.7   Karnaugh Maps

A Karnaugh map or simply a K-map is a diagram showing the graphical form of a truth table. Since there is no specific set of rules for minimizing a Boolean function using identities, it is difficult to know whether the minimum expression is obtained. The K-map provides a systematic procedure for simplifying Boolean functions of typically up to five variables. However, a computer program using a tabular method such as the Quine-McCluskey algorithm can be used to minimize Boolean functions. Note that the theory behind K-map is based on Gray code and the identity, $A + \bar{A} = 1$.

The K-map is a diagram containing squares with each square representing one of the minterms of the Boolean function. For example, the K-map of two variables (A, B) contains four squares. The four minterms $\bar{A}\bar{B}$, $\bar{A}B$, $A\bar{B}$, and AB are represented by each square. Similarly, there are 8 squares for three variables, 16 squares for four variables, and 32 squares for five variables. Since any Boolean function can be expressed in terms of minterms, the K-map can be used to visually represent a Boolean function.

The K-map is drawn in such a way that there is only a one-bit change from one square to the next (Gray code). Squares can be combined in groups of 2^n where $n = 0,1,2,3,4,5$, and the Boolean function can be minimized by following certain rules. This minimum expression will reduce the total number of gates for implementation. Thus, the cost of building the logic circuit is reduced.

3.7.1 Two-Variable K-map

Figure 3.26 shows the K-map for two variables.

Since there are four minterms with two variables, four squares are required to represent them. This is depicted in the map of Figure 3.27(a). Each square represents a minterm. Figure 3.27(b) shows the K-map for two variables. Since each variable (A or B) has a value of 0 or 1 in the K-map of Figure 3.27(b), the 0 and 1 shown on the left of the map corresponds to 'A' while the 0 and 1 on the top are assigned to the variable 'B'. The squares containing minterms with one variable change are called *adjacent* squares. A square is adjacent to another square placed horizontally or vertically next to it. For example, consider the minterms m_0 and m_1. Since $m_0 = \bar{A}\bar{B}$ and $m_1 = \bar{A}B$, there is a one variable change ($\bar{B}$ in m_0 and B in m_1, $\bar{A}$ is the same in both squares). Therefore, m_0 and m_1 are adjacent squares. Similarly, other adjacent squares in the map include m_0 and m_2, or m_1 and m_3. $m_0(\bar{A}\bar{B})$ and $m_3(AB)$ are not adjacent squares since both variables change from 0's to 1's. The adjacent squares can be combined to eliminate one of the variables. This is based on the Boolean identities or $A + \bar{A} = 1$ or $B + \bar{B} = 1$.

The adjacent squares can also be identified by considering the map as a book. By closing the book at the middle vertical line, m_0 and m_2 will respectively be placed on m_1 and m_3. Thus, m_0 and m_1 are adjacent; squares m_2 and m_3 are also adjacent. Similarly, by closing the map at the middle horizontal line, m_0 will fall on m_2 while m_1 will be placed on m_3. Thus, m_0 and m_2 or m_1 and m_3 are adjacent squares.

Now, let us consider a Boolean function, $F = \Sigma m(0,1)$. Figure 3.28 shows that the function F containing two minterms m_0 and m_1 are identified by placing 1's in the corresponding squares of the map. In order to minimize the function F, the two squares can be combined as shown since they are adjacent. The map is then inspected for common variables looking at the squares vertically and horizontally. Since $A = 0$ is common to both squares, $F = \bar{A}$. This can be proven analytically by using Boolean identities as follows:

$$F = \sum m(0, 1) = \bar{A}\bar{B} + \bar{A}B$$

$$= \bar{A}(\bar{B} + B) = \bar{A} \text{ (since } \bar{B} + B = 1)$$

In a two-variable K-map, adjacent squares can be combined in groups of 2 or 4.

Next, consider $F = \Sigma m(0,2,3)$. The K-map is shown in Figure 3.29. 1's are placed in the squares defined by the minterms m_0, m_2, and m_3. By combining the adjacent squares m_0 with m_2 and m_2 with m_3, the common terms can be determined to simplify the function F. For example, by inspecting m_0 and m_2 vertically and horizontally, the term $\bar{B}$ is the common term. On the other hand, by looking at m_2 and m_3 horizontally and vertically, variable A is the common term. The minimized form of the function F can be obtained by logically ORing these common terms. Therefore,

$$F = A + \overline{B}.$$

Note that the function $F = 1$ for $F = \Sigma m(0, 1, 2, 3)$ for in which all squares (minterms) in the K-map are 1's.

3.7.2   Three-Variable K-map

Figure 3.30 shows the K-map for three variables. Figure 3.30(a) shows a map with three literals in each square. There are eight minterms (m_0, m_1, ..., m_7) for three variables. Figure 3.30(b) shows these minterms — one for each square in the K-map.

Like the two-variable K-map, a square in a three-variable K-map is adjacent to the squares placed horizontally or vertically next to it. Consider the minterms m_1, m_2, m_3, and m_7. For example, m_3 is adjacent *to* m_1, m_2, and m_7; m_1 is adjacent to m_3; m_2 is adjacent to m_3; m_7 is adjacent to m_3. But, m_7 is adjacent neither to m_1 nor to m_2; m_1 is not adjacent to m_2 and vice versa.

Like the two-variable map, the K-map can be considered as a book. The adjacent squares can also be determined by closing the book at the middle horizontal

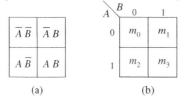

(a)   (b)

FIGURE 3.27   Two-variable K-map.

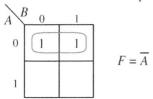

FIGURE 3.28   K-Map for $F = \Sigma m(0,1)$.

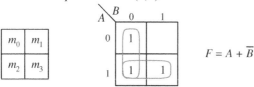

FIGURE 3.29   K-Map for $F = \Sigma m(0,2,3)$.

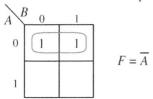

(a)   (b)

FIGURE 3.30   Three-variable K-map.

and vertical lines. For example, closing the book at the middle horizontal line, the adjacent pair of squares are m_0 and m_4, m_1 and m_5, m_3 and m_7, m_2 and m_6. On the other hand, closing the book at the middle vertical line, the adjacent pair of squares are m_0 and m_2, m_1 and m_3, m_4 and m_6, m_5 and m_7.

For a three variable K-map, adjacent squares can be combined in powers of 2: 1 (2^0), 2 (2^1), 4 (2^2) and 8 (2^3). The Boolean function is one when all eight squares are 1's. It is desirable to combine as many squares as possible. Note that grouping two adjacent squares in an n-variable (n=2^p) K-map will provide a product term of (n-p) literals. This means that, for a three-variable K-map, grouping two (2^1) adjacent squares will provide a product term of two literals and combining four (2^2) adjacent squares will provide a product term of one literal. The following examples illustrate this.

Example 3.3

Simplify the Boolean function
$$f(A, B, C) = \Sigma\, m(0, 2, 3, 4, 6, 7)$$
using a K-map.

Solution

Figure 3.31 shows the K-map along with the grouping of adjacent squares. First, a 1 is placed in the K-map for each minterm that represents the function. Next, the adjacent squares are identified by squares next to each other. Therefore, m_2, m_3, m_6, and m_7 can be combined as a group of adjacent squares. The common term for this grouping is B. Note that combining four (2^2) squares provides the result with only one literal, B. Next, by folding the K-map at the middle vertical line, adjacent squares m_0, m_2, m_4, and m_6 can be identified. Combining them together will provide the single common term $\overline{C}$. Therefore,
$$f = B + \overline{C}$$
This result can be verified analytically by using the identities as follows:
$$f = \Sigma\, m(0, 2, 3, 4, 6, 7)$$
$$= \overline{A}\,\overline{B}\,\overline{C} + \overline{A}B\overline{C} + \overline{A}BC + A\overline{B}\,\overline{C} + AB\overline{C} + ABC$$
$$= \overline{B}\,\overline{C}(A + \overline{A}) + B\overline{C}(\overline{A} + A) + BC(\overline{A} + A)$$
$$= \overline{B}\,\overline{C} + B\overline{C} + BC$$
$$= \overline{C}(\overline{B} + B) + BC$$
$$= \overline{C} + BC$$
$$= (B + \overline{C})(C + \overline{C}) = B + \overline{C} \qquad \text{(using the Distributive Law)}$$

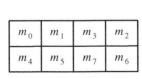
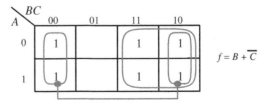

FIGURE 3.31   K-map for $f(A, B, C) = \Sigma\, m(0, 2, 3, 4, 6, 7)$.

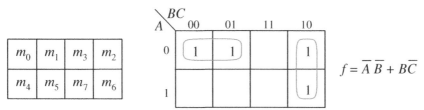

FIGURE 3.32   K-map for $f(A, B, C) = \Sigma\, m(0, 1, 2, 6)$.

Example 3.4

Simplify the Boolean function

$$f(A, B, C) = \Sigma\, m(0, 1, 2, 6)$$

using a K-map.

Solution

Figure 3.32 shows the K-map along with the grouping of adjacent squares.
From the K-map, grouping adjacent squares and logically ORing common product terms, $f = \overline{A}\,\overline{B} + B\overline{C}$

Example 3.5

(a)   Simplify the Boolean function
$F(A,B,C) = \overline{A}\,\overline{B}\,\overline{C} + A\overline{B}\overline{C} + \overline{B}C$
using a K-map.

(b)   Simplify the Boolean function for $F(A,B,C)$ given
$\overline{F}(A,B,C) = \Sigma m(0, 1, 4, 5)$
using a K-map.

Solution

(a)   The function contains three variables, A, B, and C, and is not expressed in minterm form. The first step is to express the function in terms of minterms as follows:

$$F(A,B,C) = \overline{A}\,\overline{B}\,\overline{C} + A\overline{B}\,\overline{C} + \overline{B}C(A + \overline{A})$$

$$= \overline{A}\,\overline{B}\,\overline{C} + A\overline{B}\,\overline{C} + A\overline{B}C + \overline{A}\,\overline{B}C$$

$$= \Sigma\, m(0,1,4,5)$$

Figure 3.33(a) shows the K-map. Note that the four (2^2) adjacent squares are grouped to provide a single literal by eliminating the other literals. Therefore, $F(A,B,C) = \overline{B}$. Although $F(A,B,C)$ is not expressed in minterm form, one can usually identify the squares with 1's in the K-map for the function $F(A,B,C) = \overline{A}\,\overline{B}\,\overline{C} + A\,\overline{B}C + \overline{B}C$ by inspection. This will avoid the lengthy process of converting such functions into minterm form.

(b)   Figure 3.33 (b) shows the K-map along with the grouping of adjacent squares. From the K-map, $\overline{F}(A,B,C) = \overline{B}$. Hence, $F(A,B,C) = B$

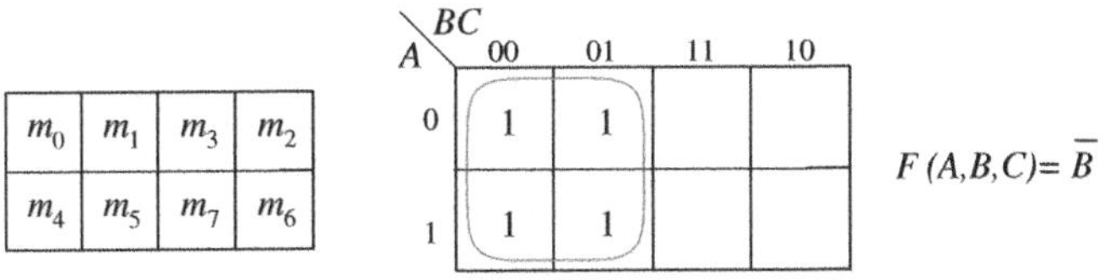

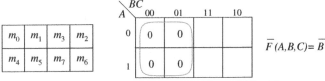

FIGURE 3.33   (a) K-map for $F(A,B,C) = \overline{A}\,\overline{B}\,\overline{C} + A\overline{B}\overline{C} + \overline{B}C$.

FIGURE 3.33   (b) K-map for $\overline{F}(A,B,C) = \overline{A}\,\overline{B}\,\overline{C} + A\overline{B}\overline{C} + \overline{B}C$.

3.7.3   Four-Variable K-map

A four-variable K-map, depicted in Figure 3.34, contains 16 squares because there are 16 minterms. Figure 3.34(a) includes four literals in each square. Figure 3.34(b) lists each minterm in its respective square. As before, a square is adjacent to the squares placed horizontally or vertically next to it. For example, in Figure 3.34(b) m_7 is adjacent to m_3, m_5, m_6, and m_{15}. Also, by closing the K-map at the middle vertical line, the adjacent pairs of squares are m_3 and m_1, m_2 and m_0, m_4 and m_6, m_{12} and m_{14}, m_8 and m_{10}, and so on. On the other hand, closing it at the middle horizontal line will provide the following adjacent squares: m_0 and m_8, m_1 and m_9, m_3 and m_{11}, m_2 and m_{10}, and so on.

For a four-variable K-map, adjacent squares can be grouped in powers of 2: 1 (2^0), 2 (2^1), 4 (2^2), 8 (2^3), and 16 (2^4). The Boolean function is one when all 16 minterms are 1. Combining two adjacent squares will provide a product term of three literals; four adjacent squares will provide a product term of two literals; eight adjacent squares will yield a product term of one literal.

Example 3.6

(a)   Simplify the Boolean function

$$f(A,B,C,D) = \Sigma m(0, 1, 2, 3, 8, 9, 10, 11, 12, 13, 14, 15)$$

using a K-map.

(b)   Simplify the Boolean function for $f(A,B,C,D)$ using a K-map given

$$\overline{f}(A,B,C,D) = \Sigma m(0, 1, 2, 3, 8, 9, 10, 11, 12, 13, 14, 15)$$

CD \ AB	00	01	11	10
00	$\overline{A}\,\overline{B}\,\overline{C}\,\overline{D}$	$\overline{A}\,\overline{B}\,\overline{C}\,D$	$\overline{A}\,\overline{B}\,C\,D$	$\overline{A}\,\overline{B}\,C\,\overline{D}$
01	$\overline{A}\,B\,\overline{C}\,\overline{D}$	$\overline{A}\,B\,\overline{C}\,D$	$\overline{A}\,B\,C\,D$	$\overline{A}\,B\,C\,\overline{D}$
11	$A\,B\,\overline{C}\,\overline{D}$	$A\,B\,\overline{C}\,D$	$A\,B\,C\,D$	$A\,B\,C\,\overline{D}$
10	$A\,\overline{B}\,\overline{C}\,\overline{D}$	$A\,\overline{B}\,\overline{C}\,D$	$A\,\overline{B}\,C\,D$	$A\,\overline{B}\,C\,\overline{D}$

(a)

CD \ AB	00	01	11	10
00	m_0	m_1	m_3	m_2
01	m_4	m_5	m_7	m_6
11	m_{12}	m_{13}	m_{15}	m_{14}
10	m_8	m_9	m_{11}	m_{10}

(b)

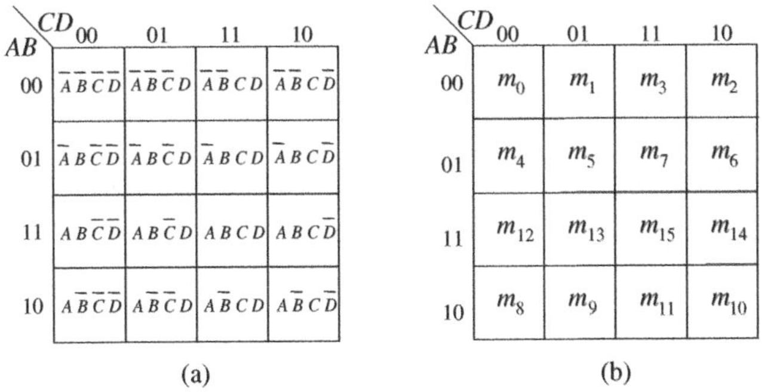

FIGURE 3.34 Four-variable K-map.

Solution

(a)   Figure 3.35(a) shows the K-map. The 8 adjacent squares combined in the bottom two rows yield the common product term of one literal, A. Because the top row is adjacent to the bottom row, combining the minterms in these two rows will provide a common product term of a single literal, $\overline{B}$. Therefore, by ORing these two terms, the minimized form of the function, $f = A + \overline{B}$ is obtained.

(b)   Figure 3.35 (b) shows the K-map. Since the equation for f (A,B,C,D) is given, the specified minterms are marked as 0's in the K-map. After grouping, $\overline{f}$ $(A,B,C,D) = A + \overline{B}$ is obtained. Hence,

$$f(A,B,C,D)= \overline{\overline{f}(A,B,C,D)} = \overline{A + \overline{B}} \text{ . Hence, using DeMorgan's theorem, } f(A,B,C,D) = \overline{A}B.$$

Example 3.7

Simplify the Boolean function $F(A,B,C,D) = \Sigma\, m(0, 2, 4, 5, 6, 8, 10)$ using a K-map.

Solution

Figure 3.36 shows the K-map. The common product term obtained by grouping the adjacent squares m_0, m_2, m_4, and m_6 will contain $\overline{A}\overline{D}$. The common product term obtained by grouping the adjacent squares m_0, m_2, m_8, and m_{10} will be $\overline{B}\overline{D}$. Combining the adjacent squares m_4 and m_5 will provide the common term $\overline{A}B\overline{C}$. ORing these common product terms will yield the minimum function, $F(A, B, C, D) = \overline{A}\overline{D} + \overline{B}\overline{D} + \overline{A}B\overline{C}$.

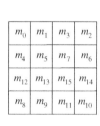
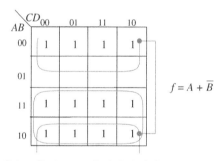

FIGURE 3.35 (a)   K-map for $f(A,B,C,D) = \Sigma m(0, 1, 2, 3, 8, 9, 10, 11, 12, 13,14, 15)$.

FIGURE 3.35 (b)   K-map for $\overline{f}(A,B,C,D) = \Sigma m(0, 1, 2, 3, 8, 9, 10, 11, 12, 13, 14, 15)$.

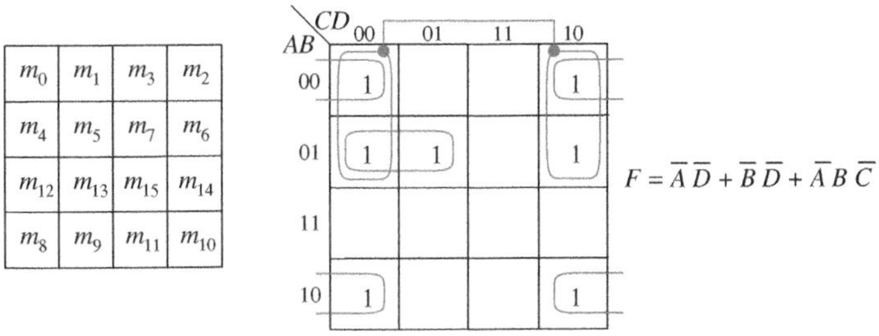

FIGURE 3.36   K-map for $F(A, B, C, D) = \Sigma\, m(0, 2, 4, 5, 6, 8, 10)$.

Example 3.8

Simplify the Boolean Function, $F = \overline{A}\,\overline{B}\,\overline{C} + \overline{A}B\,\overline{C} + \overline{A}B\,\overline{D} + \overline{A}\,\overline{B}\,CD$ using a K-map.

Solution

Figure 3.37 shows the K-map. In the figure, the function F can be expressed in terms of minterms as follows:

$$F = \overline{A}\,\overline{B}\,\overline{C}(D + \overline{D}) + \overline{A}B\overline{C}(D + \overline{D}) + \overline{A}B\overline{D}(C + \overline{C}) + \overline{A}\,\overline{B}CD$$

$$= \overline{A}\,\overline{B}\,\overline{C}D + \overline{A}\,\overline{B}\,\overline{C}\,\overline{D} + \overline{A}B\overline{C}D + \overline{A}B\overline{C}\,\overline{D} + \overline{A}B\overline{D}C + \overline{A}B\overline{D}\,\overline{C} + \overline{A}\,\overline{B}CD$$

$$= m_1 + m_0 + m_5 + m_4 + m_6 + m_4 + m_2$$

$$= m_1 + m_0 + m_5 + m_4 + m_6 + m_2 \quad \text{because } m_4 + m_4 = m_4$$

Rearranging the terms: $F = m_0 + m_1 + m_2 + m_4 + m_5 + m_6$

Therefore, $F = \Sigma\, m(0, 1, 2, 4, 5, 6)$

These minterms are marked as 1 in the K-map. The adjacent squares are grouped as shown. The minimum form of the function, $F = \overline{A}\overline{C} + \overline{A}\overline{D}$.

3.7.4   Prime Implicants

A prime implicant is the product term obtained as a result of grouping the maximum number of allowable adjacent squares in a K-map. The prime implicant is called *essential* if it is the only term covering the minterms. A prime implicant is called *nonessential* if another prime implicant covers the same minterms. The simplified expression for a function can be determined using the K-map as follows:

i)   Determine all the essential prime implicants.

ii)   Express the minimum form of the function by logically ORing the essential prime implicants obtained in i) along with other prime implicants that may be required to cover any remaining minterms not covered by the essential prime implicants.

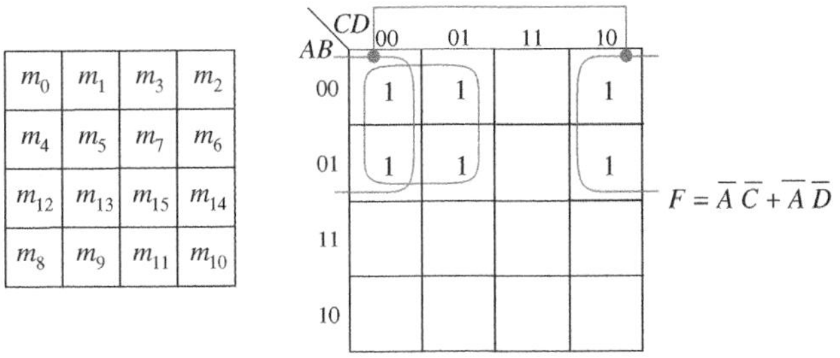

FIGURE 3.37   K-map for $F = \overline{A}\,\overline{B}\,\overline{C} + \overline{A}B\overline{C} + \overline{A}B\overline{D} + \overline{A}\overline{B}CD$.

Example 3.9

Find the prime implicants from the K-map of Figure 3.38 and then determine the simplified expression for the function.

Solution

The essential prime implicants are AB, $\overline{A}\,\overline{B}$ because minterms m_0 and m_1 can only be covered by the term $\overline{A}\,\overline{B}$ and minterms m_6 and m_7 can only be covered by the term AB.

The terms AC and $\overline{B}C$ are nonessential prime implicants because minterm m_5 can be combined with either m_1 or m_7. The term AC can be obtained by combining m_5 with m_7 whereas the term $\overline{B}C$ is obtained by combining m_5 with m_1. The function can be expressed in two simplified forms as follows:

$$f = \overline{A}\,\overline{B} + AB + AC$$
$$\text{or}$$
$$f = \overline{A}\,\overline{B} + AB + \overline{B}C$$

Example 3.10

Find the essential prime implicants from the K-map of Figure 3.39 and then find the simplified expression for the function.

Solution

The prime implicants can be obtained as follows:

1. By combining minterms m_5, m_7, m_{13}, and m_{15}, the prime implicant BD is obtained.
2. By combining minterms m_8, m_{10}, m_{12}, and m_{14}, the prime implicant $A\overline{D}$ is obtained.
3. By combining minterms m_{12}, m_{13}, m_{14}, and m_{15}, the prime implicant AB is obtained.

The terms BD and $A\overline{D}$ are essential prime implicants whereas AB is a nonessential prime implicant because minterms m_5 and m_7 can only be covered by the term BD and minterms

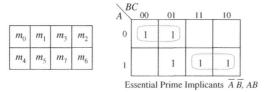

FIGURE 3.38   K-map for Example 3.9.

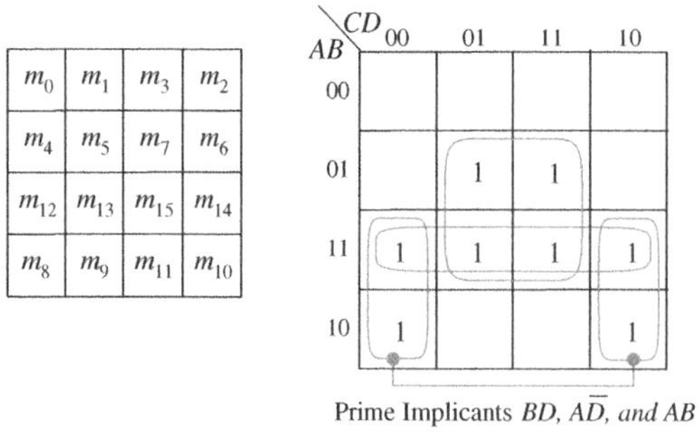

FIGURE 3.39   K-map for Example 3.10.

m_8 and m_{10} can only be covered by the term $A\overline{D}$. However, minterms m_{12}, m_{13}, m_{14}, and m_{15} can be covered by these two prime implicants (BD and $A\overline{D}$). Therefore, the term AB is not an essential prime implicant. Because all minterms are covered by the essential prime implicants, BD and $A\overline{D}$, the term AB is not required to simplify the function. Therefore,

$$f = BD + A\overline{D}$$

Example 3.11

Find the prime implicants and then simplify the function using a K-map.

$$f = \Sigma m(2, 4, 5, 8, 9, 13)$$

Solution

Figure 3.40 shows the K-map. The essential prime implicants are $\overline{A}\,\overline{B}C\overline{D}$, $\overline{A}B\overline{C}$, and $A\overline{B}\overline{C}$, and because minterms m_4 and m_5 can only be covered by the term $\overline{A}\,B\,\overline{C}$, minterms m_8 and m_9 can only be covered by the term $A\overline{B}\overline{C}$, and minterm m_2 can only be covered by the term $\overline{A}\,\overline{B}C\overline{D}$.

Minterm m_{13} can be combined with either m_5 or m_9. Combining m_{13} with m_5 will yield the term $B\overline{C}D$; combining m_{13} with m_9 will provide the term $A\,\overline{C}\,D$. Therefore, minterm m_{13} can be covered by either $B\overline{C}D$ or $A\overline{C}D$. Therefore, $B\overline{C}D$ and $A\overline{C}D$ are nonessential prime implicants. Hence, the function has two simplified forms:

$$f = \overline{A}\,\overline{B}C\overline{D} + \overline{A}B\overline{C} + A\overline{B}\,\overline{C} + B\overline{C}D$$
$$\text{or}$$
$$f = \overline{A}\,\overline{B}C\overline{D} + \overline{A}B\overline{C} + A\overline{B}\,\overline{C} + A\overline{C}D$$

3.7.5 Expressing a Boolean function in Product-of-sums (POS) form using a K-map

So far, the simplified Boolean functions derived from the K-map were expressed in sum-of-products (SOP) form. This section will describe the procedure for obtaining the simplified Boolean function in product-of-sums (POS) form.

So far, the minterms of a function, f in the K-map are represented by 1's. This means that the empty squares in the K-map can be identified as 0's. Hence, combining the appropriate adjacent squares for 0's will provide the simplified expression of the complement of the function ($\overline{f}$). By taking the complement of $\overline{f}$ and then using DeMorgan's theorem, the simplified expression for the function, f, can be obtained. The following example illustrates this.

FIGURE 3.40   K-map for $f = \Sigma\, m(2, 4, 5, 8, 9, 13)$.

Example 3.12

(a) Simplify the Boolean function, $f(A,B,C,D) = \Sigma m(0, 1, 4, 5, 6, 7, 8, 9, 14, 15)$ in product-of-sums (POS) form using a K-map.

(b) Simplify the Boolean function

$$f(A,B,C,D) = (\overline{A} + B + C + D)(A + B + \overline{C} + D)(A + B + C + D)(\overline{A} + B + \overline{C} + D)$$

using a K-map.

Solution

(a) Figure 3.41(a) shows the K-map. Combining the 0's, a simplified expression for the complement of the function can be obtained as follows:

$$\overline{f}(A,B,C,D) = \overline{B}C + AB\overline{C}$$

By DeMorgan's Theorem,

$$f(A,B,C,D) = \overline{\overline{f}}(A,B,C,D) = \overline{(\overline{B}C + AB\overline{C})} = \left(\overline{\overline{B}C}\right)\cdot\left(\overline{AB\overline{C}}\right) = (B+\overline{C})\cdot(\overline{A}+\overline{B}+C)$$

(b) The example in (a) illustrates the procedure for obtaining a simplified expression for *f* in product-of-sums (POS). The procedure is similar for simplifying a function expressed in product-of-sums (maxterms) form.

   To represent a function expressed in product-of-sums in the K-map, the complement of the function must first be obtained. The squares will then be identified as 0's for the minterms of the complement of the function. For example, consider the following function expressed in maxterm form:

$$f(A,B,C,D) = (\overline{A} + B + C + D)(A + B + \overline{C} + D)(A + B + C + D)(\overline{A} + B + \overline{C} + D)$$

   From Section 3.6, the function, $\overline{f}(A,B,C,D)$ can be obtained as follows:

$$\overline{f}(A, B,C,D) = A\overline{B}\,\overline{C}\,\overline{D} + \overline{A}\overline{B}C\overline{D} + \overline{A}\,\overline{B}\,\overline{C}\,\overline{D} + A\overline{B}C\overline{D}$$
$$= \Sigma m(0, 2, 8, 10)$$

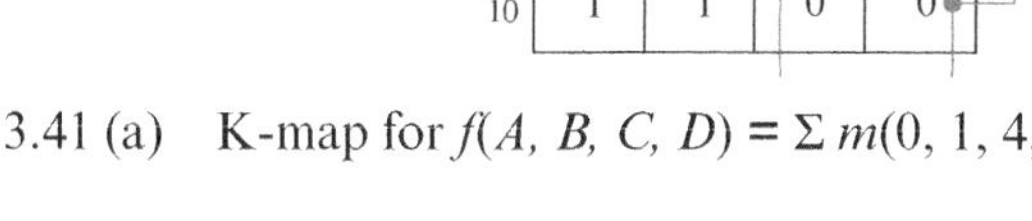

FIGURE 3.41 (a)   K-map for $f(A, B, C, D) = \Sigma m(0, 1, 4, 5, 6, 7, 8, 9, 14, 15)$.

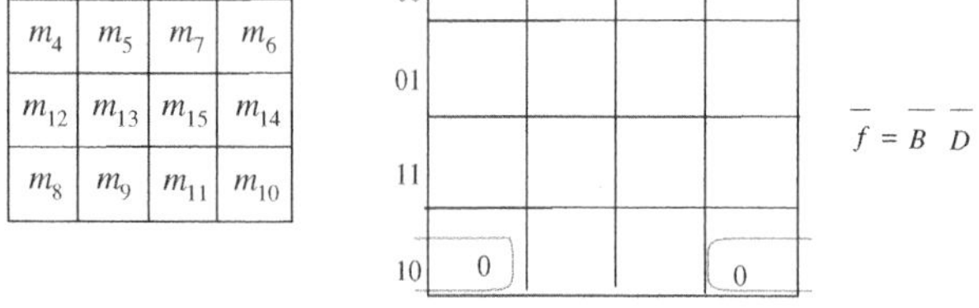

FIGURE 3.41 (b)   K-map for $= f(A,B,C,D) = \Sigma m(0, 2, 8, 10)$.

Placing 0's in the K-map for m_0, m_2, m_8, and m_{10} and then grouping adjacent 0's will provide the simplified expression for f. This is shown in Figure 3.41(b). Finally, the function, f, can be obtained by complementing the function, $\bar{f}$ and then applying DeMorgan's theorem.

From K-map of Figure 3.41(b), $\bar{f}(A, B, C, D) = \bar{B}\,\bar{D}$

Hence, $f(A,B,C,D) = \overline{\bar{f}A,B,C,D} = \overline{\bar{B}\bar{D}} = B + D$

3.7.6 Don't Care Conditions

So far, the squares of a K-map are marked with 1's for the minterms of a function. The other squares are assumed to be 0's. This is not always true, because there may be situations in which the function is not defined for all combinations of the variables. Such functions having undefined outputs for certain combinations of literals are called *incompletely specified functions*. One does not normally care about the value of the function for undefined minterms. Therefore, the undefined minterms of a function are called *don't care conditions*. Simply put, the don't care conditions are situations in which one or more literals in a minterm can never happen, resulting in nonoccurence of the minterm.

As an example, BCD numbers include ten digits (0 through 9) and are defined by four bits (0000_2 through 1001_2). However, one can represent binary numbers from 0000_2 through 1111_2 using four bits. This means that the binary combinations 1010_2 through 1111_2 (10_{10} through 15_{10}) can never occur in BCD. Therefore, these six combinations (1010_2 through 1111_2) are don't care conditions in BCD. The functions for these six combinations of the four literals are unspecified. The don't care condition is represented by the symbol X. This means that the symbol X will be placed inside a square in the K-map for which the function is unspecified. The don't care minterms can be used to simplify a function. The function can be minimized by assigning 1's or 0's for X's in the K-map while determining adjacent squares. These assigned values of X's can then be grouped with 1's or 0's in the K-map, depending on the combination that provides the minimum expression. Note that a don't care condition may not be required if it does not help in minimizing the function.

To understand the concept of don't care conditions, the circuit of Figure 3.42 is provided. In the figure, suppose that the digital circuit is required to turn the LED ON when the BCD inputs (W, X, Y, Z) are 0000_2 through 0110_2 and turn the LED OFF when the BCD inputs (W, X, Y, Z) are 0111_2 through 1001_2. However, when the inputs (W,X,Y,Z) are invalid BCD inputs 1010_2 through 1111_2, it is not specified in the problem whether the output LED will be turned ON or OFF. Hence, the output is undefined and we do not care whether the output LED is ON or OFF. In such a situation, we say that the output LED (L) will be don't cares (X) for invalid BCD values for W,X,Y, and Z. The truth table for this problem is shown in Table 3.2.

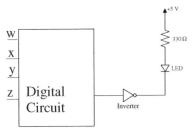

FIGURE 3.42 Figure illustrating the concept of Don't care conditions.

From the above truth table,

$$L = \overline{W}\,\overline{X}\,\overline{Y}\,\overline{Z} + \overline{W}\,\overline{X}\,\overline{Y}\,Z + \overline{W}\,\overline{X}\,Y\,\overline{Z} + \overline{W}\,\overline{X}\,YZ + \overline{W}X\,\overline{Y}\,\overline{Z} + \overline{W}\,X\,\overline{Y}Z + \overline{W}XY\,\overline{Z}$$

Hence, $L = \Sigma m(0, 1, 2, 3, 4, 5, 6)$. L can be minimized using a K-map shown in Figure 3.43 in which the X's (don't care conditions) can be used as 1's as needed to simplify the equation for L as $L = \overline{W}\,\overline{Y} + \overline{W}\,\overline{X} + Y\overline{Z}$

This will allow one to build the circuit of Figure 3.42 with minimum cost. Note that don't care conditions are used as needed for simplification. One must first group 1's along with don't care conditions (if needed) to minimize the equation. If 1's can be grouped without don't care conditions, then don't care conditions are not needed for simplification. The same rule applies while grouping 0's.

Now, suppose we re-define the problem of Figure 3.42 as follows: Turn the LED ON for valid BCD inputs (WXYZ = 0000_2 through 1001_2) and turn the LED OFF for invalid BCD inputs (WXYZ = 1010_2 through 1111_2). In this case, there are no don't care conditions since LED outputs for invalid BCD inputs are not undefined; they are defined as 0's. The LED outputs for the valid BCD inputs in the truth table will be 1's while the LED outputs for the invalid BCD inputs in the truth table will be

TABLE 3.2   Truth Table for Figure 3.42 with don't care conditions

INPUTS	LED Output
WXYZ	L
0000	1
0001	1
0010	1
0011	1
0100	1
0101	1
0110	1
0111	0
1000	0
1001	0
1010	X
1011	X
1100	X
1101	X
1110	X
1111	X

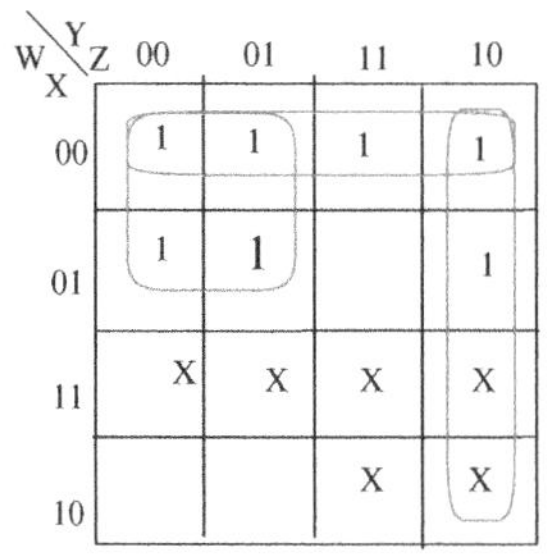

FIGURE 3.43   K-map for the Truth table of Table 3.2.

0's. The truth table is shown in Table 3.3. As before, from the truth table, L = Σm(0, 1, 2, 3, 4, 5, 6, 7, 8, 9). Figure 3.44 shows the K-map. The minimum equation for L (LED output) can be obtained in terms of the inputs (W,X,Y,Z) from the K-map as:
$L = \overline{W} + \overline{X}\ \overline{Y}$

Example 3.13

Simplify the Boolean function

$f(A,B,C) = \Sigma\, m(0, 3, 4, 6, 7)$

using a K-map. The don't care conditions, $d(A,B,C) = \Sigma m(2, 5)$

Solution

Figure 3.45 shows the K-map. To find the minimum equation for $f(A, B, C)$, one must always group the adjacent squares with 1's first, and use the don't care condition (s)

TABLE 3.3   Truth Table for figure 3.42 without don't care conditions

WXYZ	L
0000	1
0001	1
0010	1
0011	1
0100	1
0101	1
0110	1
0111	1
1000	1
1001	1
1010	0
1011	0
1100	0
1101	0
1110	0
1111	0

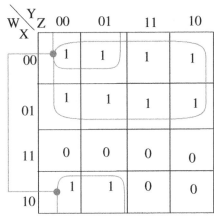

FIGURE 3.44   K-map for $L = \Sigma\, m(0, 1, 2, 3, 4, 5, 6, 7, 8, 9)$.

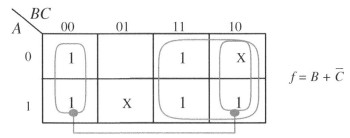

FIGURE 3.45   K-map for $f(A, B, C) = \Sigma\, m(0, 3, 4, 6, 7)$ with don't care conditions, $d(A, B, C) = \Sigma\, m(2, 5)$.

m_0	m_1	m_3	m_2
m_4	m_5	m_7	m_6
m_{12}	m_{13}	m_{15}	m_{14}
m_8	m_9	m_{11}	m_{10}

<table>
<tr><td>AB \ CD</td><td>00</td><td>01</td><td>11</td><td>10</td></tr>
<tr><td>00</td><td>1</td><td>X</td><td>0</td><td>1</td></tr>
<tr><td>01</td><td>X</td><td>1</td><td>X</td><td>X</td></tr>
<tr><td>11</td><td>1</td><td>0</td><td>X</td><td>0</td></tr>
<tr><td>10</td><td>1</td><td>0</td><td>0</td><td>1</td></tr>
</table>

FIGURE 3.46   K-map for Example 3.14 by combining 1's.

only if needed to find the most minimum solution. One must not use the don't care condition(s) if not needed to simplify. In this example, the don't care minterm, m_2 is used to obtain the minimum equation for $f(A, B, C)$, while the don't care minterm, m_5 is not required.

Hence, $f(A, B, C) = B + \overline{C}$

Example 3.14

Simplify the function $f(A, B, C, D) = \Sigma m(0, 2, 5, 8, 10, 12)$ using a K-map. Assume that the minterms $m_1, m_4, m_6, m_7,$ and m_{15} can never occur.

Solution

The don't care conditions are

$$d(A, B, C, D) = \Sigma m(1, 4, 6, 7, 15)$$

Figure 3.46 shows the K-map. By assigning $X = 1$ and combining 1's as shown, f can be expressed in sum-of-products form as follows:

$$f = \overline{C}\,\overline{D} + \overline{A}B + \overline{B}\,\overline{D}$$

On the other hand, by assigning $X = 0$ and combining 0's as shown in Figure 3.47, can be obtained as a product-of-sums. Thus,

$$\overline{f} = CD + AD + BC$$

$$f = \overline{\overline{f}} = \overline{CD + AD + BC}$$

$$= (\overline{CD})(\overline{AD})(\overline{BC})$$

$$= (\overline{C} + \overline{D})(\overline{A} + \overline{D})(\overline{B} + \overline{C})$$

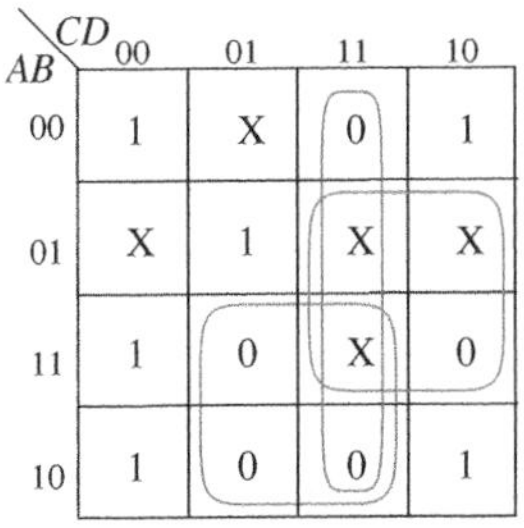

FIGURE 3.47 K-map for Example 3.14 by combining 0's.

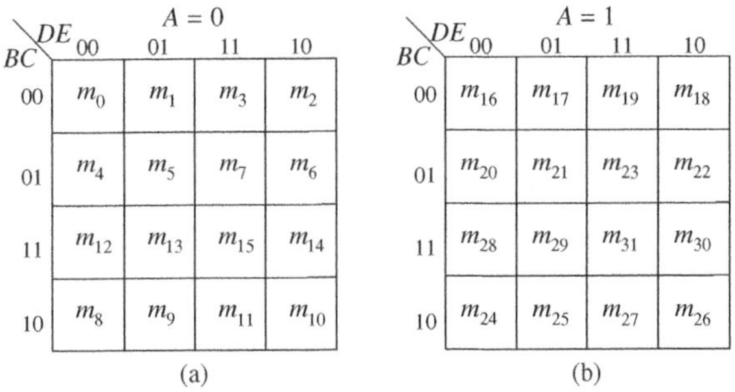

(a)   (b)

FIGURE 3.48 Five-Variable K-map.

3.7.7 Five-Variable K-map

Figure 3.47 shows a five-variable K-map. The five-variable K-map contains 32 squares. It contains two four-variable maps for $BCDE$ with $A = 0$ in one of the two maps and $A = 1$ in the other. The value of a minterm in each map can be determined by the decimal value of the five literals.

For example, minterm m_{14} from Figure 3.48(a) can be expressed in terms of the five literals as $\overline{A}BCD\overline{E}$. On the other hand, minterm m_{26} can be expressed in terms of the five literals from Figure 3.48(b) as $AB\overline{C}D\overline{E}$.

When simplifying a function, each K-map can first be considered as an individual four-variable map with $A = 0$ or $A = 1$. Combining of adjacent squares will be identical to typical four-variable maps. Next, the adjacent squares between the two K-maps can be determined by placing the map in Figure 3.48(a) on top of the map in Figure 3.48(b). Two squares are adjacent when a square in Figure 3.48(a) falls on the square in Figure 3.48(b) and vice versa. For example, minterm m_0 is adjacent to minterm m_{16}, minterm m_1 is adjacent to minterm m_{17}, and so on.

Example 3.15

Simplify the function
$$f(A, B, C, D, E) = \Sigma m(3, 7, 10, 11, 14, 15, 19, 23)$$
using a K-map.

Solution

Figure 3.48 shows the K-map.
$$f = \overline{A}BD + \overline{B}DE$$
To find the adjacent squares, the K-maps are first considered individually. From Figure 3.49(a), combining minterms m_{10}, m_{11}, m_{14}, and m_{15} will yield the product term $\overline{A}BD$. Minterms m_{19} and m_{23} are in the K-map of Figure 3.49(b). However, they are

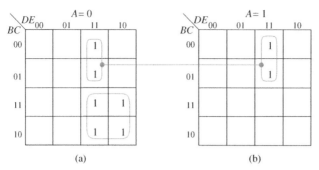

(a)   (b)

FIGURE 3.49   K-map for Example 3.15.

adjacent to minterms m_3 and m_7 in Figure 3.49(a). Combining m_3, m_7, m_{19}, and m_{23} together, the product term $\overline{B}DE$ can be obtained. Literals A or $\overline{A}$ are not included here because adjacent squares belong to both $A = 0$ and $A = 1$.

Therefore, the minimum form of f is

$$f = \overline{A}BD + \overline{B}DE$$

3.8   Quine–McCluskey Method

When the number of variables in a K-map is more than five, it becomes impractical to use K-maps in order to minimize a function. A tabular method known as *Quine–McCluskey* can be used. A computer program is usually written for the Quine–McCluskey method. One uses this program to simplify a function with more than five variables.

Like the K-map, the Quine–McCluskey method first finds all prime implicants of the function. A minimum number of prime implicants is then selected that defines the function. In order to understand the Quine–McCluskey method, an example will be provided using tables and manual check-off procedures. Although a computer program rather than manual approach is normally used by logic designers, a simple manual example is presented here so that the method can be easily understood.

The Quine–McCluskey method first tabulates the minterms that define the function. The following example illustrates how a Boolean function is minimized using the Quine–McCluskey method.

Example 3.16

In Example 3.7, $F(A, B, C, D) = \Sigma m(0, 2, 4, 5, 6, 8, 10)$ is simplified using a K-map. The minimum form is $F = \overline{A}\,\overline{D} + B\overline{D} + \overline{A}B\overline{C}$. Verify this result using the Quine–McCluskey method.

Solution

First arrange the binary representation of the minterms as shown in Table 3.4.
In the table, the minterms are grouped according to the number of 1's contained in their binary representations. For example, consider column (i). Because minterms m_2, m_4, and m_8 contain one 1, they are grouped together. On the other hand, minterms m_5, m_6, and m_{10} contain two 1's, so they are grouped together.

Next, consider column (ii). Any two minterms that vary by one bit in column (i) are grouped together in column (ii). Starting from the top row, proceeding to the bottom row, and comparing the binary representation of each minterm in column (i),

TABLE 3.4   Simplifying $F = \Sigma m(0, 2, 4, 5, 6, 8, 10)$ Using the Quine–McCluskey Method

	(i)					(ii)					(iii)			
Minterm	A	B	C	D			A	B	C	D			A	B	C	D
0	0	0	0	0	✓	0,2	0	0	–	0	✓	0,2,4,6	0	–	–	0
2	0	0	1	0	✓	0,4	0	–	0	0	✓	0,2,8,10	–	0	–	0
4	0	1	0	0	✓	0,8	–	0	0	0	✓	0,4,2,6	0	–	–	0
8	1	0	0	0	✓	2,6	0	–	1	0	✓	0,8,2,10	–	0	–	0
5	0	1	0	1	✓	2,10	–	0	1	0	✓					
6	0	1	1	0	✓	4,5	0	1	0	–						
10	1	0	1	0	✓	4,6	0	1	–	0	✓					
						8,10	1	0	–	0	✓					

pairs of minterms having only a one-variable change are grouped together in column (ii) with the variable bit replaced by the symbol –. For example, comparing $m_0 = 0000$ with $m_2 = 0010$, there is a one-variable change in bit position 1. This is shown in column (ii) by placing – in bit position 1 with the other three bits unchanged. Therefore, the top row of column (ii) contains 00–0. The procedure is repeated until all minterms are compared from top to bottom for one unmatched bit and are represented by replacing this bit position with – and other bits unchanged. A ü is placed on the right-hand side to indicate that this minterm is compared with all others and its pair with one bit change is found. If a minterm does not have another minterm with one bit change, no check mark is placed on its right. This means that the prime implicant will contain four literals and will be included in the simplified from of the function F. In column (i), for each minterm, which has a corresponding pair with one bit change is identified. These pairs are listed in column (ii).

Finally, consider column (iii). Each minterm pair in column (ii) is compared to the next, starting from the top, to find another pair with one bit change; for example m_0, $m_2 = 00$–0 and m_4, $m_6 = 01$–0. For this case, bit position 2 does not match. This bit position is replaced by – in the top row of column (iii). Therefore, in column (iii), the top row groups these four minterms 0, 2, 4, 6 with $ABCD$ as $0 – – 0$. Similarly, all other pairs in column (ii) are compared from top to bottom for one bit change and are listed accordingly in column (iii) if an unmatched bit is found. A check mark is placed in the right of column (ii) if an unmatched bit is found between two pairs. Note that minterms 4 and 5 do not have any other pair in the list of column (ii) having one unmatched bit. Therefore, this pair is not checked on the right and must be included in the simplified form of F as a prime implicant containing three variables. The two rows of column (iii) (0,2,4,6 and 0,4,2,6) are the same and contain $0 – – 0$. Therefore, this term should be considered once. Similarly, the groups 0,2,8,10 and 0,8,2,10 containing -0-0 should be considered ones. In column (iii), there are no more groups that exist with one unmatched bit.

The comparison process stops. The prime implicants will be the unchecked terms $\overline{A}B\overline{C}$ (from column (ii)) along with, $\overline{A}\,\overline{D}$ and $\overline{B}\overline{D}$ [from column (iii)]. Thus, the simplified form for F is $F = \overline{A}\,\overline{D} + \overline{B}\,\overline{D} + \overline{A}B\overline{C}$
This agrees with the result of Example 3.7.

3.9   Implementation of Digital Circuits with NAND, NOR, and Exclusive- OR/ Exclusive-NOR Gates

This section first covers implementation of logic circuits using NAND and NOR gates. These gates are extensively used for designing digital circuits. The NAND and NOR gates are called *universal gates* because any digital circuit can be implemented with them. These gates are , therefore, more commonly used than AND and OR gates.

Next, identities associated with Exclusive-OR operation are covered. Finally, design of parity generation and checking circuits using Exclusive-NOR gate is provided.

3.9.1 NAND Gate Implementation

Any logic operation can be implemented using NAND gates. Figure 3.50 shows how NOT, AND, OR, and Invert-OR operations can be implemented with NAND gates. Note that Invert-OR operation is required for sum-of-products (SOP) implementation. This will be shown later in this section.

A Boolean function can be implemented using NAND gates by first obtaining the simplified expression of the function in terms of AND, OR, and NOT operations. The function can then be converted to NAND-only logic. A function expressed in sum-of-products (SOP) form can be readily implemented using only NAND gates.

Example 3.17

Implement the simplified function $F = \overline{XY + XZ}$ using NAND gates.

Solution

First implement the function using AND, OR, and NOT gates as follows:

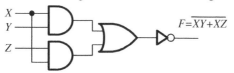

Now convert the AND, OR, and NOT gates to NAND gates as follows:

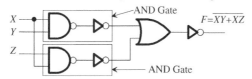

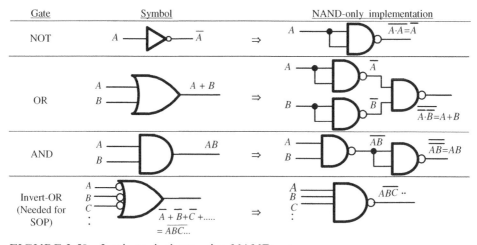

FIGURE 3.50   Logic equivalents using NAND gates.

The NOT gates can be represented as bubbles at the inputs of the OR gate as shown below.

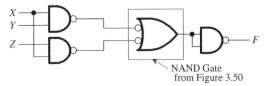

NAND Gate
from Figure 3.50

Therefore, the function $F = \overline{X\,Y} + \overline{XZ}$ can be implemented using only NAND gates as follows:

This is a three-level implementation since 3 gate delays are required to obtain the output F. This circuit may not be cost effective. But it may be convenient to build the circuit with only NAND gates without any other logic gates.

Example 3.18

Simplify the following Boolean function using a K-map:

$$f(A, B, C, D) = \Sigma m(0, 4, 8, 11, 12, 15)$$

Draw a schematic using NAND gates.

Solution

In order to draw a schematic using only NAND gates, we must express $f(A, B, C, D)$ in SOP form by grouping adjacent 1's in the K-map as shown in Figure 3.51.From the K-map,

$$f(A, B, C, D) = \overline{C}\overline{D} + ACD$$

Figure 3.52 shows the logic diagram using AND, OR and NOT gates.

Note that the logic circuit has four gate delays. Figure 3.53 shows the various steps for converting the schematic of Figure 3.52 to a a logic diagram with NAND gates. In Figure 3.53(a), each AND gate of Figure 3.52 is represented by an AND gate with two inverters at the output. For example, consider AND gate 1 of Figure 3.52. The AND gate and an inverter are used to form the NAND gate shown in the top row of Figure 3.53(b) with an inverter (indicated by a bubble at the OR gate input). AND gate 3 is represented in the same way as AND gate 1 in Figure 3.53(b).

Finally, in Figure 3.53(c), the OR gate with the bubbles at the input in Figure 3.53(b) is replaced by a NAND gate. Thus, the NAND gate implementation in Figure 3.53(c) is obtained.

3.9.2 NOR Gate Implementation

As with NAND gates, any logic operation can be implemented using NOR gates. Figure 3.54 shows how NOT, AND, OR, and Invert-AND operations can be implemented with NOR gates. Note that Invert-AND operation is required for product of sums (POS) implementation.

A Boolean function can be implemented using NOR gates by first obtaining the simplified expression of the function in terms of AND-OR- NOT logic operations.

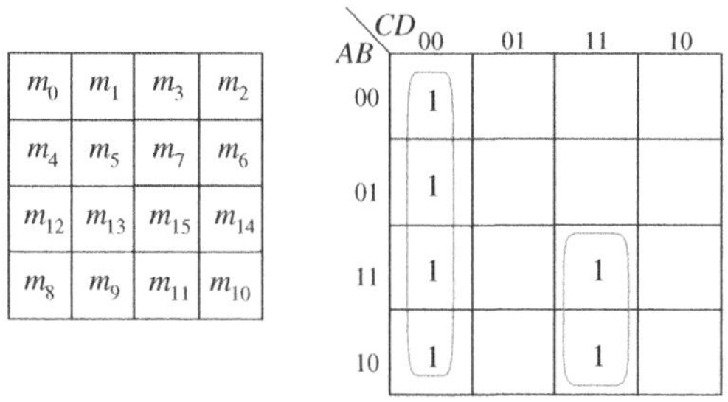

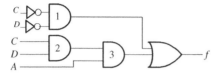

m_0	m_1	m_3	m_2
m_4	m_5	m_7	m_6
m_{12}	m_{13}	m_{15}	m_{14}
m_8	m_9	m_{11}	m_{10}

FIGURE 3.51   K-map for Example 3.18.

FIGURE 3.52   Logic diagram for implementation of $f = \overline{C}\overline{D} + ACD$.

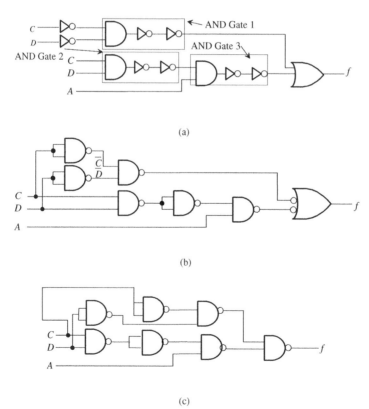

(a)

(b)

(c)

FIGURE 3.53   Steps for NAND gate implementation of Figure 3.52.

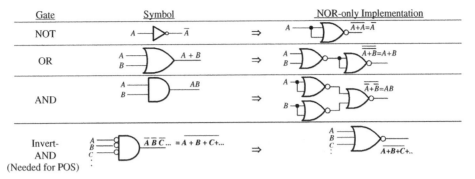

FIGURE 3.54   Logic equivalents using NOR gates.

The function can then be converted to NOR-only logic. A function expressed in product-of-sums (POS) form can be readily implemented using only NOR gates.

Example 3.19

Implement the following function using NOR gates:

$$f = \bar{a}\,(b+c)\,(a+d)$$

Solution

Figure 3.55 shows the AND-OR implementation of the logic equation. Figure 3.56 shows the NOR implementation.

Example 3.20

Simplify the function for $f\,(A,B,C,D)$ given $\bar{f}\,(A,\,B,C,D) = \Sigma m(3, 9, 11, 13, 14)$ using a K-map. Assume that the don't care conditions are $d(A,\,B,C,D) = \Sigma m(1, 4, 6, 7, 15)$. Draw a schematic using only NOR gates.

Solution

In order to draw a schematic using only NOR gates, we must express $f(A,\,B,C,D)$ in POS form by grouping adjacent 0's in the K-map.
Figure 3.57 shows the K-map. By assigning X = 0 and combining adjacent 0's as shown, $f(A,\,B,C,D)$ can be expressed in product-of-sums (POS) form as follows: $\bar{f} = CD + AD + BC$. Hence,

$$f = \bar{\bar{f}} = \overline{CD + AD + BC}$$

$$= (\overline{CD})(\overline{AD})(\overline{BC})$$

$$= (\bar{C} + \bar{D})(\bar{A} + \bar{D})(\bar{B} + \bar{C})$$

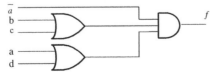

FIGURE 3.55   AND-OR implementation of Example 3.19.

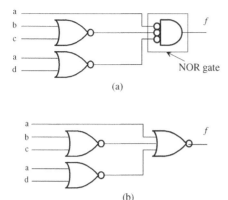

(a)

(b)

FIGURE 3.56   NOR implementation of Example 3.19.

Figure 3.57   K-map for Example 3.20.

Figure 3.58(a) shows the logic diagram using AND, OR and NOT gates. Figure 3.58(b) and 3.58(c) show the various steps for converting the schematic of Figure 3.58(a) to a logic diagram with NOR gates.

3.9.3   XOR / XNOR Implementations

As mentioned before, the Exclusive-OR operation between two variables A and B can be expressed as

$$A \oplus B = A\overline{B} + \overline{A}B.$$

The Exclusive-NOR or equivalence operation between A and B can be expressed as

$$A \odot B = \overline{A \oplus B} = AB + \overline{A}\,\overline{B}.$$

The following identities are applicable to the Exclusive-OR operation:
i) $A \oplus 0 = A \cdot 1 + \overline{A} \cdot 0 = A$
ii) $A \oplus 1 = A \cdot 0 + \overline{A} \cdot 1 = \overline{A}$
iii) $A \oplus A = A \cdot \overline{A} + \overline{A} \cdot A = 0$
iv) $A \oplus \overline{A} = A \cdot A + \overline{A} \cdot \overline{A} = A + \overline{A} = 1$

Finally, Exclusive-OR is commutative and associative:

$A \oplus B = B \oplus A$
$(A \oplus B) \oplus C = A \oplus (B \oplus C)$
$= A \oplus B \oplus C$

The Exclusive-NOR operation among three or more variables is called an *even function* because the Exclusive-NOR operation among three or more variables includes product terms in which each term contains an even number of 1's. For example, consider Exclusive-NORing three variables as follows:

$$f = \overline{A \oplus B \oplus C} = \overline{(A\overline{B} + \overline{A}B) \oplus C}$$

Let $D = A\overline{B} + \overline{A}B$. Then $\overline{D} = \overline{A\overline{B} + \overline{A}B} = AB + \overline{A}\,\overline{B}$.

Hence, $f = \overline{D \oplus C}$

$$= DC + \overline{D}\,\overline{C}$$

$$= (A\overline{B} + \overline{A}B)C + \overline{(A\overline{B} + \overline{A}B)}\,\overline{C}$$

$$= (A\overline{B} + \overline{A}B)C + (AB + \overline{A}\,\overline{B})\overline{C}$$

Hence, $f = A\overline{B}C + \overline{A}BC + ABC + \overline{A}\,\overline{B}\,\overline{C}$

Note that in this equation, $f = 1$ when one or more product terms in the equation are 1. However, by inspection, the binary equivalents of the right-hand side of the equation are 101, 011, 110, and 000. That is, the function is expressed as

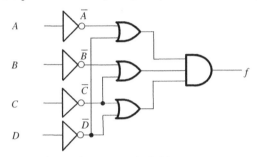

FIGS 3.58 (a)   Figure for Example 3.20.

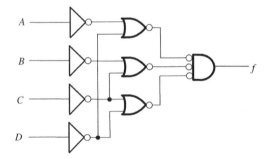

FIGS 3.58 (b)   Figure for Example 3.20.

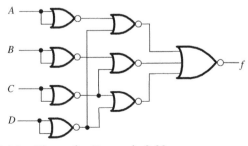

FIGS 3.58 (c)   Figure for Example 3.20.

the logical sum (OR) of product terms containing even numbers of ones. Therefore, the function is called an *even function*. Similarly, it can be shown that Exclusive-OR operation among three or more variables is an odd function.

Exclusive-OR or Exclusive-NOR operation can be used for error detection and correction using parity during data transmission. Note that parity can be classified as either odd or even. The parity is defined by the number of 1's contained in a string of data bits. When the data contains an odd number of 1's, the data is said to have "odd parity"; On the other hand, the data has "even parity" when the number of 1's is even. To illustrate how parity is used as an error check bit during data transmission, consider Figure 3.59.

Suppose that Computer X is required to transmit a 3-bit message to Computer Y. To ensure that data is transmitted properly, an extra bit called the *parity bit* can be added by the transmitting Computer X before sending the data. In other words, Computer X generates the parity bit depending on whether odd or even parity is used during the transmission. Suppose that odd parity is used. The odd parity bit for the three-bit message will be as follows:

Message			Odd Parity Bit
A	B	C	P
0	0	0	1
0	0	1	0
0	1	0	0
0	1	1	1
1	0	0	0
1	0		1
1	1	0	1
1	1	1	0

Here $P = 1$ when the 3-bit message ABC contains an even number of 1's. Thus, the parity bit will ensure that the 3-bit message contains an odd number of 1's before transmission. $P = 1$ when the message contains an even number of 1's. Therefore, P is an even function. Thus,

FIGURE 3.59　Parity generation and checking.

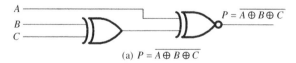

(a) $P = \overline{A \oplus B \oplus C}$

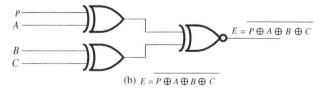

(b) $E = \overline{P \oplus A \oplus B \oplus C}$

FIGURE 3.60　Implementation of parity generation and checking using XOR / XNOR gates.

$$P = \overline{A \oplus B \oplus C}$$

The transmitting Computer X generates this parity bit. Computer X then transmits 4-bit information (a 3-bit message along with the parity bit) to Computer Y. Computer Y receives this 4-bit information and checks to see whether each 4-bit data item contains an odd number of 1's (odd parity). If the parity is odd, Computer Y accepts the 3-bit message; otherwise the computer sends the 4-bit information back to Computer X for retransmission. Note that Computer Y checks the parity of the transmitted data using the equation

$$E = \overline{P \oplus A \oplus B \oplus C}$$

Here the error $E = 1$ if the four bits have an even number of ones (even parity). That is, at least one of the four bits is changed during transmission. On the other hand, the error bit, $E = 0$ if the 4-bit data has an odd number of ones. Figure 3.60 shows the implementation of the parity bit, $P = \overline{A \oplus B \oplus C}$, and the error bit, $E = \overline{P \oplus A \oplus B \oplus C}$.

Questions and Problems

3.1   Perform the following operations. Include your answers in hexadecimal.
        $A6_{16}$ OR 31_{16}; $F7A_{16}$ AND $D80_{16}$; $36_{16} \oplus 2A_{16}$

3.2   Given $A = 1001_2$, $B = 1101_2$, find: A OR B; $B \wedge A$; $\overline{A}$; $A \oplus A$.

3.3   Perform the following operation: $A7_{16} \oplus FF_{16}$.
        What is the relationship of the result to $A7_{16}$?

3.4   Draw a logic diagram to implement $F = ABCDE$ using only 3-input AND gates.

3.5   Draw a logic diagram using two-input AND and OR gates to implement the
        following function $F = P(P + Q)(P + Q + R)(P + Q + R + S)$ without any
        simplification; then analyze the logic circuit to verify that $F = P$.

3.6   Prove the following identities algebraically and by means of truth tables:

        (a) $(A + B)(\overline{A + B}) = 0$

        (b) $A + \overline{A}B = A + B$

        (c) $XY + \overline{X}\,\overline{Y} + X\overline{Y} + \overline{X}Y = 1$

        (d) $(\overline{A + \overline{A}B}) = \overline{A}\,\overline{B}$

        (e) $(\overline{X} + Y)(X + \overline{Y}) = \overline{X \oplus Y}$

        (f) $\overline{B}\,\overline{C} + ABC + \overline{A}\,\overline{C} = \overline{C \oplus (AB)}$

3.7   Simplify each of the following Boolean expressions as much as possible using
        identities:

        (a) $XY + (1 \oplus X) + X\overline{Z} + X\overline{Y} + XZ$
        (b) $AB\overline{C} + AB\overline{C}D + AB\overline{D}$
        (c) $BC + \overline{A}BC\overline{D} + \overline{A}BCD + \overline{A}BCD$
        (d) $(\overline{X} + \overline{Y})(\overline{X}\overline{Y}) + ZX\underline{Y} + X\overline{Z}Y$
        (e) $(A + \overline{B})(A + B)(\overline{AB})$
        (f) $(A + \overline{A}B)((\overline{A} + B + C)(\overline{C})((\overline{A} + B)$
        (g) $x\,\overline{z} + \overline{x}yz + x\,\overline{y}$
        (h) $y\,\overline{z} + x\,y\,\overline{z} + xz + y$
        (i) $(A + \overline{A}B)(A + B) + \overline{B}$
        (j) $(\overline{X} + \overline{Y}) + (X \oplus (X \oplus XY))$
        (k) $(X \oplus (\overline{X}Y \oplus (X \oplus X)))$

3.8   Using DeMorgan's theorem, draw logic diagrams for $AB\overline{C} + \overline{A}\overline{B} + BC$
        (a)   Using only AND gates and inverters.
        (b)   Using only OR gates and inverters.
        You may use two-input and three-input AND and OR gates for (a) and (b).

3.9   Using truth tables, express each one of the following functions and their
        complements in terms of sum of minterms and product of maxterms:
        (a) $F = ABC + \overline{A}BD + \overline{A}\,\overline{B}\,\overline{C} + AC\overline{D}$
        (b) $F = (W + X + Y)(W\overline{X} + Y)$

3.10 Express each of the following expressions in terms of minterms and maxterms.

(a) $F = B\overline{C} + \overline{A}B + B(A + C)$
(b) $F = (A + \overline{B} + C)(\overline{A} + B)$

3.11 Minimize each of the following functions using a K-map:

(a) $F(A, B, C) = \Sigma\, m(0, 1, 4, 5)$
(b) $F(A, B, C) = \Sigma\, m(0, 1, 2, 3, 6)$
(c) $F(X, Y, Z) = \Sigma\, m(0, 2, 4, 6)$

3.12 Minimize each of the following expressions for F using a K-map.

(a) $F(A,B,C) = \overline{B}\,\overline{C} + ABC + AB\overline{C}$
(b) $F(A,B,C) = \overline{A}B\overline{C} + BC$
(c) $F(A,B,C) = \overline{A}C + A(B\overline{C} + B\overline{C})$

3.13 Simplify each of the following functions for F using a K-map.

(a) $F(W, X, Y, Z) = \Sigma\, m(0, 1, 4, 5, 8, 9)$
(b) $F(A,B,C,D) = \Sigma\, m(0, 2, 8, 10, 12, 14)$
(c) $\overline{F}(A,B,C,D) = \Sigma\, m(2, 4, 5, 6, 7, 10, 14)$
(d) $F(W, X, Y, Z) = \Sigma\, m(2, 3, 6, 7, 8, 9, 12, 13)$
(e) $\overline{F}(W, X, Y, Z) = \Sigma\, m(0, 2, 4, 6, 8, 10, 12, 14)$
(f) $F(W, X, Y, Z) = \Sigma\, m(1, 3, 5, 7, 9, 11, 13, 15)$

3.14 Minimize each of the following expressions for $\overline{F}$ using a K-map in sums-of-product form:

(a) $F(W, X, Y, Z) = \overline{W}\overline{X}YZ + WYZ$
(b) $F = \overline{A}\,\overline{B}\,\overline{C}\,\overline{D} + \overline{A}CD + ABCD$
(c) $F = (\overline{A} + \overline{B} + C + \overline{D})(\overline{A} + B + C + \overline{D})(A + \overline{B} + C + \overline{D})$

3.15 Find essential prime implicants and then minimize each of the following functions for F using a K-map:

(a) $F(A, B, C, D) = \Sigma\, m(3, 4, 5, 7, 11, 12, 15)$
(b) $F(W, X, Y, Z) = \Sigma\, m(2, 3, 6, 7, 8, 9, 12, 13, 15)$

3.16 Minimize each of the following functions for f using a K-map and don't care conditions, d.

(a) $f(A, B, C) = \Sigma\, m(1, 2, 4, 7)$
    $d(A, B, C) = \Sigma\, m(5, 6)$
(b) $f(X, Y, Z) = \Sigma\, m(2, 6)$
    $d(X, Y, Z) = \Sigma\, m(0, 1, 3, 4, 5, 7)$
(c) $f(A, B, C, D) = \Sigma\, m(0, 2, 3, 11)$
    $d(A, B, C, D) = \Sigma\, m(1, 8, 9, 10)$
(d) $f(A, B, C, D) = \Sigma\, m(4, 5, 10, 11)$
    $d(A, B, C, D) = \Sigma\, m(12, 13, 14, 15)$

3.17 Minimize the following expression using the Quine–McCluskey method. Verify the results using a K-map. Draw logic diagrams using NAND gates. Assume true and complemented inputs.

$$F(A, B, C, D) = \Sigma\, m(0, 1, 4, 5, 8, 12)$$

3.18   Minimize the following expression using a K-map:
$F = AB + \overline{A}\,\overline{B}\,\overline{C}\,\overline{D} + C\overline{D} + \overline{A}\,\overline{B}\,C\,D$
and then draw schematics using:
(a)   NAND gates.
(b)   NOR gates.

3.19   Minimize the following function $F(A, B, C, D) = \Sigma\,m(6, 7, 8, 9)$ using a K-map assuming that the condition $AB = 11_2$ can never occur. Draw schematics using:
(a)   NAND gates
(b)   NOR gates

3.20   Simplify each of the following functions for F using a K-map. Draw a logic diagram for each using NAND-only and NOR-only gates.
(a) $F(W, X, Y, Z) = \Sigma\,m(0, 2, 4, 8, 12)$
(b) $F(W, X, Y, Z) = \Sigma\,m(0, 2, 4, 7, 11, 12, 13)$
(c) $F(A, B, C, D) = \Sigma\,m(1, 3, 5, 6, 8, 9, 10, 14, 15)$
(d) $\overline{F}(A, B, C, D) = \Sigma\,m(0, 1, 4, 5, 12, 13)$
         Don't Cares, d $(A, B, C, D) = \Sigma\,m(2, 6, 14)$
(e) $F(A, B, C, D) = \Sigma\,m(3, 7, 8, 9, 12, 15)$
         Don't Cares, d $(A, B, C, D) = \Sigma\,m(2, 6, 14)$
(f) $F(W, X, Y, Z) = \Sigma\,m(4, 6, 8, 12)$
         Don't Cares, $d\,(W, X, Y, Z) = \Sigma\,m(11, 14)$

3.21   It is desired to compare two 4-bit numbers for equality. If the two numbers are equal, the circuit will generate an output of 1. Draw a logic circuit using a minimum number of gates of your choice.

3.22   Show analytically that $A \oplus (A \oplus B) = B$.

3.23   Show that the Boolean function, $f = A \oplus B \oplus AB$ between two variables, A *and B*, can be implemented using a single two-input gate.

3.24   Design a parity generation circuit for a 5-bit data (4-bit message with an even parity bit) to be transmitted by computer X. The receiving computer Y will generate an error bit, $E = 1$, if the 5-bit data received has an odd parity; otherwise, $E = 0$. Draw logic diagrams for both parity generation and checking using XOR gates.

3.25   Draw a logic diagram for a two-input (A,B) Exclusive-OR operation using only four two-input (A,B) NAND gates. Assume that complemented inputs $\overline{A}$ and $\overline{B}$ are not available.

3.26   Determine by inspection whether the function, F in each of the following is odd or even, and comment on the result:
(a) $\overline{F} = A \oplus B \oplus C$
(b) $F = A \oplus B \oplus C$

4

COMBINATIONAL LOGIC

4.1 Basic Concepts

Digital logic circuits can be classified into two types: combinational and sequential. A combinational circuit is designed using logic gates in which application of inputs generates the outputs at any time. An example of a combinational circuit is an adder, which produces the result of addition as output upon application of the two numbers to be added as inputs.

A sequential circuit, on the other hand, is designed using logic gates and memory elements known as *flip-flops*. Note that the flip-flop is a one-bit memory. A sequential circuit generates the circuit outputs based on the present inputs and the outputs (states) of the memory elements. The sequential circuit is basically a combinational circuit with memory. Note that a combinational circuit does not require any memory (flip-flops),whereas sequential circuits require flip-flops to remember the present states. A counter is a typical example of a sequential circuit. To illustrate the sequential circuit, suppose that it is desired to count in the sequence 0, 1, 2,3, 0, 1,... and repeat. In binary, the sequence is 00, 01, 10, 11, 00, 01, ..., and so on. This means that a two-bit memory using two flip-flops is required for storing the two bits of the counter because each flip-flop stores one bit. Let us call these flip-flops with outputs A and B. Note that initially $A = 0$ and $B = 0$. The flip-flop changes outputs upon application of a clock pulse. With appropriate inputs to the flip-flops and then applying the clock pulse, the flip-flops change the states (outputs) to $A = 0$, $B = 1$. Thus, the count to 1 can be obtained. The flip-flops store (remember) this count. Upon application of appropriate inputs along with the clock, the flip-flops will change the status to $A = 1$, $B = 0$; thus, the count to 2 is obtained. The flip-flops remember (store) the present value of the count at the outputs until a common clock pulse is applied to the flip-flops. The inputs to the flip-flops are manipulated by a combinational circuit based on A and B as inputs. For example, consider $A = 1$, $B = 0$. The inputs to the flip-flops are determined in such a way that the flip-flops change the states at the clock pulse to $A = 1$, $B = 1$; thus, the count to 3 is obtained. The process is repeated.

In this chapter, analysis and design of typical combinational circuits will be covered.

4.2 Analysis of a Combinational Logic Circuit

Analyzing a combinational circuit means that the circuit is given, the output equation (s) in terms of the input (s) and the truth table need to be determined. Hence, the combinational logic circuit can be analyzed by (i) first, identifying the number of inputs and outputs, (ii) expressing the output functions in terms of the inputs, and (iii) determining the truth table for the logic diagram.

Fundamentals of Digital Logic and Microcontrollers, Sixth Edition. M. Rafiquzzaman.
© 2014 John Wiley & Sons, Inc. Published 2014 by John Wiley & Sons, Inc.

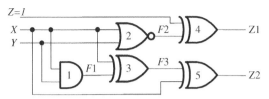

FIGURE 4.1   Analysis of a combinational logic circuit.

As an example, consider the combinational circuit in Figure 4.1 There are three inputs (X, Y, and Z) and two outputs ($Z1$ and $Z2$) in the circuit. Let us now express the outputs $F1$ and $F2$ in terms of the inputs. The output $F1$ of the AND gate #1 is $F1 = XY$. The output $F2$ of NOR gate #2 can be expressed as $F2 = \overline{X+Y}$. The output of the XOR gate #3 is

$$F3 = X \oplus F1 = (X \oplus XY)$$

Because one of the inputs of the XOR gate #4 is 1, its output is inverted. Therefore,

$$Z1 = \overline{F2} = X + Y$$

Finally,

$$Z2 = X \oplus F3 = X \oplus (X \oplus XY)$$

Therefore,

$$Z2 = X \oplus (X(1 \oplus Y))$$

$$= X \oplus X\bar{Y} \text{ since } 1 \oplus Y = \bar{Y}$$

$$= X(1 \oplus \bar{Y})$$

$$= XY \text{ since } 1 \oplus \bar{Y} = Y$$

Another way of determining $Z2$ is provided below:

$$Z2 = X \oplus F3 = X \oplus (X \oplus XY) = X \oplus X \oplus XY = 0 \oplus (XY) = XY$$

The truth table shown in Table 4.1 can be obtained by using the logic equations for $Z1$ and $Z2$.

4.3   Design of a Combinational Circuit

Design of a combinational circuit is basically a reverse process of analyzing the circuit. Note that the term *design* means that a designer needs to come up with a logic circuit following certain criteria such as minimum cost and highest speed. In this book, logic circuits are designed based on minimum cost. That is why minimization techniques are used to reduce the number of logic gates in a circuit. This will provide lower implementation cost. In summary, a combinational circuit can be designed using three steps as follows:

1) **Truth Table:** If the truth table is given, go to step 2; else, determine the inputs and outputs from problem definition and then derive the truth table.

TABLE 4.1   Truth table for Figure 4.1 with input, $Z = 1$

Inputs		Outputs	
X	Y	$Z1$	$Z2$
0	0	0	0
0	1	1	0
1	0	1	0
1	1	1	1

TABLE 4.2   Truth Table for **F**

A	B	C	F
0	0	0	0
0	0	1	1
0	1	0	1
0	1	1	1
1	0	0	1
1	0	1	1
1	1	0	1
1	1	1	0

2) **Minimization:** Minimize the number of inputs (literals) in order to express the outputs. This reduces the number of gates and thus the implementation cost. In order to accomplish this, first use K-map(s). Note that the number of K-maps is determined by the number of outputs. If the output(s) cannot be simplified using K-map(s), then obtain output equation(s) from the truth table directly, and then simplify using Boolean identities. Simplification is also sometimes possible using the K-map(s) first and then using Boolean identities.

3) **Logic diagram or schematic:** Draw the logic diagram or schematic using logic gates. Note that since the terms *Logic diagram* and *Schematic* mean the same thing, they will be used interchangeably throughout this book.

In order to illustrate the design procedure, consider the following example. Suppose it is desired to design a combinational circuit with three inputs (A, B, and C) and one output F. The output F is one if A, B, and C are not equal ($A \neq B \neq C$); $F = 0$ otherwise. The three steps for designing the circuit is provided next:

Step 1 Truth Table: Since the truth table is not given, the number of inputs and outputs are determined. There are three inputs (A, B, and C) and one output, F. The truth table is then obtained from problem definition and is shown in Table 4.2.

Step 2 Minimization: Output, F in the truth table of Table 4.2 is simplified using a K-map (Figure 4.2(a)). The output equation is obtained as $F = A\overline{B} + \overline{A}C + B\overline{C}$

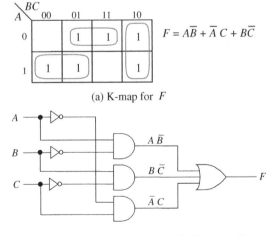

(a) K-map for *F*

(b)   Logic Diagram for the output, *F*

FIGURE 4.2   K-map and the logic diagram for **F**.

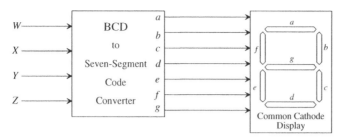

FIGURE 4.3   BCD to seven-segment code converter.

Step 3 Logic diagram or Schematic: Using equations of Step 2, the schematic is drawn as shown in Figure 4.2 (b).

4.4   Multiple-Output Combinational Circuits

A combinational circuit may have more than one output. In such a situation, each output must be expressed as a function of the inputs. A digital circuit called the *code converter* is an example of multiple-output circuits. A code converter transforms information from one binary code to another. As an example, consider the BCD to seven-segment code converter shown in Figure 4.3. The code converter in the figure can be designed to translate a BCD input digit (*W*, *X*, *Y*, and *Z*) to its corresponding seven-segment code for displaying the decimal digit. Note that seven-segment displays (both common cathode display and common anode display) are covered in Chapter 1.

The inputs *W*, *X*, *Y*, and *Z* can be entered into the code converter via four switches as was discussed in Chapter 1. A combinational circuit can be designed for the code converter that will translate each digit entered using four bits into appropriate seven output bits (one bit for each segment) of the display.

In this case, the code converter has four inputs and seven outputs. This code converter is commonly known as a *BCD to seven-segment decoder*. With four bits (*W, X, Y,* and *Z*), there are sixteen combinations (0000_2 through 1111_2) of 0's and 1's. BCD allows only 10 (0000_2 through 1001_2) of these 16 combinations, so the invalid numbers (1010_2 through 1111_2) will never occur for BCD and can be considered as don't cares in K-maps because it does not matter what the seven outputs (*a* through *g*) are for these invalid combinations.

The 7447 (TTL) is a commercially available BCD to 7-segment decoder/driver chip. It is designed for driving a common-anode display. A LOW output will light a segment while a HIGH output will turn it OFF. For normal operation, the LT (Lamp test) and BI/RBO (Blanking Input / Ripple Blanking Input) must be open or connected to HIGH. The 7448 chip, on the other hand, is designed for driving a common-cathode display.

TABLE 4.3   Truth table for converting decimal digits (since common- cathode, a 1 will turn a segment ON and a 0 will turn it OFF)

Decimal Digit to be Displayed	BCD Input Bits				Seven-Segment Output Bits						
	W	*X*	*Y*	*Z*	*a*	*b*	*c*	*d*	*e*	*f*	*g*
2	0	0	1	0	1	1	0	1	1	0	1
4	0	1	0	0	0	1	1	0	0	1	1
9	1	0	0	1	1	1	1	0	0	1	1

To illustrate the design of a BCD to seven-segment decoder, consider designing a code converter for displaying the decimal digits 2, 4, and 9, and turning the display OFF for other valid BCD digits, using the diagram shown in Figure 4.3. First, it is obvious that the BCD to seven-segment decoder has four inputs and seven outputs. Three design steps are provided in the following.

Step 1: Truth Table Table 4.3 shows the truth table.

Step 2: Minimization Figures 4.4 i) through iv) show the K-maps. Note that Seven-Segment output bits (a through g) with the same column values will need one K-map. For example, b and g with column values 111 will have the same K-map, c and f with column values 011 will have the same K-map, and finally, d and e with column values 100 will have the same K-map.

Step 3: Schematic Figure 4.4 v) shows the schematic. Note that the valid BCD digits that are not to be displayed (0, 1, 3, 5, 6, 7, 8), in this example, the combinational circuit for the code converter will generate 0's for the seven output bits (a through g) to turn them OFF. However, these seven bits will be don't-cares in the K-map for the invalid BCD digits 10 through 15.

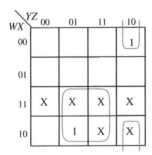

i) K-map for a: $a = WZ + \bar{X}Y\bar{Z}$

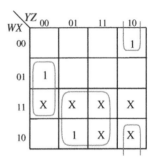

ii) K-map for b, g : $b, g = X\,\bar{Y}\,\bar{Z} + WZ + \bar{X}Y\bar{Z}$

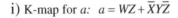

iii) K-map for c, f: $c, f = X\,\bar{Y}\,\bar{Z} + WZ$

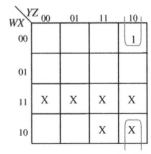

iv) K-map for d, e: $d, e = \bar{X}Y\bar{Z}$

FIGURE 4.4   i) through iv) K-maps

Example 4.1

Design a digital circuit that will convert the BCD codes for the decimal digits (0 through 9) to their Gray codes.

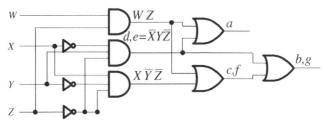

v) Schematic

FIGURE 4.4v)   BCD to seven-segment decoder design for displaying decimal digits
                2, 4, and 9.

Solution

Because both Gray code and BCD code are represented by four bits, for each decimal digit,
there are four inputs and four outputs. Note that 4-bit binary combination will provide 16
(2^4) combinations of 1's and 0's. Because only ten of these combinations (0000 through
1001) are allowed in BCD, the invalid combinations 1010 through 1111 can never occur in
BCD. Therefore, these six binary inputs are considered as don't cares. This means that it
does not matter what binary values are assumed by $f_3 f_2 f_1 f_0$ for $WXYZ = 1010$ through 1111.

Step 1: Truth Table   Table 4.4 shows the truth table.

Step 2: Minimization   Figures 4.5 a) through d) show the K-maps.

Step 3: Schematic   Figure 4.5 e) shows the schematic

4.5   Typical Combinational Circuits

This section describes typical combinational circuits. Topics include comparators,
decoders, encoders, multiplexers, demultiplexers, adders, and subtractors. These
digital components are implemented in MSI chips. Note that MSI (Medium-Scale
Integration) is defined in Chapter 1.

4.5.1   Comparators

The digital comparator is a widely used combinational system. Figure 4.6 shows a two-bit
digital comparator, which provides the result of comparing two two-bit unsigned numbers
as follows:

TABLE 4.4   Truth Table for Example 4.1

Decimal	Input BCD Code				Output Gray Code			
Digit	W	X	Y	Z	f_3	f_2	f_1	f_0
0	0	0	0	0	0	0	0	0
1	0	0	0	1	0	0	0	1
2	0	0	1	0	0	0	1	1
3	0	0	1	1	0	0	1	0
4	0	1	0	0	0	1	1	0
5	0	1	0	1	0	1	1	1
6	0	1	1	0	0	1	0	1
7	0	1	1	1	0	1	0	0
8	1	0	0	0	1	1	0	0
9	1	0	0	1	1	1	0	1

Input Comparison	Outputs		
	G	E	L
$A > B$	1	0	0
$A < B$	0	0	1
$A = B$	0	1	0

The design steps for the two-bit comparator are provided below.

Step 1: Truth Table: Table 4.5 provides the truth table for the two-bit comparator.

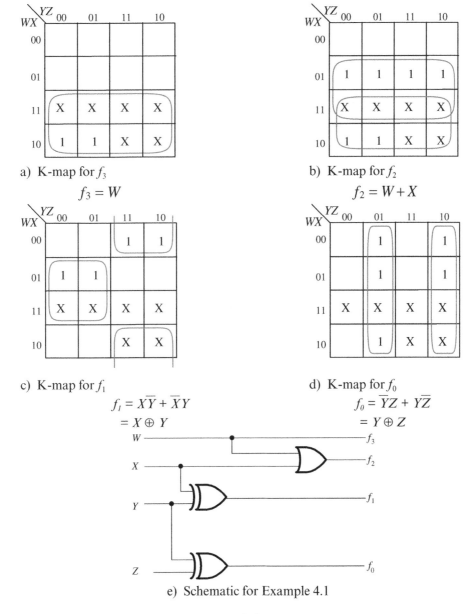

a) K-map for f_3

$$f_3 = W$$

b) K-map for f_2

$$f_2 = W + X$$

c) K-map for f_1

$$f_1 = X\overline{Y} + \overline{X}Y$$
$$= X \oplus Y$$

d) K-map for f_0

$$f_0 = \overline{Y}Z + Y\overline{Z}$$
$$= Y \oplus Z$$

e) Schematic for Example 4.1

FIGURE 4.5   K-maps and Logic Circuit for Example 4.1.

Step 2: Minimization: Figure 4.7(a) shows the K-maps.

Step 3: Schematic: Figure 4.7(b) shows the schematic.

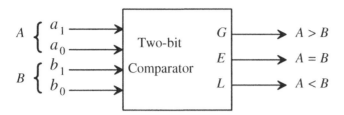

FIGURE 4.6   Block diagram of a two-bit comparator.

TABLE 4.5   Truth Table for the two-bit Comparator

Inputs				Outputs		
a_1	a_0	b_1	b_0	G	E	L
0	0	0	0	0	1	0
0	0	0	1	0	0	1
0	0	1	0	0	0	1
0	0	1	1	0	0	1
0	1	0	0	1	0	0
0	1	0	1	0	1	0
0	1	1	0	0	0	1
0	1	1	1	0	0	1
1	0	0	0	1	0	0
1	0	0	1	1	0	0
1	0	1	0	0	1	0
1	0	1	1	0	0	1
1	1	0	0	1	0	0
1	1	0	1	1	0	0
1	1	1	0	1	0	0
1	1	1	1	0	1	0

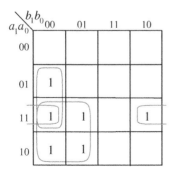

$$G = a_1\overline{b_1} + a_0\overline{b_1}\,\overline{b_0} + a_1a_0\overline{b_0}$$

i) K-map for G

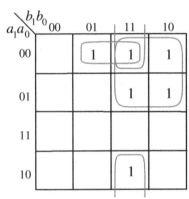

$$E = \overline{a_1}\,\overline{a_0}\,\overline{b_1}\,\overline{b_0} + \overline{a_1}a_0\overline{b_1}b_0 + a_1a_0b_1b_0 + a_1\overline{a_0}b_1\overline{b_0}$$

$$= \overline{a_1}\,\overline{b_1}(\overline{a_0}\,\overline{b_0} + a_0b_0) + a_1b_1(a_0b_0 + \overline{a_0}\,\overline{b_0})$$

$$= (a_0b_0 + \overline{a_0}\,\overline{b_0})(a_1b_1 + \overline{a_1}\,\overline{b_1})$$

$$= (a_0 \odot b_0)(a_1 \odot b_1)$$

ii) K-map for E

Note that the symbol $\odot$ means XNOR or equivalence operation as described in Section 3.2.4 of Chapter 3.

$$L = \overline{a_1}b_1 + \overline{a_0}b_1b_0 + \overline{a_1}\,\overline{a_0}b_0$$

iii) K-map for L

FIGURE 4.7(a)   K-maps for the two-bit comparator.

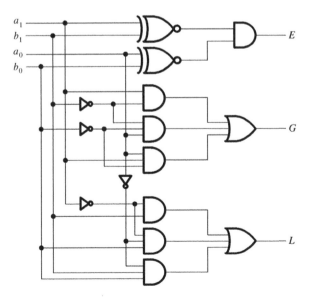

FIGURE 4.7(b)   Design of a two-bit comparator.

4.5.2 Decoders

An n-bit binary number provides 2^n minterms or maxterms. For example, a two-bit binary number will generate 4 (2^2) minterms or maxterms. A decoder is a combinational circuit, when enabled, selects one of 2^n minterms or maxterms at the output based on 'n' input combinations. However, a decoder sometimes may have less than 2^n outputs. For example, the BCD to seven-segment decoder has 4 inputs and 7 outputs rather than 16 (2^4) outputs. Note that a decoder may have a HIGH enable (for example, E), or a LOW enable (for example, $\overline{E}$) or multiple enables based on the design.

As an example, the block diagram of a typical 2-to-4 decoder generating minterms is shown in Figure 4.8(a). This decoder generates a minterm when enabled by a HIGH. Table 4.6(a) provides the truth table. In the truth table, the symbol X is the don't care condition, which can be 0 or 1. Also, $E = 0$ disables the decoder. On the other hand, the decoder is enabled when $E = 1$. For example, when $E = 1$, $x_1 = 0$, $x_0 = 0$, and the output d_0 is one while the other outputs d_1, d_2, and d_3 are 0's, and we say that the decoder has generated the minterm m_0.

As another example, the block diagram of a 2-to-4 decoder generating maxterms is shown in Figure 4.8(b). This decoder generates a maxterm when enabled by a HIGH. Table 4.6(b) provides the truth table. In the truth table, the symbol X is the don't care condition, which can be 0 or 1. Also, $E = 0$ disables the decoder. On the other hand, the decoder is enabled when $E = 1$. For example, when $E = 1$, $x_1 = 1$, $x_0 = 1$, then the output $\overline{d_3}$ is 0 while the other outputs $\overline{d_0}$, $\overline{d_1}$, and $\overline{d_2}$ are 1's, and we say that the decoder has generated the maxterm $\overline{m_3}$.

Next, the design of the decoder of Figure 4.8(a) will be provided. The three design steps are as follows:

<u>**Step 1:**</u> **Truth Table:** Table 4.6(a) provides the truth table for the 2-to-4 decoder.

<u>**Step 2:**</u> **Minimization:** By placing the minterms from the truth table on K-maps, minimization is not possible. Hence, output equations are obtained directly from the truth table. The equations are:

$$d_0 = E\overline{x_1}\,\overline{x_0} \qquad d_1 = E\overline{x_1}x_0 \qquad d_2 = Ex_1\overline{x_0} \qquad \text{and } d_3 = Ex_1x_0$$

<u>**Step 3:**</u> **Schematic:** Figure 4.9 shows the schematic.

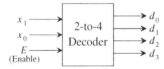

(a) 2-to-4 decoder generating minterms

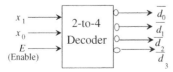

(b) 2-to-4 decoder generating maxterms

FIGURE 4.8   Block diagram of the 2-to-4 decoder.

In general, for n inputs, the n-to-2^n decoder when enabled selects one of 2^n minterms or maxterms at the output based on the input combinations. The decoder actually provides binary to decimal conversion operation. Large decoders can be designed using small decoders as the building blocks. For example, a 4-to-16 decoder can be designed using five 2-to-4 decoders as shown in Figure 4.10, and the truth table for the 4-to-16 decoder is provided in Table 4.7.

Commercially available decoders are normally built using NAND gates rather than AND gates because it is less expensive to produce the selected decoder output in its complement form. Also, most commercial decoders contain one or more enable inputs to control the circuit operation. An example of the commercial decoder is the 74HC138 or the 74LS138. This is a 3-to-8 decoder with three enable lines G_1, $\overline{G_{2A}}$, and $\overline{G_{2B}}$. When $G_1 = H$, $\overline{G_{2A}} = L$ and $\overline{G_{2B}} = L$, the decoder is enabled. The decoder has three inputs, C (Most Significant Bit), B (Next), and A (Least Significant Bit), and eight outputs $\overline{Y_0}$, $\overline{Y_1}$, $\overline{Y_2}$, ..., $\overline{Y_7}$. As an example, when $CBA = 001$ and the decoder enabled, the selected output line $\overline{Y_1}$ (line 1) goes to LOW while the other output lines stay HIGH. Hence, the 74138 is a 3-to-8 decoder and generates maxterms. Decoders along with logic gate(s) can be used to implement Boolean function(s). In order to

TABLE 4.6 (a)   Truth Table of the 2-to-4 Decoder generating minterms for Figure 4.8(a)

Inputs			Outputs			
E	x_1	x_0	d_0	d_1	d_2	d_3
0	X	X	0	0	0	0
1	0	0	1	0	0	0
1	0	1	0	1	0	0
1	1	0	0	0	1	0
1	1	1	0	0	0	1

TABLE 4.6 (b)   Truth Table of the 2-to-4 Decoder generating maxterms for Figure 4.8(b)

Inputs			Outputs			
E	x_1	x_0	$\overline{d_0}$	$\overline{d_1}$	$\overline{d_2}$	$\overline{d_3}$
0	X	X	1	1	1	1
1	0	0	0	1	1	1
1	0	1	1	0	1	1
1	1	0	1	1	0	1
1	1	1	1	1	1	0

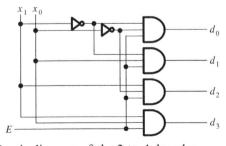

FIGURE 4.9   Logic diagram of the 2-to-4 decoder.

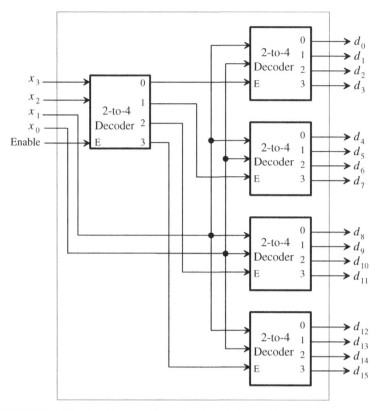

FIGURE 4.10   Implementation of a 4-to-16 decoder using
                2-to-4 decoders.

accomplish this, the Boolean function(s) must be expressed in terms of minterms or maxterms form. Example 4.2 illustrates this.

Example 4.2

Draw a schematic to implement F (W,X) = W using:
(a) a 2-to-4 decoder and a minimum number of logic gates. Assume that the decoder outputs a HIGH on the selected line when enabled by a LOW.
(b) a 2-to-4 decoder and a minimum number of logic gates. Assume that the decoder outputs a LOW on the selected line when enabled by a HIGH.

Solution

(a) In order to implement this, the Boolean function F (W,X) must be expressed in minterm form. Hence, $F(W,X) = W (X + \overline{X}) = WX + W\overline{X} = m_3 + m_2$ Hence, $F(W,X) = \Sigma m(2, 3)$
Figure 4.11(a) shows the schematic.

TABLE 4.7   Truth Table of the 4-to-16 decoder generating minterms for Figure 4.10

INPUTS					OUTPUTS															
E	x_3	x_2	x_1	x_0	d_0	d_1	d_2	d_3	d_4	d_5	d_6	d_7	d_8	d_9	d_{10}	d_{11}	d_{12}	d_{13}	d_{14}	d_{15}
0	X	X	X	X	0	0	0	0	0	0	0	0	0	0	0	0	0	0	0	0
1	0	0	0	0	1	0	0	0	0	0	0	0	0	0	0	0	0	0	0	0
1	0	0	0	1	0	1	0	0	0	0	0	0	0	0	0	0	0	0	0	0
1	0	0	1	0	0	0	1	0	0	0	0	0	0	0	0	0	0	0	0	0
1	0	0	1	1	0	0	0	1	0	0	0	0	0	0	0	0	0	0	0	0
1	0	1	0	0	0	0	0	0	1	0	0	0	0	0	0	0	0	0	0	0
1	0	1	0	1	0	0	0	0	0	1	0	0	0	0	0	0	0	0	0	0
1	0	1	1	0	0	0	0	0	0	0	1	0	0	0	0	0	0	0	0	0
1	0	1	1	1	0	0	0	0	0	0	0	1	0	0	0	0	0	0	0	0
1	1	0	0	0	0	0	0	0	0	0	0	0	1	0	0	0	0	0	0	0
1	1	0	0	1	0	0	0	0	0	0	0	0	0	1	0	0	0	0	0	0
1	1	0	1	0	0	0	0	0	0	0	0	0	0	0	1	0	0	0	0	0
1	1	0	1	1	0	0	0	0	0	0	0	0	0	0	0	1	0	0	0	0
1	1	1	0	0	0	0	0	0	0	0	0	0	0	0	0	0	1	0	0	0
1	1	1	0	1	0	0	0	0	0	0	0	0	0	0	0	0	0	1	0	0
1	1	1	1	0	0	0	0	0	0	0	0	0	0	0	0	0	0	0	1	0
1	1	1	1	1	0	0	0	0	0	0	0	0	0	0	0	0	0	0	0	1

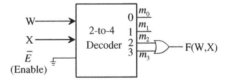

(a) Schematic for Example 4.2(a)

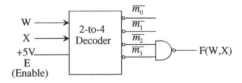

(b) Schematic for Example 4.2(b)

FIGURE 4.11   Schematics for Example 4.2.

(b) For this case, the Boolean function F (W,X) must be expressed in maxterm form. Hence, using $F(W,X) = m_3 + m_2$ from part (a),

$$F(W,X) = \overline{\overline{\overline{F}}}(W,X) = \overline{\overline{m_2 + m_3}} = \overline{\overline{m_2} \bullet \overline{m_3}}$$

Figure 4.11(b) shows the schematic.

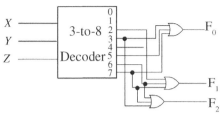

FIGURE 4.12   Schematic for Example 4.3.

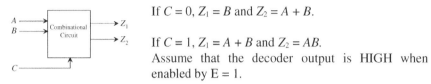

If $C = 0$, $Z_1 = B$ and $Z_2 = A + B$.

If $C = 1$, $Z_1 = A + B$ and $Z_2 = AB$.
Assume that the decoder output is HIGH when enabled by E = 1.

FIGURE 4.13   Figure for Example 4.4.

Example 4.3

A combinational circuit is specified by the following equations:
$F_0 (X,Y,Z) = \overline{X} YZ + XZ$, $F_1 (X,Y,Z) = Y (X + \overline{Z})$, $F_2 = X Y + \overline{X} Y Z$. Draw
a logic diagram using a 3-to-8 decoder and external gates. Assume that the decoder outputs a HIGH on the selected line.

Solution

In order to implement the equations for F, using a decoder, these equations must first be expressed in minterm form as follows:
$F_0 (X,Y,Z) = \overline{X} YZ + XZ (Y + \overline{Y}) = \overline{X} YZ + XYZ +X \overline{Y} Z = \Sigma m(3, 5, 7)$
$F_1 (X,Y,Z) = Y(X+\overline{Z}) = XY + Y\overline{Z} = XY(Z+\overline{Z}) + Y\overline{Z}(X+\overline{X}) = XYZ + XY\overline{Z} + XY\overline{Z} + \overline{X}Y\overline{Z} = \Sigma m(2,6,7)$
$F_2 = X Y + \overline{X} Y Z = XY (Z + \overline{Z}) + \overline{X} Y Z = XYZ + XY \overline{Z} + \overline{X} Y Z = \Sigma m(3, 6, 7)$

Figure 4.12 shows the schematic.

Example 4.4

Design a combinational circuit using a decoder and logic gates to implement the functions depicted in Figure 4.13.

Solution

The truth table is shown in Table 4.8.
From the truth table,

$$Z_1 = m \Sigma (2, 3, 5, 6, 7)$$
$$Z_2 = m \Sigma (1, 2, 3, 7)$$

The logic diagram is shown in Figure 4.14.

4.5.3   Encoders

An encoder is a combinational circuit that performs the reverse operation of a decoder. An encoder has a maximum of 2^n inputs and n outputs. Figure 4.15 shows the block diagram of a 4-to-2 encoder. Table 4.9 provides the truth table of the 4-to-2 encoder.

From the truth table, it can be concluded that an encoder actually performs decimal-to-binary conversion. In the encoder defined by Table 4.9, it is assumed that only

TABLE 4.8   Truth Table for Example 4.4

	Inputs		Outputs	
C	B	A	Z_1	Z_2
0	0	0	0	0
0	0	1	0	1
0	1	0	1	1
0	1	1	1	1
1	0	0	0	0
1	0	1	1	0
1	1	0	1	0
1	1	1	1	1

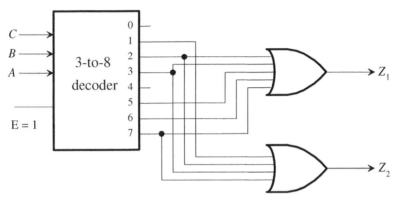

FIGURE 4.14   Implementation of Example 4.4 using a decoder and OR gates.

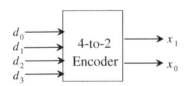

FIGURE 4.15   Block diagram of a 4-to-2 encoder.

TABLE 4.9   Truth Table of the 4-to-2 Encoder

	Inputs			Outputs	
d_0	d_1	d_2	d_3	x_1	x_0
1	0	0	0	0	0
0	1	0	0	0	1
0	0	1	0	1	0
0	0	0	1	1	1

TABLE 4.10   Truth Table of the 4-to-2 Priority Encoder

	Inputs			Outputs	
d_3	d_2	d_1	d_0	x_1	x_0
1	y	y	y	1	1
0	1	y	y	1	0
0	0	1	y	0	1
0	0	0	1	0	0

$y = 0$ or 1

one of the four inputs can be HIGH at anytime. If more than one input is 1 at the same time, an undefined output is generated. For example, if d_1 and d_2 are 1 at the same time, both x_0 and x_1 are 1. This represents binary 3 rather than 1 or 2. Therefore, in an encoder in which more than one input can be active simultaneously, a priority scheme must be implemented in the inputs to ensure that only one input will be encoded at the output.

A 4-to-2 priority encoder will be designed next. Suppose that it is assumed that inputs with higher subscripts have higher priorities. This means that d_3 has the highest priority and d_0 has the lowest priority. Therefore, if d_0 and d_1 become one simultaneously, the output will be 01 for d_1.
Three design steps are provided below:

Step 1: Truth Table: Table 4.10 shows the truth table of the 4-to-2 priority encoder.

Step 2: Minimization: Figures 4.16a(i) and 4.16a(ii) show the K-maps. Note that $d_3 d_2 d_1 d_0 = 0000$ is a don't care condition in K-maps since it does not occur in the truth table.

Step 3: Schematic: Figure 4.16a(iii) shows the schematic.

Note that chips are limited by the number of pins. Priority encoders can sometimes be used to expand the number of input pins for handling more input signals. For example, the Motorola/Freescale 68000 16-bit 64-pin microprocessor provides three interrupt input pins. Using a 74148 (8-to-3 priority encoder) chip, the interrupt capability of the 68000 can be expanded to provide 8 additional inputs.

The 74148 is a commercially available 8-to-3 priority encoder. Figure 4.16b(i) shows the 74148 functional block diagram along the truth table in Figure 4.16b(ii).

The operation of the 74148 shown in Figure 4.16b(i) will now be explained. Suppose that $\overline{EI} = 0$, $\overline{I0} = 0$, and $\overline{I1}$ through $\overline{I7}$ are 1's, then from the truth table of 4.16b(ii), $\overline{GS} = 0$, $\overline{EO} = 1$, $\overline{A2}\,\overline{A1}\,\overline{A0} = 111$. Note that the 74148 generates active low ouputs.

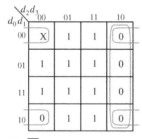

(i) K-map for $\overline{x_0}$

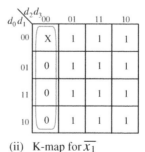

(ii) K-map for $\overline{x_1}$

$$\overline{x_0} = \overline{d_1}\,\overline{d_3} + d_2\overline{d_3}$$

$$x_0 = (d_1 + d_3)(\overline{d_2} + d_3)$$

$$\overline{x_1} = \overline{d_2}\,\overline{d_3}$$

$$x_1 = d_2 + d_3$$

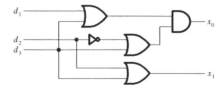

(iii) Logic diagram

FIGURE 4.16a   K-maps and logic diagram of a 4-to-2 priority encoder.

Figure 4.16b(iii) shows the schematic of a 16-to-4 priority encoder obtained by connecting two 74148's along with three NAND gates. The configuration generates active high outputs.

For example, consider Figure 4.16b(iii). Suppose $\overline{I15} = 0$ and $\overline{I8}$ through $\overline{I14}$ are 1's in the 74148#2. Also, suppose $\overline{I0}$ through $\overline{I7}$ of the 74148#1 are 1's. Since $\overline{EI}$ of the

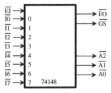

(i) 74148 Functional block Diagram

INPUTS									OUTPUTS				
$\overline{EI}$	$\overline{I0}$	$\overline{I1}$	$\overline{I2}$	$\overline{I3}$	$\overline{I4}$	$\overline{I5}$	$\overline{I6}$	$\overline{I7}$	$\overline{A2}$	$\overline{A1}$	$\overline{A0}$	$\overline{GS}$	$\overline{EO}$
1	X	X	X	X	X	X	X	X	1	1	1	1	1
0	1	1	1	1	1	1	1	1	1	1	1	1	0
0	X	X	X	X	X	X	X	0	0	0	0	0	1
0	X	X	X	X	X	X	0	1	0	0	1	0	1
0	X	X	X	X	X	0	1	1	0	1	0	0	1
0	X	X	X	X	0	1	1	1	0	1	1	0	1
0	X	X	X	0	1	1	1	1	1	0	0	0	1
0	X	X	0	1	1	1	1	1	1	0	1	0	1
0	X	0	1	1	1	1	1	1	1	1	0	0	1
0	0	1	1	1	1	1	1	1	1	1	1	0	1

(ii) 74148 Truth Table [X can be 0 or 1]

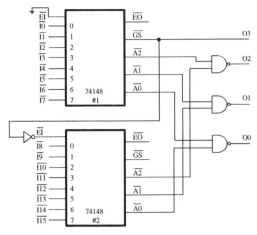

(iii) 16-to-4 Priority ender using two 74148's

FIGURE 4.16b 74148 Block Diagram, Truth Table, and expansion.

74148#1 is connected to ground, outputs $\overline{A2}$, $\overline{A1}$, $\overline{A0}$, $\overline{GS}$ are 1's. Note that the output $\overline{GS}$ of the 74148#1 is inverted and connected to $\overline{EI}$ of the 74148#2 . Hence, the priority encoder 71148#2 will be enabled. Since $\overline{I15}$ (line 7) of the 74148#2 is 0 , the outputs $\overline{A2}$ $\overline{A1}$ $\overline{A0}$ of the 74148#2 will be 000 (inverted 111). As mentioned before, the 74148

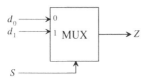

FIGURE 4.17   Block diagram of a 2-to-1 multiplexer.

TABLE 4.11   Truth Table of the 2-to-1 Multiplexer

S	d_0	d_1	Z
0	0	0	0
0	0	1	0
0	1	0	1
0	1	1	1
1	0	0	0
1	0	1	1
1	1	0	0
1	1	1	1

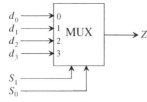

(a)   K-map for the 2-to-1 MUX

(b)   Logic diagram of the 2-to-1 MUX

FIGURE 4.18   K-map and Logic diagrm of the 2-to-1 MUX.

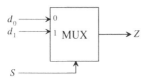

FIGURE 4.19   Block-diagram Representation of a Four-input Multiplexer.

TABLE 4.12   Truth table of the four input multiplexer

S_1	S_0	Z
0	0	d_0
0	1	d_1
1	0	d_2
1	1	d_3

produces active low outputs. Hence, in Figure 4.16b(iii) , $\overline{A2}\ \overline{A1}\ \overline{A0}$ of the 74148#2 are 000 while $\overline{A2}\ \overline{A1}\ \overline{A0}$ of the 74148#1 are 111. Since $\overline{A2}\ \overline{A1}\ \overline{A0}$ of the 74148#2 are NANDed with $\overline{A2}\ \overline{A1}\ \overline{A0}$ of the 74148#1 , outputs O2 O1 O0 will be 111. Also, O3 = $\overline{GS}$ of the 74148#1 = 1. Hence, O3 O2 O1 O0 = 1111 generating acive HIGH when $\overline{I15}$ = 0 at the output of the 16-to-4 encoder.

4.5.4   Multiplexers

A multiplexer (abbreviated as MUX) is a combinational circuit that selects one of n input lines and provides it on the output. Thus, the multiplexer has several inputs and only one output. One or more select lines identify or address one of several inputs and provides it on the output line. Figure 4.17 shows the block diagram of a 2-to-1 multiplexer. The two inputs can be selected by one select line, S. When $S = 0$, input line 0 (d_0) will be presented as the output. On the other hand, when $S = 1$, input line 1 (d_1) will be produced at the output.

The 2-to-1 MUX of Figure 4.17 is designed next. Three design steps are provided below:

Step 1: Truth Table: Table 4.11 shows the truth table of the multiplexer.

Step 2: Minimization: Figures 4.18(a) shows the K-map. From the K-map,
$$Z = \overline{S}d_0 + Sd_1.$$

Step 3: Schematic: Figure 4.18(b) shows the schematic.

In general, a multiplexer with n select lines can select one of 2^n data inputs. Hence, multiplexers are sometimes referred to as *data selectors*.

A large multiplexer can be implemented using a small multiplexer as the building block. For example, consider the block diagram and the truth table of a 4-to-1 multiplexer shown in Figure 4.19 and Table 4.12 respectively. The 4-input multiplexer can be implemented using three 2-to-1 multiplexers as shown in Figure 4.20.

In Figure 4.20, the select line S_0 is applied as input to the multiplexers MUX 0 and MUX 1. This means that $Z_0 = d_0$ or d_1 and $Z_1 = d_2$ or d_3, depending on whether $S_0 = 0$ or 1. The select line S_1 is given as input to the multiplexer MUX 2. This implies that $Z = Z_0$ if S_1 = 0; otherwise $Z = Z_1$. In this arrangement if $S_1 S_0 = 11$, then $Z = d_3$ because $S_0 = 1$ implies that $Z_0 = d_1$ and $Z_1 = d_3$ because $S_1 = 1$, the MUX 2 selects the data input Z_1, and thus $Z = d_3$. The other entries of the truth table of Table 4.12 can be verified in a similar manner.

Multiplexers can be used to implement Boolean equations. For example, consider realizing $f(x,y,z)= x\overline{z} + yz$ using a 4-to-1 multiplexer. First, the Boolean equation for $f(x,y,z)$ is expressed in minterm form as follows: $f(x,y,z)=x\overline{z}\,(y+\overline{y}) + yz\,(x + \overline{x})= xy\overline{z} + x\,\overline{y}\,\overline{z} + xyz + \overline{x}\,yz$. The next step is to use two of the three variables (x,y,z) as select inputs. Suppose

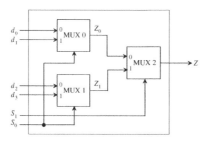

FIGURE 4.20   Implementation of a Four-Input Multiplexer Using Only Two-input Multiplexers.

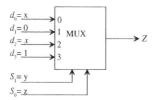

FIGURE 4.21   Implementation of a Boolean equation using a 4-to-1 multiplexer.

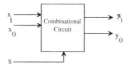

FIGURE 4.22   Figure for Example 4.5.

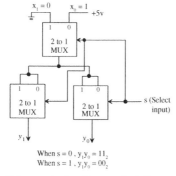

FIGURE 4.23   Schematic for Example 4.5.

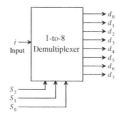

FIGURE 4.24   1-to-8 demultiplexer.

y and z are arbitrarily chosen as select inputs. The four combinations $(\bar{y}\,\bar{z},y\bar{z},\bar{y}z,yz)$ of the select inputs, y and z are then required to be factored out of minterm form for $f(x,y,z)$ to determine the inputs to the 4-to-1 multiplexer as follows: $f\,(x,y,z)=\bar{y}\,\bar{z}(x) + \bar{y}z\,(0) + y\bar{z}(x)$ $+yz\,(x+\bar{x})= \bar{y}\,\bar{z}(x) + \bar{y}z\,(0) + y\bar{z}(x) + yz\,(1)$. Hence, the above equation for $f(x,y,z)$ can be implemented using the 4-to-1 multiplexer of Figure 4.19 as follows: $S_1 = y, S_0 = z, d_0 = x$, $d_1=0, d_2=x, d_3=1$. Figure 4.21 shows the implementation.

Next, consider implementing $f(a,b,c) = \Sigma m\,(0,2,3,7)$ using the 4-to-1 multiplexer of Figure 4.19. The first step is to group the truth table in pairs as follows:

```
a b c   f
0 0 0   1
0 0 1   0        f= c̄
-------------
0 1 0   1
0 1 1   1        f=1
-------------
1 0 0   0
1 0 1   0        f=0
----------------
1 1 0   0
1 1 1   1        f=c
----------------
```

Next, select 'a' and 'b' as Select inputs. Note that in each pair of the table, 'f' can be 0 or 1 or c or $\bar{c}$. For example, when ab=00, output f= $\bar{c}$ because f=1 when c=0, and f=0 when c=1. Hence, the 4-to-1 multiplexer of Figure 4.19 can be connected as follows: $S_1 =a, S_0 = b, d_0 = \bar{c}, d_1 =1, d_2 = 0, d_3 = c$.

Example 4. 5

Draw a schematic to implement a combinational circuit that will select one of two inputs $(x_1\,x_0)$, and then perform the following functions based upon a select input, 's' depicted in the figure 4.22. Connect $x_1\,x_0$ to appropriate levels such that : when s = 0, then two-bit output $(y_1\,y_0) =11_2$; on the other hand, when s = 1, then two-bit output $(y_1\,y_0) = 00_2$. Use only 2-to-1 multiplexers for the combinational circuit. Do not use any logic gates.

Solution

Figure 4.23 shows the schematic.

4.5.5   Demultiplexers

The demultiplexer is a combinational circuit that performs the reverse operation of a multiplexer. The demultiplexer has only one input and several outputs. One of the outputs is selected by the combination of 0's and 1's of the select inputs. These inputs determine one of the output lines to be selected; data from the input line is then transferred to the selected output line. Figure 4.24 shows the block diagram of a 1-to-8 demultiplexer. Suppose that i = 1 and $S_2S_1S_0 = 010$; output line d_2 will be selected and a 1 will be output on d_2.

4.5.6   Binary / BCD Adders and Binary Subtractors

When two bits x and y are added, a sum bit and a carry bit are generated. A combinational circuit that adds two bits is called a *half-adder*. Figure 4.25 shows a block diagram of the half-adder. The half-adder is designed next. Three design steps are provided next:

FIGURE 4.25   Block Diagram of a Half-Adder.

TABLE 4.13   Truth Table of the Half-Adder

| Inputs | | Outputs | | Decimal |
x	y	C	S	Value
0	0	0	0	0
0	1	0	1	1
1	0	0	1	1
1	1	1	0	2

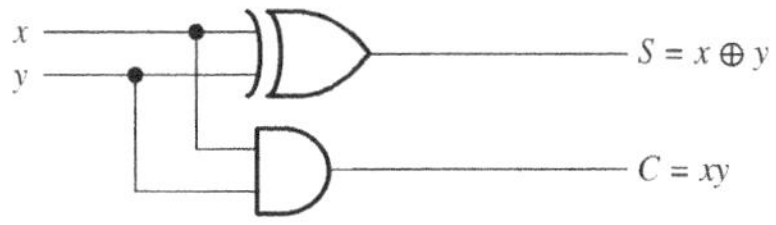

FIGURE 4.26   Logic diagram of the half-adder.

Step 1: Truth Table: Table 4.13 shows the truth table of the half-adder.

Step 2: Minimization: Output equations cannot be minimized using K-maps. Hence, output equations are obtained directly from the truth table as follows:
$S = \bar{x}y + x\bar{y} = x \oplus y$, $C = xy$

Step 3: Schematic: Figure 4.26 shows the logic diagram of the half-adder.
Next, consider addition of two 4-bit numbers as follows:

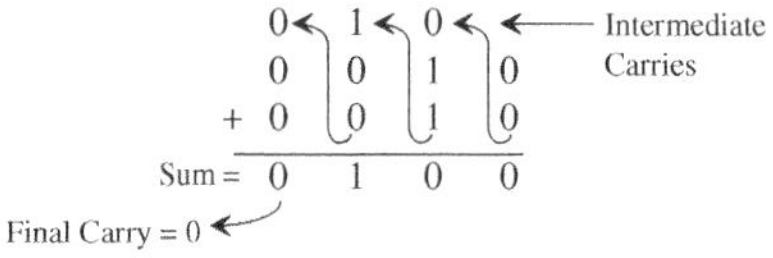

This addition of two bits will generate a sum bit and a carry bit. The carry may be 0 or 1. Also, there will be no previous carry while adding the least significant bits (bit 0) of the two numbers. This means that two bits need to be added for bit 0 of the two numbers. On the other hand, addition of three bits (two bits of the two numbers and a previous carry, which may be 0 or 1) is required for all the subsequent bits. Hence, an adder is needed to add three bits. A combinational circuit that adds three bits, generating a sum bit and a carry bit (which may be 0 or 1), is called a *full adder*. Figure 4.27 shows the block diagram of a full adder. The full adder adds three bits, x, y, and z, and generates a sum bit (S) and a carry bit (C). The full adder is designed next. Three design steps are provided below.

Step 1: Truth Table: Table 4.14 shows the truth table of the full adder.

FIGURE 4.27   Block diagram of a full adder.

TABLE 4.14 Truth Table of the Full Adder

Inputs			Outputs		Decimal
x	y	z	C	S	Value
0	0	0	0	0	0
0	0	1	0	1	1
0	1	0	0	1	1
0	1	1	1	0	2
1	0	0	0	1	1
1	0	1	1	0	2
1	1	0	1	0	2
1	1	1	1	1	3

Step 2: Minimization: The sum, S cannot be minimized using a K-map. Hence, the equation for the sum, S is obtained directly from the truth table and then simplified using Boolean identities as follows:

From the truth table, $S = \bar{x}\,\bar{y}z + \bar{x}y\bar{z} + x\,\bar{y}\,\bar{z} + xyz = (\bar{x}y + x\,\bar{y})\,\bar{z} + (xy + \bar{x}\,\bar{y})z$
Let $w = \bar{x}y + x\,\bar{y}$ then $\bar{w} = xy + \bar{x}\,\bar{y}$ Hence, $S = \bar{w}\,z + w\,\bar{z} = w \oplus z = x \oplus y \oplus z$
Also, from the truth table, $C = \bar{x}yz + x\bar{y}z + xy\bar{z} + xyz = (\bar{x}y + x\bar{y})z + xy\,(z + \bar{z}) = wz + xy$
    where $w = (\bar{x}y + x\bar{y}) = x \oplus y$ Hence, $C = (x \oplus y)z + xy$.

Another form of Carry can be written as follows:

$C = \bar{x}yz + x\bar{y}z + xy\bar{z} + xyz = \bar{x}yz + x\bar{y}z + xy\bar{z} + xyz + xyz + xyz$ (Adding redundant terms xyz)
 $= yz\,(\bar{x} + x) + xz\,(y + \bar{y}) + xy\,(z + \bar{z}) = yz + xz + xy$

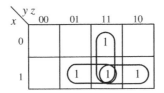

(a) K-map for C of the full adder

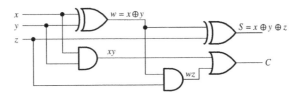

(b) Schematic of the full adder

FIGURE 4.28 K-map for C and the schematic of the full adder.

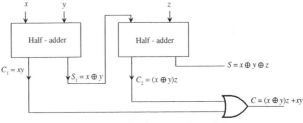

FIGURE 4.29 Full addr using two half-adder and an OR gate.

This form of C can also be obtained using the K-map of figure 4.28(a) as follows:
$C = yz + xz + xy$

Step 3: **Schematic:** Figure 4.28(b) shows the schematic of the full adder.

Note that the names *half-adder* and *full adder* are based on the fact that two half-adders are required to obtain a full adder. This can be obtained as follows. One of the two half-adders with inputs, x and y will generate the sum, $S_1 = x \oplus y$ and the carry, $C_1 = xy$. The sum (S_1) output can be connected to one of the inputs of the second half-adder with z as the other input. Thus, the sum output (S) and the carry output (C_2) of the second half-adder will be $S = x \oplus y \oplus z$ and $C_2 = (x \oplus y)z$. The carry outputs of the two half-adders can be logically ORed to provide the carry (C) of the full adder as $C = (x \oplus y)z + xy$. Therefore, two half-adders and a two-input OR gate can be used to obtain a full adder. Figure 4.29 depicts this.

A 4-bit binary adder (also called *Ripple Carry Adder*) for adding two 4-bit numbers $x_3x_2x_1x_0$ and $y_3y_2y_1y_0$ can be implemented using one half-adder and three full adders as shown in Figure 4.30. A full adder adds two bits if one of its inputs is grounded. This means that the half-adder in Figure 4.30 can be replaced by a full-adder with its input C_{in} grounded. Figure 4.31 shows implementation of a 4-bit binary adder using four full adders.

Next, design of a BCD adder for adding two single BCD digits will be provided. From Chapter 2, addition of two BCD digits is correct if the binary sum is less than or equal to 1001_2 (9 in decimal). A binary sum greater than 1001_2 results into an invalid BCD sum; adding 0110_2 to an invalid BCD sum provides the correct sum with an output carry of 1. Furthermore, addition of two BCD digits (each digit having a maximum value of 9) along with carry will provide a sum not exceeding 19 in decimal (10011_2). A BCD adder can be designed by implementing required corrections in the result for decimal numbers from 10 through 19 (1010_2 through 10011_2). Therefore, a correction is necessary for the following:

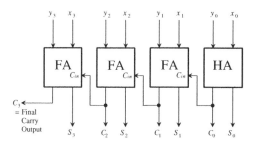

FIGURE 4.30   4-bit binary adder using one half-adder and three full adders.

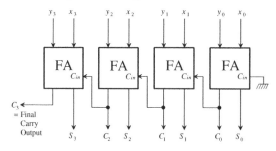

FIGURE 4.31   Four-bit binary adder using full adders.

i) If the binary sum is greater than or equal to decimal 16 (This will generate a carry of one)

ii) If the binary sum is 1010_2 through 1111_2

For example, consider adding packed BCD numbers 99 and 38:

	1		
99	1001	1001	BCD for 99
+38	0011	1000	BCD for 38
137	1101	0001	invalid sum
	+0110	+0110	add 6 for correction
0001	0011	0111	
1	3	7	← correct answer 137

This means that a circuit can be designed to indicate that BCD correction (addition of 0110 to the sum) is necessary after binary addition of two single BCD digits. The circuit will generate:

i) an output (C_{11}) of 1 if the binary sum is $S_3S_2S_1S_0 = 1010_2$ through 1111_2.

or

ii) a carry (C_0) of 1 if the binary sum produces a carry ; this means that the binary sum is greater than or equal to decimal 16.

For case i), using the K-map of Figure 4.32,

$C_{11} = S_1S_3 + S_2S_3$

Hence, $C_{11} = S_1S_3 + S_2S_3 = S_3 (S_1 + S_2)$.

Combining cases i) and ii), the output of the circuit, $C_1 = C_0 + S_3 (S_1 + S_2)$. This is implemented in the Figure 4.33.

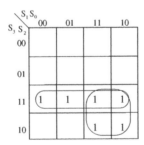

FIGURE 4.32   K-map for C_{11}.

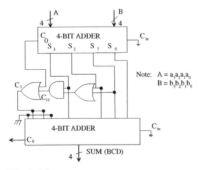

FIGURE 4.33   BCD Adder.

Note that C_1 is the output of the logic circuit which can be 0 or 1 based on the inputs C_0 or C_{11} of the circuit. When $C_1 = 0$, then zeros are added to $S_3S_2S_1S_0$. This situation occurs when $S_3S_2S_1S_0$ is less than or equal to 1001_2. However, when $C_1 = 1$, then the binary number 0110 is added to $S_3S_2S_1S_0$ using the second 4-bit adder. This situation occurs when $S_3S_2S_1S_0$ is 1010_2 through 1111_2 or if a carry is generated. The carry output from the second 4-bit adder can be discarded. Note that BCD parallel adder for adding n BCD digits can be obtained using n BCD adders by connecting the output carry of each low BCD adder to C_{in} of the next BCD adder.

Next, half-subtractor and full-subtractor will be discussed. Similar to half-adder and full-adder, there are half-subtractor and full-subtractor. Using half- and full-subtractors,

TABLE 4.15   Truth table for the Half-subtractor

x (minuend)	y (subtrahend)	B (borrow)	R (result)
0	0	0	0
0	1	1	1
1	0	0	1
1	1	0	0

TABLE 4.16   Truth table for the full-subtractor

x	y	z	B (Borrow)	R (Result)
0	0	0	0	0
0	0	1	1	1
0	1	0	1	1
0	1	1	1	0
1	0	0	0	1
1	0	1	0	0
1	1	0	0	0
1	1	1	1	1

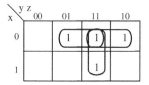

FIGURE 4.34   K-map for B for full-subtactor.

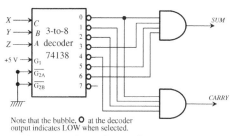

Note that the bubble, ◯ at the decoder output indicates LOW when selected.

FIGURE 4.35(a)   Implementation of a Full-adder Using a 74138 Decoder and Two 4-input AND Gates.

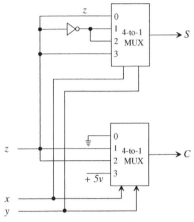

FIGURE 4.35(b)   Full adder using MUX's.

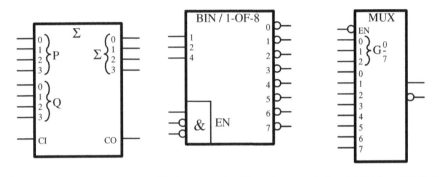

a) 4-bit Binary Adder
    (74LS283 or 74HC283)

b) 3-to-8 Decoder (74LS138 or
    74HC138)

c) 8-to-1 Multiplexer (74LS151
    or 74HC151) (providing both
    true and complemented
    outputs)

FIGURE 4.36   IEEE Symbols.

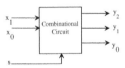

FIGURE 4.37   Figure for Example 4.6.

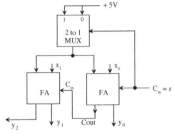

FIGURE 4.38   Schematic for Example 4.6.

subtraction operation can be implemented with logic circuits in a direct manner. A half-subtractor is a combinational circuit that subtracts two bits generating a result (R) bit and a borrow (B) bit. The truth table for the half-subtractor is provided in Table 4.15.

The borrow (B) is 0 if x is greater than or equal to y; B = 1 if x is less than y. From the truth table, $R = \bar{x} y + x \bar{y} = x \oplus y$ and $B = \bar{x} y$.

A full-subtractor is a combinational circuit that performs the operation among three bits x - y - z generating a result bit (R) and a borrow bit (B). The truth table for the full-subtractor is provided in Table 4.16.

The truth table for R (Table 4.16) is the same as the truth table for S of the full adder (Table 4.14). Hence, $R = x \oplus y \oplus z$. Also, using the K-map of Figure 4.34, $B = \bar{x} y + \bar{x} z + yz$.

It is advantageous to implement addition and subtraction with full-adders since both operations can be obtained using a single logic circuit. Next, consider implementing a full-adder using a decoder, and then multiplexers.

Because any Boolean function can be expressed in terms of minterms or maxterms, a decoder can be used to implement the function or its complement. Since the 74138 generates a LOW (maxterm) on the selected output line, the complement of a Boolean function can be obtained by logically ANDing the appropriate minterms. For example, consider the truth table of the full adder listed in Table 4.14. The inverted sum and the inverted carry can be expressed in terms of minterms as follows:

$$\overline{SUM} = \Sigma\, m(0,3,5,6)$$
$$\overline{CARRY} = \Sigma\, m(0,1,2,4)$$

Hence, using DeMorgan's theorem

$$SUM = \overline{m_0} \bullet \overline{m_3} \bullet \overline{m_5} \bullet \overline{m_6} \text{ and } CARRY = \overline{m_0} \bullet \overline{m_1} \bullet \overline{m_2} \bullet \overline{m_4}$$

Figure 4.35a shows the implementation of a full adder using a 74138 decoder (*C=X*, *B=Y, A=Z*) and two 4-input AND gates. Note that the 74138 in the Manufacturer's data sheet uses the symbols *C, B, A* as three inputs to the decoder with *C* as the most significant bit and *A* as the least significant bit.

Next, consider implementing the full adder using a minimum number of 4-to-1 multilexers. After grouping Table 4.14 of the full-adder in pairs, the modified table is shown below:

Inputs			Outputs		
x	*y*	*z*	*C*	*S*	
0	0	0	0	0	
0	0	1	0	1	$C = 0, S = z$
0	1	0	0	1	
0	1	1	1	0	$C = z, S = \bar{z}$
1	0	0	0	1	
1	0	1	1	0	$C = z, S = \bar{z}$
1	1	0	1	0	
1	1	1	1	1	$C = 1, S = z$

Since the full adder has two outputs (C,S), two 4-to-1 multiplxers are needed. The schematic is shown in figure 4.35(b).

4.6   IEEE Standard Symbols

IEEE has developed standard graphic symbols for commonly used digital components such as adders, decoders, and multiplexers. These are depicted in Figure 4.36.

Example 4.6

Draw a schematic to implement a combinational circuit that will input a two-bit unsigned number $(x_1 x_0)$, and then perform the following functions based upon a select input, 's' as depicted in the figure 4.37.

Connect $x_1 x_0$ to appropriate levels such that when s =0, then three-bit output $(y_2y_1y_0)$ = x_1x_0+3; on the other hand, when s = 1, then three-bit output $(y_2y_1y_0)$ = $x_1x_0 + 4$. Use a minimum number of full adders and multiplexers. Do not use any logic gates.

Solution

Since the combinational circuit needs to perform two functions, one 2-to-1 mux is needed. Also, for two-bit arithmetic operations by the combinational circuit, two full adders are needed. Hence, by manipulating the inputs of the MUX, the schematic shown in Figure 4.38 will accomplish this.

Example 4.7

Design combinational circuits using full adders and multiplexers as building blocks to implement (a) a 4-bit adder/subtractor; add when S =0 and subtract when S = 1. (b) multiply a 4-bit unsigned number by 2 when S=0 and transfer zero to output when S = 1.

Solution

(a) The subtraction x - y of two binary numbers can be performed using twos complement arithmetic. As discussed before, x - $y = x$ + (ones complement of y) + 1.

A 4-bit adder/subtractor is shown in Figure 4.39(a).

The adder/subtractor in Figure 4.39(a) utilizes four MUX's. Each MUX has one select line (S) and is capable of selecting one of two lines, y_n or $\overline{y_n}$.

The 4-bit adder/subtractor of Figure 4.39(a) either adds two 4-bit numbers and performs $(x_3x_2x_1x_0)$ ADD $(y_3y_2y_1y_0)$ when S = 0 or performs the subtraction operation $(x_3x_2x_1x_0)$ MINUS $(y_3y_2y_1y_0)$ for S = 1. The select bit S can be implemented by a switch. When S = 0, each MUX outputs the true value of y_n (n = 0 through 3) to the corresponding inputs of the full adder FA_n (n = 0 through 3). Because S = 0 (C_{in} for FA_0 = 0), the four full adders perform the desired 4-bit addition. When S = 1 (C_{in} for FA_0 = 1), each MUX

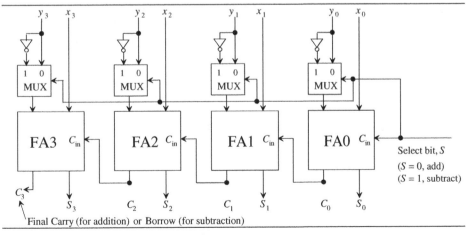

FIGURE 4.39   (a) 4-bit Adder / Subtractor.

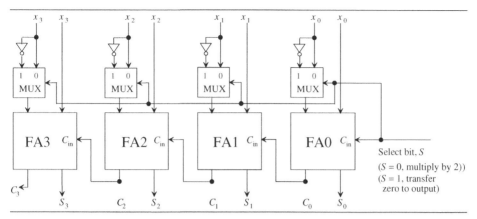

Figure 4.39(b)   Solution to Part (b).

generates the ones complement of y_n at the corresponding inputs of the full adder FA_n. Because $S = C_{in} = 1$, the four full adders provide the following operation: $(x_3 x_2 x_1 x_0)$ - $(y_3 y_2 y_1 y_0) = (x_3 x_2 x_1 x_0) + (\overline{y_3}\,\overline{y_2}\,\overline{y_1}\,\overline{y_0}) + 1$

(b) Assume 4-bit output $S_3 S_2 S_1 S_0$. Figure 4.39 (b) shows the implementation.

4.7   Read-Only Memories (ROMs)

Read-only memory, commonly called *ROM*, is a nonvolatile memory (meaning that it retains information in case of power is switched off) that provides read-only access to the stored data. A block-diagram representation of a ROM is shown in Figure 4.40. The total capacity of this ROM is $2^n \times m$ bits. Whenever an n-bit address is placed on the address line, the m-bit information stored in this address will appear on the data lines. The m-bit output generated by the ROM is also called a *word*.

For example, a 1K × 8 -bit (1024 × 8) ROM chip contains 10 address pins ($2^{10} = 1024 = 1K$) and 8 data pins. Therefore, $n = 10$ and $m = 8$. On the other hand, an 8K × 8 -bit (8192 × 8) ROM chip includes 13 address pins ($2^{13} = 8192 = 8K$) and 8 data pins. Thus, $n = 13$ and $m = 8$.

A ROM is an LSI (Large Scale Integration) chip that is normally designed using an array of MOS transistors. A ROM is a combinational circuit. Internally, a ROM contains a decoder and OR gates; this is illustrated in Figure 4.41. Although diodes are not used for fabricating ROMs, a diode-based Mask ROM is shown in Figure 4.42 for simplicity, and just for illustrative purposes.

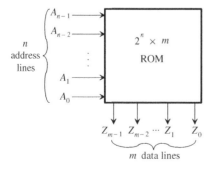

FIGURE 4.40   Block-diagram representation of a ROM.

A typical 3-input diode OR gate is shown in Figure 4.43. Resistor R pulls the output down to a LOW level as long as all the inputs are LOW. However, if any input is connected to a high voltage source (3 to 5 volts), the output is pulled HIGH to within one diode drop of the input. Thus, the circuit operates as an OR gate. To illustrate the operation of a ROM, consider the 2×4-bit ROM of Figure 4.42. In this system , when $A_1A_0 = 00$, the decoder output line 0 will be HIGH. This causes the diodes D_{00} and D_{01} to conduct, and thus the output $Z = Z_3Z_2Z_1Z_0 = 0011$. Similarly, when $A_1A_0 = 01$, the decoder output line 1 goes high, diode D_{10} conducts, and the output will be $Z = Z_3Z_2Z_1Z_0 = 0100$. Table 4.17 shows the truth table.

Figure 4.44 shows the subcategories of ROMs. ROMs are designed using MOS technology. ROM implementation offers a cost-effective solution for building circuits to perform useful tasks such as square root and transcendental function computations.

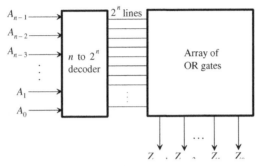

FIGURE 4.41   Internal Structure of a ROM.

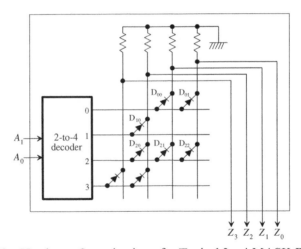

FIGURE 4.42   Hardware Organization of a Typical 2 x 4 MASK ROM.

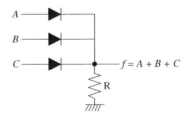

FIGURE 4.43   Diiode-OR Gate.

TABLE 4.17   Truth Table implemented by the ROM of Figure 4.40

A_1	A_0	Z_3	Z_2	Z_1	Z_0
0	0	0	0	1	1
0	1	0	1	0	0
1	0	0	1	1	1
1	1	1	1	0	0

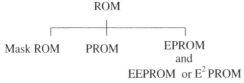

FIGURE 4.44   Subcategories of ROMs.

Mask ROM are programmed by placing the switching devices such as MOS transistors at the appropriate intersection points of the row and column lines. Hence, in a mask ROM, the contents of the ROM are initialized by the manufacturer at the time of its production. This means that this approach is well suited for producing a standard circuit such as a bar-code generator. Because these types of ROMs are mass-produced, their costs are also very low. However, a mask ROM cannot be reconfigured by a user. That is, a user cannot alter its contents.

Occasionally, a user may wish to develop a specific ROM-based circuit as demanded by the application. In this case, a ROM that allows a user to initialize its contents is required. A ROM with such a flexibility is known as a *PROM* (programmable ROM). In this device, the manufacturer places a switching element along with a fusible link at each intersection. This implies that all ROM cells are initialized with a 1. If a user desires to store a zero in a particular cell, the fuse is blown at that point. This activity is called *programming*, and it may be accomplished by passing electrical impulses. It should be pointed out that in such a ROM a user can program the ROM only once. That is, it is not possible to reprogram a PROM once the fuse is blown.

When a new product is developed, it may be necessary for the designer to modify the contents of the ROM. A ROM with this capability is referred to as an *EPROM* (erasable programmable ROM). Usually, the contents of this memory are erased by exposing the ROM chip to ultraviolet light. Typical erase times vary between 10 and 30 minutes. After erasure, the ROM may be reprogrammed by passing voltage pulses at the special inputs. The 2764 chip is a typical example of an EPROM. It is a 28-pin 8K × 8 chip contained in a dual in-line package (DIP). It has 13 address input pins and 8 data output pins. Note that the 2764 needs 13 ($2^{13} = 8192$) pins to address 8192 (8K) locations.

The growth in IC technology allowed the production of another type of ROM whose contents may be erased using electrical impulses. These memory devices are customarily referred to as *electrically erasable PROMs* (EEPROMs or E^2PROMs). The main advantage of an EEPROM is that its contents can be changed without removing the chip from the circuit board.

4.8   Programmable Logic Devices (PLDs)

A programmable logic device (PLD) is a generic name for an IC (Integrated Circuit) chip capable of being programmed by the user after it is manufactured. It is programmed by blowing fuses. The theory behind PLD is based on the concept of SOP (Sum-Of-Products). Since a Boolean function can be expressed in SOP form, a PLD chip contains an array of AND gates and OR gates. There are two types of PLDs. They are identified by the location of the fuses on the AND-OR array as shown in Figure 4.45.

The PROM was discussed in the last section. A PROM contains a decoder (number of fixed AND gates) and programmable OR gates. PLDs evolved from PROMs. There are two types of PLDs. They are PLA and PAL.

The PLA (Programmable Logic Array) was developed before PAL. The PLA includes several AND and OR gates, both of which are programmable. The PAL (Programmable Array Logic), on the other hand, includes programmable AND gates and fixed OR gates. The PLA is very flexible in the sense that the necessary AND terms can be logically ORed to provide the desired Boolean functions.

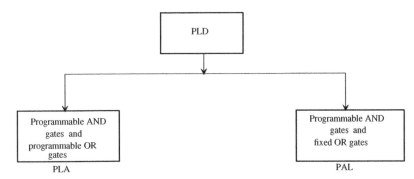

FIGURE 4.45　Types of PLD.

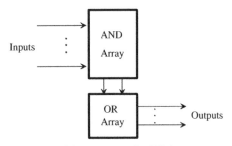

FIGURE 4.46　Internal Structure of a PLA.

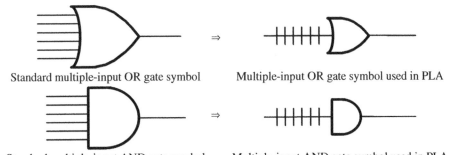

FIGURE 4.47　Multiple input AND and OR Gate Symbols for PLA.

The internal structure of a typical PLA is shown in Figure 4.46. The AND array of this system generates the required product terms, and the OR array is used to OR the product terms generated by the AND array. As in the case of the ROM, these gate arrays can be realized using MOS transistors.

Some PLAs provide inverted outputs by XORing each output with one. Finally, product terms can be shared in PLAs.

In order to illustrate a PLA, a special AND gate or OR gate symbol with multiple inputs shown in Figure 4.47 will be utilized in the following examples.

Next, consider implementing the following equations using a PLA:

$$Z_0 = \overline{A}\overline{B} + BC, \; Z_1 = \overline{A}\overline{B} + AC, \; Z_2 = \overline{A}\overline{C} + BC$$

Figure 4.48 shows the PLA with programmable links after programming. The schematic in Figure 4.48 is shown in Figure 4.49 in which the programmable links are

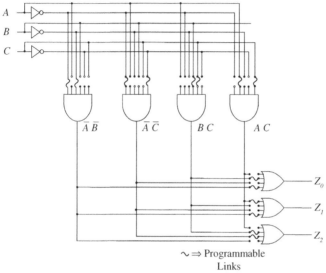

FIGURE 4.48   PLA with programmable links.

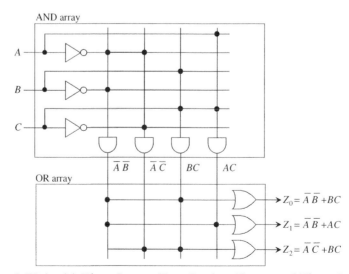

FIGURE 4.49   A PLA with Three Inputs, Four Product Terms, and Three Outputs.

replaced by dot symbols. Note that each AND gate shown in Figure 4.49 is with a single horizontal input line. A dot on the horizontal line indicates that an input connection to an AND gate is created by programming (shown by a programmable link in Figure 4.48). The dots at the OR gate input connections are created in a similar manner.

This PLA has three inputs, A, B, and C. The AND generates from product terms $\overline{A}\overline{B}$, $\overline{A}\overline{C}$, BC, and AC. These product terms are logically summed up in the OR array, and the outputs Z_0, Z_1, and Z_2 are generated. The use of PLAs is very cost-effective when the number of inputs in a combinational circuit realized by a ROM is very high and all input combinations are not used. For example, consider the following multiple output functions:

$$W = AE + BC$$
$$X = CD + FE$$
$$Y = FG + HI$$

To implement these Boolean functions in a ROM, a 512×3 array is needed because there are nine inputs (A through I) ($2^9 = 512$) and three outputs (W, X, Y), but the same functions can be realized in a PLA using six product terms, nine inputs, and three outputs, as shown in Figure 4.50. Therefore, a considerable savings in hardware can be achieved with PLAs.

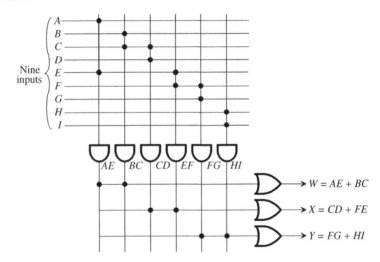

FIGURE 4.50   A PLA with Nine Inputs, Six Product Terms, and Three Outputs.

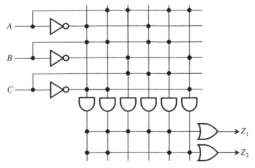

FIGURE 4.51   PLA Implementation of Example 4.8.

Example 4.8

Implement Example 4.4 using PLAs.

Solution

From Example 4.4,

$$Z_1(C, B, A) = \sum m(2, 3, 5, 6, 7)$$
$$= \overline{C}B\overline{A} + \overline{C}BA + C\overline{B}A + CB\overline{A} + CBA$$

$$Z_2(C, B, A) = \sum m(1, 2, 3, 7)$$
$$= \overline{C}\,\overline{B}A + \overline{C}B\overline{A} + \overline{C}BA + CBA$$

Figure 4.51 shows the PLA implementation.

4.9   Commercially Available Field Programmable Devices (FPDs)

Both mask programmable and field programmable PLAs are available. Mask programmable PLAs are similar to mask ROMs in the sense that they are programmed during fabrication. Field programmable PLAs (FPLAs), on the other hand, can be programmed by the user with a computer-aided design (CAD) program to select a minimum number of product terms to express the Boolean functions.

There are three types of commercially available Field Programmable Devices (FPDs). These are Simple PLD (SPLD), Complex PLD (CPLD), and Field Programmable Gate Array (FPGA). SPLDs can handle tens to hundreds of gates, and cannot be used for larger digital-design applications. Therefore, CPLD (complex PLD) chips are designed by the manufacturers such as Altera and Xlinix to accomplish this. CPLDs can handle thousands of gates. Both SPLD and CPLD are nonvolatile; they retain programs in case power is switched off. The original FPGA is SRAM-based and can handle millions of gates, and is volatile. That is, programs are lost in case power is switched off. However, flash memory-based novolatile FPGAs are currently available.

The SPLD is typically based on PAL architecture, and uses EPROM technology to implement the switches. Note that PAL is a registered trademark of Advanced Micro Devices, Inc. (AMD). PALs were introduced by Monolithic Memories (a division of AMD) in 1970. The PAL chips are usually identified by a two-digit number followed by a letter and then one or two digits. The first two-digit number specifies the number of inputs whereas the last one or two digits define the number of outputs. The fixed number of AND gates are connected to either an OR or a NOR gate. The letter H indicates that the output gates are OR gates; the letter L is used when the outputs are NOR gates; the letter C is used when the outputs include both OR and NOR gates. Note that OR outputs generate active HIGH whereas NORs provide active LOW outputs. On the other hand, OR-NOR gates include both active HIGH and active LOW outputs.

For example, the PAL16L8 is a 20-pin chip with a maximum of 16 inputs, up to 8 outputs, one power pin, and one ground pin. The 16L8 contains 10 non shared inputs, six inputs that are shared by six outputs, and two non shared outputs. Figure 4.52 shows the pin diagram of the PAL16L8. Note that PEEL (Programmable Electrically Erasable Logic) devices or Erasable PLDs such as 18CV8 or 16V8 are available for instant reprogramming just like an EEPROM. These devices utilize CMOS EEPROM technology. These erasable PLDs use electronic switches (MOS transistors) rather than fuses so that they are erasable and reprogrammable like EEPROMs.

A typical CPLD contains several PLDs. Each PLD containing AND and OR gates with EEPROM or EPROM or Flash memory (designed using EEPROM and EPROM) to implement the programmable switches along with all the interconnections in the same chip. Altera Corporation's EPM7032LC44-6 (36 user I/O pins) is an example of a typical CPLD.

IC manufacturers such as Altera and Xlinix also took a different approach for handling larger applications. They devised FPGA (Field Programmable Gate Array) chips which can be programmed at the user's location. A typical FPGA chip contains several smaller individual logic blocks (memory, multiplexers, gates, and flip-flops) along with all interconnections in a single chip. The FPGA does not use EEPROM technology to implement the switches; the programming information is stored in memory and the flip-flops are needed to implement sequential circuits. The memory is normally

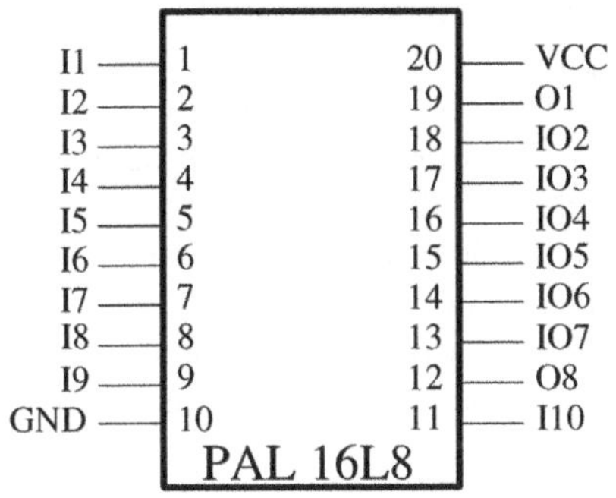

I1 1 20 VCC
I2 2 19 O1
I3 3 18 IO2
I4 4 17 IO3
I5 5 16 IO4
I6 6 15 IO5
I7 7 14 IO6
I8 8 13 IO7
I9 9 12 O8
GND 10 11 I10
PAL 16L8

FIGURE 4.52 Pinout for PAL 16L8.

TABLE 4.18 Truth Table to be inplemented in the SRAM of Figure 4.53

Inputs		Output
x	y	$z = x \oplus y$
0	0	0
0	1	1
1	0	1
1	1	0

4 x 1-bit SRAM
READ = 1
Two-bit address
x
y
00 0
01 1
10 1
11 0
z

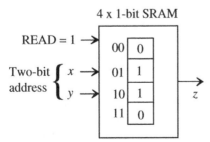

FIGURE 4.53 A Simplifed block diagram implementing XOR truth table of Table 4.18.

programmed to store a look-up table (truth table) containing the combinational circuit functions for the logic block. Note that the concept of FPGA is based on the fact that a truth table (hence, a combinational logic circuit) can be implemented in memory.

For example, consider a simplified memory implementing an XOR gate ($z = x \oplus y$) with two inputs (x,y) and one output (z) of Figure 4.53. Let us implement the truth table of Table 4.18 in the memory of Figure 4.53. The memory can be programmed with the truth table such that the contents of memory address 00 as 0, address 01 as 1, address 10 as 1, and address 11 as 0 and thus, the XOR truth table can be implemented in the SRAM.

In summary, an FPGA Chip basically contains two blocks. They are lookup tables and switch matrices. The lookup table block is contained in memory. This block along with mux's and flip-flops is called a Configurable Logic Block (CLB). An FPGA chip contains several CLBs which are surrounded by a ring of switch matrices containing multiplexers. The switch matrices allow the user to program the connections among lookup tables in the CLBs. The switch matrices contain I/O interface blocks which connect the CLB signals to FPGA pins.

FPGAs are very popular these days because of the flexibility of implementing users' complex design via software onto a single chip. In addition, these programmed FPGA chips can then be interfaced readily to other chips on the board which may be required to perform the functions of a specific application. Altera Corporation's EPF10K10PLCC (84 user I/O pins) is an example of a typical FPGA chip. Finally, products can be developed using FPGA from conceptual design via prototype to production in a very short time since debugging software is faster than debugging hardware.

4.10   HARDWARE DESCRIPTION LANGUAGE (HDL)

Hardware Description Languages (HDLs) such as Verilog or VHDL along with CAD (Computer-aided design) tools, allow FPGAs to be programmed with millions of gates in a short time. A CAD system contains a number of programs that are used to design a logic circuit. Some of these tools are discussed in the following.

A "Schematic Capture" tool utilizes a schematic editor program for entering, modifying and drawing a schematic on a computer screen using menus, keyboard commands, and the mouse. As mentioned before, the word "schematic" means a digital logic circuit in which logic gates along with their interconnections are shown. The schematic capture tool maintains an internal library consisting of graphic symbols for gates. The designer can place these gates on the screen from the library , and can then connect with lines that represent wires. The "schematic capture" is useful for small design. For larger circuits, the designer normally uses a Hardware Description Language (HDL) to write codes to specify the circuit.

Synthesis is basically similar in concept to the functions provided by a compiler. The compiler generates the object code. Synthesis, on the other hand, generates a database with instructions on how to integrate the hardware specified by the HDL code. Using "Synthesis" software of the CAD system, the designer can produce a circuit either from a schematic or from the HDL code. Using either approach, the Synthesis tool generates a set of logic expressions describing the functions required to obtain the circuit. These initial logic expressions are not in minimum form because they are basically devised by the designer using the CAD tools. For larger circuits, it may be difficult for the designer to manually come up with a better circuit. The designer may be able to use the "Synthesis" CAD tool to obtain a better circuit automatically based on the designer's original circuit.

A "Functional Simulator" is a CAD tool that is used to verify the correct operation of the circuit being designed. The " Simulator" can be used for simulation that takes into consideration the specified inputs and produces the desired outputs. Timing diagrams are used to represent the simulation results. The designer can check these timing diagrams to verify that the circuit performs according to the specifications.

Verilog and VHDL are two popular HDLs for designing both combinational and sequential circuits. Note that Verilog and VHDL are not programming languages like C. Rather, they are used to describe a logic circuit; they do not compute anything. An overview of both Verilog and VHDL will be provided next.

Verilog (developed by Design Automation in 1984 and later acquired by Cadence Design Systems) is a hardware description language. Verilog is not an acronym. Verilog syntax is based mostly on C and some Pascal. Note that Cadence Design Systems made Verilog available to the public during the early 90's. It was then accepted as an IEEE (Institute of Electrical and Electronics Engineers) standard (number 1364) in 1995.

VHDL is an acronym for VHSIC Hardware Description Language. VHSIC stands for Very High Speed Integrated Circuits. The design of VHDL evolved from the United States Department of Defense (DOD) VHSIC program. VHDL is based on the Ada programming language. The design of VHDL started in 1983 and after going through several versions was formally accepted as an IEEE standard number 1076 in 1987.

Since undergraduate students are usually familiar with C programming language, it is easier for them to learn Verilog compared to VHDL. Hence, Verilog will be covered in this book.

In order to design systems using Verilog, two levels of abstractions or their combinations are used. These are Structural modeling and Behavioral modeling. Structural modeling can be used to describe a schematic or a logic diagram (gates and interconnections) of a system. This type of modeling makes the designer's task easy for hardware implementation. Structural modeling is used for small systems.

"Hierarchical" structural modeling can be used by the designer to decompose a large digital system into smaller blocks or modules. The designer can define a block that is used repeatedly. This common block can be used by other blocks in the Verilog code to accomplish the desired task.

The Behavioral modeling, on the other hand, is used to describe a system in terms of what it does and how it behaves rather than in terms of its components and their interconnections. Boolean expressions are used to accomplish this. Behavioral modeling is used for larger systems.

Behavioral modeling is typically used to describe sequential circuits, although it can also be used to describe combinational circuits. The flow of data in Behavioral modeling can be represented via concurrent or sequential statements. Concurrent statements are executed in parallel as soon as data is available at the inputs while sequential statements are executed in the order that they are written. Behavioral modeling uses either sequential statements or concurrent statements. The first method is useful in describing complex digital systems. When behavioral model is described by concurrent statements, it is called *dataflow modeling*. The Dataflow modeling describes a digital circuit in terms of its function and flow of data through the circuit. Dataflow modeling utilizes a number of operators that can be used on variables to obtain the outputs.

A Verilog design program can be written and simulated using software tools provided by the manufacturers including Xlinix and Altera. A Verilog code called *test bench* can be written to test a Verilog design. A test bench program allows the designer to monitor the output(s) based on the application of appropriate inputs. These outputs can then be verified for correctness. Test results can be represented in terms of both waveform and tabular form. The waveform typically contains timing diagrams to graphically show the relationship between time, inputs, and outputs.

4.11   Verilog basics

In the following, basic features associated with Verilog will be covered.

4.11.1   Verilog keywords

Verilog includes approximately 100 predefined identifiers called *keywords*. These keywords must be lower case, and specify the language constructs. Some of the keywords include `module`, `endmodule`, `input`, `output`, `not`, `or`, `nor`, `nand`, `xor`, and `xnor`. The keywords are printed in a different font for clarity although this is not a requirement of Verilog. Verilog compilers ignore blank lines, TAB and SPACE keys. Verilog keywords are reserved, and cannot be used as user-defined names.

4.11.2   A typical Verilog Segment

Verilog describes a digital system as a set of modules. A module is a basic block in Verilog which contains the statements specifying the circuit. A typical Verilog segment is given below:

```
module module name (.,..,..); // A typical Module
        <port list>
        <declarations>
        <module items>
endmodule
```

Verilog module   In the above, the module is enclosed by the keyword `module` and ended by the keyword `endmodule`.

The module name identifies a module uniquely. This means that a name or an identifier is assigned to a module to identify it. This name must start with an alpha character or an underscore (_) rather than a number. The module name can be an alpha character (upper case or lower case) followed by numbers and/or alpha characters. Identifiers are case sensitive. This means that Full_adder and full_adder are distinct variables.

The contents inside the parentheses after the module name contain the inputs and outputs called *ports*. The ports are included in parentheses with commas separating them. One may place all inputs followed by all outputs and vice versa as long as the same order is used by another module for calling this module. A semicolon (;) is used to terminate the `module` statement. The two slashes (//) shown in the above Verilog module is used before a single line comment.

Nesting of modules is not permitted in Verilog. That is, a module cannot be placed between `module` and `endmodule` of another `module`. However, modules can be instantiated within other modules. This provides hierarchical modeling of design in Verilog. The name of a Verilog module is not available outside the module unless hierarchical modeling is used. The instance names must be defined when modules are instantiated. A module instantiation statement associates the signals in the module instantiation with the ports in a module definition. There are two ways to represent the association. These are positional association, and named association. These two methods cannot be mixed. In positional association, each signal in the module instantiation is mapped by position to the corresponding signal in the module definition. In order to illustrate positional association, consider the following Verilog program:

```
module system;
      wire [3:0] d;
      subsystem f1 (d[3], d[1], d[2], d[0]);
endmodule
module subsystem (x,y,w,z);
      input x, y;
      output w, z;
endmodule
```

In the above program, the module system has an instance of the module subsystem inside it. The connections to the subsystem are made by placing the bit vectors of the identifier (d in this case) at the desired positions in the port definitions of the subsystem module. In the above, d[3] is associated with x, d[1] with y, d[2] with w, and d[0] with z. Note that the inputs and outputs inside the module subsystem should be with inputs followed by outputs such as module subsystem (x,y,w,z); as in the above example or the outputs followed by the inputs such as module subsystem (w,z,x,y);The ordering must be done properly. Therefore, in the positional association, the names of the connecting signals must be included at the appropriate positions in the module port list. Positional association is used for small systems while named association is used for large systems.

In the named association, Verilog connects external signals by the port names rather than by positions. The port connections can be specified in any order as long as the port names in the module definition precisely match the external signals. For example, the above Verilog program with positional association can be rewritten using named association as follows:

```
module system;
      wire [3:0] d;
      subsystem f1 (.w(d[0]), .x(d[3]), .y(d[2]),.z(d[1]));
endmodule
module subsystem (w, x, y, z);
      input x, y;
      output w, z;
endmodule
```

In the above, d[0] is associated with w, d[1] with z, d[2] with y, and d[3] with x. The ordering of the ports of instance f1 of subsystem module is not important because the signals are associated by names. Note that if an instance of a module contains an unconnected port, the position of the port in the instantiation is left empty. For example, consider a module representing a three-input OR gate with declaration as or (f, a, b, c); . If it is desired to keep the input at position b unconnected, then an instance of or will be or (f, a, , c); . Note that an unconnected module input is placed in high impedance state automatically, and unconnected outputs are not used.

Verilog ports Each port in the <port list> is defined by keywords input and output based on the port directions. Verilog also supports bidirectional ports which can be defined by keyword inout . A semicolon (;) is used to terminate the port statement.

Verilog declarations　The <declarations> define data objects by keywords reg or wire. A digital circuit can be described by Verilog using a schematic (Structural modeling) or by procedural statements (Behavioral modeling). Nets specify the connections between gates by using the keyword wire in structural modeling while variables are signals generated by procedural statements declared by the keyword reg in behavioral modeling. There are two types of data in Verilog. They are nets and registers. Net variables behave like wires in actual circuits while register variables store values in procedural statements. Nets mean connection between hardware elements. Nets are driven continuously by the outputs of devices they are connected to. Nets are typically declared by the keyword wire.　Net is a class of data that includes wire as one data type. The keyword　wire　is used for a net type variable while the keywords reg and　integer　are used for register type variable. Keywords reg　and wire are one-bit wide by default. To define a wider reg　or wire, the left and right bit positions are defined in square brackets separated by a colon. For example, reg　[7:0]

TABLE 4.19　Some of the Verilog operators

Verilog Operator	Operation type	Symbol
Arithmetic	Binary addition	+
	Binary subtraction	−
	Binary multiplication	*
	Binary division	/
Bitwise	One's complement	~
	AND	&
	OR	|
	XOR	∧
Reduction	AND	&
	OR	|
	XOR	∧
Logical	NOT	!
	AND	&&
	OR	||
Shift	Logical shift left	<<
	Logical shift right	>>
Relational	Less than	<
	Greater than	>
	Logical equality	==
	Logical inequality	!==
Conditional	Conditional	?:
Concatenation	Concatenation	{ }

a,b; declares two variables a and b as 8 bits with the most significant bit as bit 7 (a[7] or b[7]) and the least significant bit as bit 0 (a[0] or b[0]). The size of the `integer` is determined by the size of the host computer. If the word length of the host computer is 64-bit then the integer size is also 64-bit.

Verilog registers (defined by keyword `reg`) typically retain their values until a new value is stored. Verilog registers are different from hardware registers which need a clock. Verilog register does not require a clock. Also, Verilog register does not need a driver like the net. Values of Verilog registers can be changed anytime during simulation by replacing with another value. Finally, Verilog combines the two keywords `output` and `reg` into a single keyword called `output reg`. For example, the following two statements,

```
output a;
reg a;
```
can be combined into a single statement as
```
output reg a;
```

module items   The <module items> for structural modeling are the statements allowed in this modeling (to be discussed later) while <module items> for behavioral modeling (to be discussed later) may be initial block or always block. Verilog uses keywords `begin` and `end` like Pascal to define a block. Curley brackets { } are not used with `begin` and `end` unlike C since these brackets are used for concatenation in Verilog.

4.11.3   Verilog operators

Verilog operators are useful for logic synthesis. Table 4.19 shows a list of the Verilog operators, with complement operator as the highest priority and the conditional operator as the lowest priority. A list of precedence of Verilog operators can be obtained from the Internet. The meaning of some of these operators will be provided in the following.

The meaning of the Arithmetic operators is obvious. They perform operations using binary numbers. As an example, consider the arithmetic operation. If X, Y, and Z are 8-bit vectors, then Z = X + Y will add 8-bit X with 8-bit Y , and will store the 8-bit result in Z.

Most bitwise operators except one's complement perform bit-by-bit logic operations on two binary numbers. As an example, consider Z = X | Y where X, Y, and Z are 4-bit wide. Suppose, X = 0110, and Y = 1010. After Z = X | Y, Z = 1110 is obtained as follows:

$$
\begin{array}{ll}
 & X = 0110 \\
\text{OR} & Y = 1010 \\
\hline
 & Z = 1110
\end{array}
$$

Reduction operators perform bitwise logic operations as specified on the bits of the same word, and then produce a one-bit result. As an example, consider the reduction operation &x where x is a 4-bit number. In this case, the operation &x means x[3]&x[2]&x[1]&x[0]. Suppose x = 0111, then &x = 0&1&1&1 = 0.

Logical operators are typically used for finding true or false conditions. Consider two 8-bit numbers X and Y. Suppose X = 00000000 and Y = 11000111. Note that logical operators assign a value of 1 to a nonzero number while assign a value of 0 if the number is zero; the specified operation is then performed. Therefore, with the assumed values of X and Y, '0' is assigned to X and '1' is assigned to Y. Results using these values for the logical operators are provided below:

!X = NOT 0 = 1
!Y = NOT 1 = 0

X && Y = 0 AND 1 = 0
X ||Y = 0 OR 1 = 1

    Note that for a single bit, the operator ~ and operator ! have the same meaning. This means that if x is one-bit then ~x = !x

    Shift operations logically shift a binary number to the left or to the right by a specified number of bits (n). When the binary number is shifted to the left by the shift count (n), the upper n bits are discarded while the low-order n bits are filled with 0's. For example, consider X = 00110101. After X<<3, X is shifted to the left three positions. The upper three bits of X (001) will be shifted out, and the low-order three bits of X (101) will be filled with three 0's so that the value of X after shifting will be 10101000.

    On the other hand, when the binary number is shifted to the right by the shift count (n), the lower n bits are discarded while the high-order n bits are filled with 0's. For example, consider the same value of X =00110101. After X>>3, X is shifted to the right three positions. The low-order three bits of X (101) will be discarded, and the high-order three bits of X (001) will be filled with three 0's so that the value of X after shifting will be 00000110.

    As an example of logical operators, consider logical equality operator (==). The statement (X==Y) compares X with Y bit-by-bit. The comparison produces a true result of 1 if X equals Y (true condition); otherwise (if X is not equal to Y), the comparison produces a false result of 0. Similarly, the meaning of "Less than" and "Greater than" operators can be explained. Relational operators can be used in 'if-else' and 'for' constructs in the same manner as the C-language. These operators are used for determining true or false conditions. A value of '1' is assigned for the true condition while a '0' is assigned for the false condition.

    Conditional operator in Verilog is denoted by the symbol ? to be used with the assign keyword. For example, consider the statement, assign z = s ? x : y; . This means that if s = 1 then z = x, else z = y for s = 0. Note that in this expression, 's' is the condition, z = x is the true expression while z = y is the false expression.

    The Concatenate operator combines two or more vectors, and produces a vector of a bigger size. As an example, consider a full adder. The full adder adds three inputs bits (a, b,cin) and generates two output bits (cout, s). The Concatenate operator symbol { } can be used to concatenate two single bits (cout and s) as a two-bit number in a single statement using the keyword assign as follows: assign {cout, s} = a + b + cin;

4.11.4   Verilog Constants

Constants in Verilog are decimal integers by default. However, the syntax 'b,'d, or 'h can be used before a number to define it as binary, decimal or hexadecimal respectively. Furthermore, the total number of bits in a number can be represented by placing the number before the quote. For example, 4'b1111 and 4'hf will represent 15 in decimal. Also, the Verilog keyword parameter declares and assigns the value to be a constant. For example, parameter x = 5; will assign the value of integer 5 to x.

4.11.5   Modeling logical conditions in a circuit

To precisely model all logical conditions in a circuit, each bit in Verilog can be one of the following:1'b0, 1'b1, 1'bz (high impedance), or 1'bx (don't care). 1'b0 and 1'b1 respectively correspond to 0 and 1.Verilog includes 1'bz for the situation when the designer needs to define a high impedance state. Furthermore, Verilog includes 1'bx to specify a don't care condition. Sometimes, miswiring of gates may also result into an unknown value of the output in certain situation. For example, if the designer may make a mistake and may connect outputs of two gates together. This output may want to assume a value of either 0 or 1. This may cause physical damage to certain logic families. In order for the simulator to detect such problems, the 1'bx (don't care) definition can be used for the output.

4.11.6   Verilog `if-else` and `case-endcase` structures
These C structures are used in behavioral modeling.

The `if-else` construct   The syntax for the `if-else` construct is as follows:

```
if (cond)
        statement1;
else
        statement2;
```

This means that if the condition is true, the statement1 is executed; else (if the condition is false), statement1 is skipped, and the statement2 is executed.

The `case-endcase` construct   The if-else statement allows one to select an expression based on true or false condition. However, if there are several possible expressions, Verilog code using `if-else` becomes cumbersome. In such a situation, the `case-endcase` construct can be useful.

The syntax for the construct is as follows:

```
case (controlling parameters)
    option1: equation;
    option2: equation;
    option3: equation;
    ---
    ---
    ---
    default: equation;
endcase
```

The controlling parameters are compared with each option, the equation with respect to the matched (first) option is executed. A `default` statement should be used when the listed options do not include the controlling parameters.

4.11.7   A typical Verilog Simulator

A Verilog simulator includes a built-in system function called $time for representing simulated time. This means that $time provides a measure of actual time for the hardware to function when fabricated. $time is expressed as an integer value rather than by time units such as seconds. However, designers typically use one time unit as one nanosecond. Time control statements may be included in Behavioral Verilog. A statement will not be executed with the symbol # followed by a number until the specified number of time steps has elapsed. This allows Verilog to model propagation delays of logic gates. The symbol # when used in test benches generates a sequence of patterns at particular times that will behave like inputs to the hardware being designed. Also, if the symbol @ is used before a statement , the statement that follows will not be executed until the statement with @ is completed.

The test bench for the simulation is normally written by the designer. The test bench tests the Verilog design by applying stimulas and providing outputs during simulation. Test benches utilize procedural blocks which start with either the keywords `initial` or `always` for providing stimulas for the test circuit. Note that the `initial` block is executed once at the start of simulation, and it does not play any role in synthesis.

An example of a simple `initial` block is provided below:

```
initial
        begin
                #0
                  x=1'b0; y=1'b0; z=1'b0;
                #50
                  x=1'b0; y=1'b0; z=1'b1;
                #50
                  x=1'b0; y=1'b1; z=1'b0;
        end
```

In the above, keywords `begin` and `end` are used to define the block with the time units defined by the symbol #.

At time = 0,
x = 0, y = 0 and z = 0.
At time = 50 ns,
x = 0, y = 0 and z = 1
Finally, at time = 100 ns,
x = 0, y = 1 and z = 0.

A simple test bench has the following structure:

<module name>
<reg and wire declarations>
<Instantiate the Verilog design>
<Generate stimulus using `initial` and `always` keywords>
<Produce the outputs using $monitor for verification>
endmodule

The inputs applied to the test (design) block for simulation are declared in the stimulus block as `reg` data type. The outputs (responses) of the test block that are to be monitored and verified are declared as `wire` data type. The test block has no inputs or outputs. The stimulus block produces inputs for the test block and verifies the output of the test block. `initial` and `always` procedural blocks can be used to produce the output. The simulator can represent the output as waveforms or in tabular form using Verilog system tasks such as $monitor. The syntax for $monitor is provided below:

$monitor ("time = %d x = %2d y = %3d z = %2b",$time, x, y, z);

Verilog system task, `$monitor` can be used to display the output of the design block under test. Verilog simulator allows the output to be represented in binary (%b or %B), octal (%o or %O), decimal (%d or %D) or hexadecimal (%hor %H). $time is a built-in function that provides the simulation time. In the above $monitor statement time, x, and y are displayed in decimal while z is represented in binary. Another way to display the output is by using system task `$display`. Note that `$display` is used to display one time value of variables. In contrast $monitor displays variables whenever changes in variables occur during simulation. The syntax for `$display` is `$display("%b%d",x,y);` which will display x in binary and y in decimal.

4.12   Verilog modeling examples for combinational circuits

As mentioned in the last section, there are basically three ways to write Verilog codes. They are:
1.   Structural modeling
2.   Dataflow modeling
3.   Behavioral modeling.

Structural modeling is based on hardware schematic. As mentioned before, a "Hierarchical" structural modeling can be used by the designer to decompose a large digital system into smaller blocks or modules. The designer can define a block that is used repeatedly. This common block can be used by other blocks in the Verilog code to accomplish the desired task. Dataflow modeling, on the other hand, uses a number of logic operators that work on inputs to generate the outputs.

Finally, behavioral modeling uses mostly C language, and some pascal constructs that show the expected behavior of the circuit. Behavioral modeling uses procedural statements such as if-else statement , and represents digital circuits at a functional level, and not at the gate level. As described before, although behavioral modeling is used mostly for sequential circuits, it can also be used for combinational circuits. Several examples of describing simple combinational circuits using Verilog will be provided in this section.

4.12.1   Structural modeling

In order to describe a combinational logic circuit using structural modeling, consider writing the Verilog code for $f = a + b\,\bar{c}$ (Section 3.6).
For structural modeling, the schematic for $f = a + b\,\bar{c}$ must be drawn first. Figure 4.54 (a) shows the schematic. Using the schematic, Verilog code is written as follows:

```
module func(a, b, c, f);
    input a, b, c;          //line #1
    output f;               //line #2
    wire y0,y1;             //line #3
    not (y0, c);            //line #4
    and (y1, b, y0);        //line #5
    or (f, y1, a);          //line #6
endmodule
```

Verilog codes should be written in such a way that they are easy to read. That is why keywords `module` and `endmodule` are aligned properly. The module name "`func`" is used arbitrarily. In order to explain the above Verilog code, line #1 through line #6 are used as comments after //.

Since there are three inputs (a,b,c) shown in the schematic, line #1 defines them using the `input` keyword. Similarly, line #2 defines one output (f) using the `output` keyword.

As mentioned before, nets specify the connections between gates by using the keyword `wire` in structural modeling. Hence, keyword `wire` (line #3) is used to declare the intermediate outputs y0 and y1 shown in the schematic.

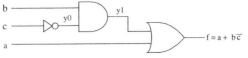

FIGURE 4. 54   (a) Schematic for $f = a + b\,\bar{c}$.

The statement in line #4, not (y0,c); uses the keyword not, and specifies 'y0' as the output and 'c' as the input. Hence, this statement means that y0 = not c.

The statement in line #5, and (y1, b, y0); uses the keyword and, and specifies 'y1' as the output with 'b' and 'y0' as inputs. Hence, this statement means y1 = b and y0.

The statement in line #6, or(f, y1, a); uses the keyword or, and specifies 'f' as the output with 'y1' and 'a' as inputs. Hence, this statement means f = a or y1. Hence, f = a + b $\bar{c}$.

Using Altera Quartus II CAD software, the timing diagram of Figure 4.54 (b) and the truth table of Figure 4.54(c) are obtained. In order to verify the result provided by the waveform, the following values are obtained: at 60ns: a = 1, b = 1, c = 0, and f = 1. The value of f can then be verified analytically after substituting the values of a, b, c, and then determining the value of f from equation, f = a + b c. Similarly, the simulation results provided by the truth table of Figure 4.54 (c) can be verified.

Waveform

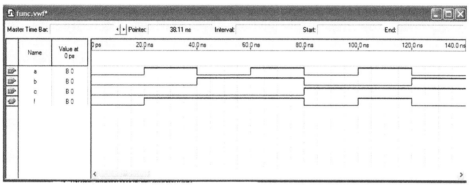

FIGURE 4.54 (b)   Waveform for the schematic of Figure 4.54 (a).

Tabular Results

INPUTS a b c;

OUTPUTS f;

UNIT ns;

RADIX HEX;

PATTERN

	a b c	f			
0.0>	0 0 0	= 0	20.0>	1 0 0	= 1
40.0>	0 1 0	= 1	60.0>	1 1 0	= 1
80.0>	0 0 1	= 0	100.0>	1 0 1	= 1
120.0>	0 1 1	= 0	140.0>	0 0 0	= 0

FIGURE 4.54 (c)   Truth table for the schematic of Figure 4.54 (a).

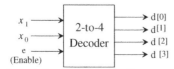

FIGURE 4.55   2-to-4 Decoder block diagram.

Example 4.9

Write a Verilog description of the 2-to-4 decoder of Figure 4.55 re-drawn from Figure 4.8(a). Use structural modeling and the schematic of Figure 4.56 (re-drawn from Figure 4.9).

Solution

Using the above schematic, the Verilog description for the decoder using structural modeling is provided below:

```
// Structural description of a 2-to-4 decoder
module decoder2to4 (x1, x0, e, d);
        input x1, x0, e;
        output [0:3] d;
        wire x11, x00;
        not
                inv1 (x11, x1),
                inv2 (x00, x0);
        and
                and1 (d[0], x11, x00,e),
                and2 (d[1], x11, x0, e),
                and3 (d[2], x1, x00, e),
                and4 (d[3], x1, x0, e);
endmodule
```

The above structural description for the 2-to-4 decoder contains three inputs (x1, x0, e), and four outputs (d[0] through d[3]). The wire declaration provides internal connections. Two NOT gates are used to obtain complements x11 and x00 of the inputs x1 and x0 respectively while the four AND gates are used for the outputs d[0] through d[3]. In the gate list such as and1 (d[0], x11, x00, e);, the output d[0] is always listed first followed by inputs x11, x00, and e. The keyword "and" is written once for all AND operators, and in this case, provides output d[0] by logically ANDing x11, x00, and e. Note that if a Verilog operation is required several times in a program such as "not" requiring twice in the above, the Verilog code can be written in two ways. The two NOT operations, in the above, are written using the keyword "not" once followed by two

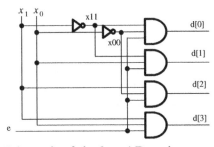

FIGURE 4.56 Schematic of the 2-to-4 Decoder.

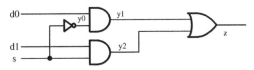

FIGURE 4.57 2-to-1 Multiplexer schematic.

different labels inv1 and inv2 separated by commas, and terminated by ;. An alternate Verilog code for the two NOT operations can be written the keyword "not" as follows:

```
not (x11, x1);
not (x00, x0);
```

Similarly, alternative codes for other logic operations in the above can be written.

Example 4.10

Write a Verilog description for the 2-to-1 multiplexer of Figure 4.57 (redrawn from Figure 4.18(b)) using structural modeling.

Solution

Using the schematic of Figure 4.57, Verilog description of the 2-to-1 MUX is written as follows:

```
//Structural modeling for 2-to-1 mux
module mux2(d0,d1,s,z);
    // I/O port declarations
output z;
input d0, d1, s;
    // Internal nets
wire y0, y1, y2;
    // Instantiate logic gate primitives
not(y0, s);
and(y1, d0, y0);
and(y2, d1, s);
or(z, y1, y2);
endmodule
```

Example 4.11

Using Figure 4.58 (re-drawn from 4.29), write a Verilog description for the full adder using two half adders and an OR gate as described in Section 4.5.6. Use Hierarchical modeling.

Solution

Assume x, y, z as three inputs and cout, sum as the two outputs of the full adder. x and y can be applied as the inputs to the first half adder generating sum, $s1 = x \oplus y$ and carry, $c1 = xy$. $s1$ can be applied as one of the inputs to the second half adder with z as the other input. The second half adder will produce a sum, $sum = x \oplus y \oplus z$ which is the desired sum of the full adder. The carry output, $c2$ of the second half adder will be $(x \oplus y) z$. $c1$ and $c2$ can be logically ORed together to provide the carry output (cout) of the full adder. This is depicted in Figure 4.29 which is redrawn as shown in Figure 4.58.

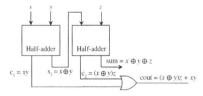

FIGURE 4.58   Full-adder using two half-adders.

Note that the schematic of the half-adder shown in Figure 4.26 is re-drawn, and shown in Figure 4.59.

This schematic of figure 4.59 is used to write the Verilog code for the half-adder. The half-adder module is then instantiated twice according to Figure 4.58 to obtain the full adder.

The Verilog description is given below:

```
// Half Adder
module half_adder (s,c,x,y);
      output s,c;
      input x,y;
      xor (s,x,y);
      and (c,x,y);
endmodule
// Full adder is obtained by instantiating half adder twice
//(Hierarchical modeling)
module full_adder (sum,cout,x,y,z);
      output sum,cout;
      input x,y,z;
      wire s1,c1,c2;
      half_adder B1(s1,c1,x,y);
      half_adder B2(sum,c2,s1,z);
      or(cout,c1,c2);
endmodule
```

Example 4.12

Using Figures 4.60 and 4.61, write a verilog description for a four-bit binary adder using Hierarchical modeling.

Solution

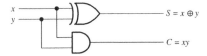

FIGURE 4.59   Half-adder schematic.

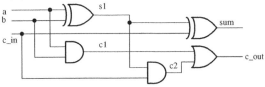

FIGURE 4.60   Full adder schematic.

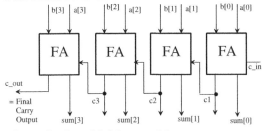

FIGURE 4.61   Figure for four-bit binary adder.

Verilog description of the full adder is written using the schematic of Figure 4.28 (b) which is redrawn for convenience,and is shown in Figure 4.60.

The full adder module is called fulladd which is then called four times according to Figure 4.31 which is redrawn, and is shown in figure 4.61.

Verilog description of the 4-bit binary adder using hierarchical modeling is provided below:

```
// Define the full_adder
module fulladd(sum, c_out, a, b, c_in);
// I/O port declarations
  output sum, c_out;
  input a, b, c_in;

// Internal nets
 wire s1, c1, c2;

// Instantiate logic gate primitives
 xor (s1, a, b);
 and (c1, a, b);
 xor (sum, s1, c_in);
 and (c2, s1, c_in);
 or (c_out, c2, c1);
endmodule

// Define a 4-bit binary adder
module fulladd4(sum, c_out, a, b, c_in);

// I/O port declarations
    output [3:0] sum;
    output c_out;
    input [3:0] a, b;
    input c_in;
// Internal nets
    wire c1, c2, c3;

// Instantiate four one-bit full adders.
    fulladd fa0(sum[0], c1, a[0], b[0], c_in);
    fulladd fa1(sum[1], c2, a[1], b[1], c1);
    fulladd fa2(sum[2], c3, a[2], b[2], c2);
    fulladd fa3(sum[3], c_out, a[3], b[3], c3);
endmodule
```

Note that in Verilog, nesting of modules is not permitted. That is, a module cannot be placed between module and endmodule of another module. However, modules can be instantiated within other modules. This provides hierarchical modeling of design in Verilog. In the above program, the full-adder is defined by instantiating primitive gates. The next module describes the 4-bit binary adder by instantiating four full-adders. The instantiation is done by using the name of the module that is instantiated with the same port names in this case.

The statement, `fulladd fa0(sum[0], c1, a[0], b[0], c_in);` in the submodule called "fulladd fa0" instantiates the full adder module called" fulladd". This submodule is given the name fa0 which is a user-defined identifier that must start with an alpha character or the underscore (_). The order in which the ports are listed inside the bracket of the submodule "fulladd fa0" is the same as those of the module" fulladd". The other three submodules with names "fa1", "fa2", and "fa3" can be explained similarly.

4.12.2   Dataflow modeling

Structural modeling can be cumbersome for writing Verilog codes for large circuits. However, Dataflow modeling can be used for describing the behavior of large circuits.

Dataflow modeling utilizes Boolean operators along with logic expressions. Dataflow modeling in Verilog allows a digital system to be designed in terms of its function. Dataflow modeling utilizes Boolean equations, and uses a number of operators that can act on inputs to produce outputs. There are approximately 30 Verilog operators. These operators can be used to synthesize digital circuits. Some of the operators are listed in the Table 4.19.

All Boolean equations are executed concurrently whenever any one of the values on the right hand side of one or more equations changes. This is accomplished using Verilog's continuous assignment statement. This statement uses the keyword `assign`. A continuous assignment statement is used to assign a value to a net. A net is not a verilog keyword. It is used to specify the output (defined by `output` or `wire` using declaration statements) of a gate. For example, consider the following assignment statement:
`assign e = (a∧b)&(~c)|d);`

Note that in the above, the keyword `assign` means that the value of 'e' is continuously updated whenever one or more values (a,b,c,d) on the right hand side of the equation change. The Boolean expression on the right hand side of the above equation is first evaluated, and the AND gate output is connected to wire e.

In order to illustrate dataflow modeling in Verilog, consider writing the Verilog code for $f = a + b\,\overline{c}$ (Section3.6). Verilog code for this logic equation was written using structural modeling in section 4.12.1.

Verilog code using dataflow modeling is provided in the following:

```
module function(a,b,c,f);
      input a,b,c;
      output f;
  assign f = a | (b&~c);
endmodule
```

The above code is self explanatory.

Example 4.13

Write a Verilog description of the 2-to-4 decoder of Figure 4.56 using dataflow modeling

Solution

The four output equations for the decoder are obtained as follows:
$d0 = e\overline{x_1}\,\overline{x_0}$, $d1 = e\overline{x_1}x_0$, $d2 = ex_1\overline{x_0}$, and $d3 = ex_1x_0$

```
module decoder2to4 (e, x1, x0, d0, d1, d2, d3);
   input e, x0, x1;
   output d0, d1, d2, d3;
   assign d0 = (e & ~x1 & ~x0);
   assign d1 = (e & ~x1 & x0);
   assign d2 = (e & x1 & ~x0);
   assign d3 = (e & x1 & x0);
endmodule
```

The above dataflow program uses Verilog keyword `assign` followed by Boolean equations using Boolean operators.

Example 4.14

Write a Verilog description for the 2-to-1 multiplexer of Figure 4.57 using dataflow modeling.

Solution

From section 4.5.4, the equation for the MUX output is $z = \overline{s}d0 + sd1$. Verilog code using dataflow modeling is provided below:

```
module mux_2to1 (s, d0, d1, z);
    input s, d0, d1;
    output z;
    assign z = (~s & d0) | (s & d1);
endmodule
```

Example 4.15

Write a Verilog description for the 2-to-1 multiplexer of Figure 4.17 using the conditional operator.

Solution

```
module mux 2to1 (s, d0, d1, z);
      input s, d0, d1;
      output z;
      assign z = s ? d1 : d0;
endmodule
```

Note: In the above, if s = 1 then z = d1 else, if s = 0 then z = d0.

Example 4.16

Write a Verilog description for a 4-bit binary adder using dataflow modeling. Use the concatenation operator.

Solution

```
// 4-bit binary adder using the concatenation operator
module adder4 (x, y, cin, s, cout);
    input [3:0] x, y;
    output cout;
    output [3:0];
    assign {cout, s} = x + y + cin;
endmodule
```

4.12.3 Behavioral modeling

Behavioral modeling is used to describe a circuit using logic expressions and C language constructs. Behavioral description uses the keyword `always` followed by one or more procedural statements. If there are more than one procedural statement, they are executed sequentially. That is, they are executed in the order they are written. If there are several statements in the `always` block, then the keywords `begin` and `end` must be used. Keywords `begin` and `end` are not required for a single statement in the `always` block. A simple example with keywords `begin` and `end` using the equations of the full adder is provided below:

```
always @ (a, b, c)
begin
    sum = a ∧ b ∧ c;
    carry = (b&c)| (a &c) | (a&b);
end
```

The outputs of the procedural statements must be declared by the keyword `reg`. Input ports cannot be declared as `reg` since they do not retain values, rather affect the changes in the external signals they are connected to. Note that a `reg` data type retains its value until a new value is assigned.

As an illustration of behavioral modeling, consider the following Verilog code written using behavioral modeling for the 2-to-1 multiplexer of Figure 4.17:

```
module mux (s, d0, d1, z);
    input s, d0, d1;
    output z;
    reg z;
    always @ (s , d0, d1)
    begin
        if (s == 0) z = d0;
        else z = d1;
    end
endmodule
```

In the above, note that output z in behavioral modeling must be declared as both `output` and `reg`. The if-else procedural statement in the always block is executed and the output z is updated if, at any time, inputs (s, d0, d1) included inside parentheses after the "always @" changes. The statement "if-else" is a conditional statement whose condition is the select input, s. When s = 0, then the output z = d0. When s = 1, then the output z = d1. It should be pointed out that no semicolon (;) is needed after the `always` statement.

Example 4.17

Write a Verilog description for a 2 to 4 decoder of Figure 4.55 with one high enable. Use behavioral modeling with `if-else` and `case-endcase` statements.

Solution

```
module decoder2to4 (e, i, d);
      output [3:0] d;
      input [1:0]i;
      input e;
      reg [3:0] d;
      always @ (i , e)
            begin
                  if (e==1)
                        case (i)
                              0: d = 4'b 0001;
                              1: d = 4'b 0010;
                              2: d = 4'b 0100;
                              3: d = 4'b 1000;
                        endcase
                  else
                        d = 4'b 0000;
            end
endmodule
```

In the above, i (two-bit input) and e (one-bit enable) are declared as inputs while vector "d " is declared as 4-bit `reg` output. The conditional statement `if-else` allows execution of the `case` statements if e=logic 1. Note that the decoder is enabled when the enable line, e equals logic 1. The logical operator `==` is used for logical equality in the if expression. If e = logic 1 , the statements (between `case` and `endcase`) are executed sequentially. The statement `if (e==1)` is executed as soon as any of the inputs after @in the `always` statement changes. The `case` statement is used for multiple branching. For example, `case(i)` determines the value of the two-bit vector, i and compares it with the values with the list of the statements. The assignment statement associated with the first value that matches is executed. Since the vector i is a two-bit vector, it can be any of the four values from 0 to 3. For example, consider the statement 2: d= 4'b0100;. If i = 10 (2 in decimal), then the `case` statement after executing 2: d= 4'b0100; will assign four-bit vector, d with the binary value 0100. This means that the line 2 of the decoder output is high while others are low. An optional default value can be used for the `case` statement. This is for assigning other values such as don't care (x) or high impedance (z). Also, in the above, if e == 0, the 4-bit output vector, d is assigned with low values. This is shown as part of the `else` statement. This means that the decoder is disabled.

Example 4.18

Write a Verilog description using behavioral modeling for a BCD to seven-segment code converter (Section 4.4) for driving a common-cathode display for displaying the decimal digits 2, 4, and 9. The converter will turn the display OFF for any other inputs.

Solution

```
module code_converter (bcd_in,seven_seg_out);
      input [3:0] bcd_in;
      output [6:0] seven_seg_out;
      reg [6:0] seven_seg_out;
        // bcd_in = abcdefg
      parameter two  = 7'b1101101;
      parameter four = 7'b0110011;
      parameter nine = 7'b1110011;
      parameter other = 7'b0000000;
      always @ (bcd_in)
      begin
            case (bcd_in)
            2: seven_seg_out = two;
            4: seven_seg_out = four;
            9: seven_seg_out = nine;
            default: seven_seg_out = other;
            endcase
      end
endmodule
```

Example 4.19

Write a Verilog description for a full-adder using 74138 decoder and gates of Figure 4.62 (re-drawn from figure 4.35).

Solution

This problem implements a full adder using a 3to8 decoder and two 4 input AND gates as shown in figure 4.62. Behavioral modeling is used for implementation of 3to8 decoder and the four-input AND gate while structural modeling is used for the interconnection of the decoder with the AND gates using the schematic of Figure 4.62 (redrawn from figure 4.35). For the 4 input AND gate, the inputs are ANDed using the bitwise AND operator "&".

//Description: Full Adder Using 3-to-8 MUX with AND gates.

// APPROACH: Behavioral for the implementation of the decoder

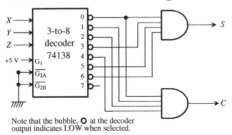

Note that the bubble, O at the decoder output indicates LOW when selected.

FIGURE 4.62   Figure for Example 4.19.

```
// and 4 input AND gates.
//Structural approach when combining the decoder and
//AND gates.
//decoder74138 3 to 8 decoder with active low outputs.
//INPUTS:   --X, Y, Z( select lines )
//          --G1, nG2A, nG2B ( enable lines)
//          Out[7:0] ( eight output lines)
//OUTPUTS: --high impendance "Z" outputs when chip not selected
//          --active low output on line selected.
//            (if chip selected)
module decoder74138 (nout, G1, nG2A, nG2B, X , Y , Z);
      output [7:0] nOut;
      input G1, nG2A, nG2B, X, Y, Z;
      reg [7:0] nOut;
      always @(G1 , nG2A , nG2B , X , Y , Z)
      begin
         if((G1, nG2A , nG2B) ==3'b100)
// chip enabled
         begin // if statement
// select conditions for select lines w/ active low outputs
            case ( { X, Y, Z})
               0: nOut[7:0] = 8'b1111_1110;
               1: nOut[7:0] = 8'b1111_1101;
               2: nOut[7:0) = 8'b1111_1011;
               3: nOut[7:0] = 8'b1111_0111;
               4: nOut[7:0] = 8'b1110_1111;
               5: nOut[7:0] = 8'b1101_1111;
               6: nOut[7:0] = 8'b1011_1111;
               7: nOut[7:0] = 8'b0111_1111;
               default: nOut [7:0] = 8'bx;   //this should
                                             // never happen
            endcase
         end // if statement
         else
         // chip disabled
            begin
            nOut [7:0] = 8'hzz;
            end
         end // always statement
endmodule
//AND4: 4 input and gate

//INPUTS: --A,B,C,D

//OUTPUTS: --Out AND output of all four inputs

module AND4 (Out,A,B,C,D);
      output Out;
```

```
        input A,B,C,D;
        reg Out;
        always@(A or B or C or D)
        begin
          Out=A & B & C & D;
        end
endmodule
```

```
//Full-Add: Full adder using 3to8 decoder 74138 and 2 four input AND gates
//INPUTS : -- X , Y , Z (X bit to add, Y bit to add , Z carry to add)
```

```
//OUTPUTS: --S = sum bit
//      --C = Carry out bit
        module Full_Add (C,S,X,Y,Z);
        output C , S;
        input X , Y , Z;
        wire [7:0] decoder_out;
```

```
// 3 to 8 decoder enabled with bits to be added as inputs
```

```
decoder74138 decoder74138_0(decoder-out [7:0],1'b1,1'b0,1'b0, X,Y,Z);
```

```
// use 4 input AND gates to do final sum and carry
```

```
    AND4 AND4_0(S,decoder_out[0],decoder_out[3],decoder_out[5],decoder_out[6]);
```

```
    AND4 AND4_1(C,decoder_out[0],decoder_out[1],decoder_out[2],decoder_out[4]);
```

```
endmodule
```

Questions and Problems

4.1   Find function F for the following circuit:

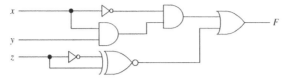

4.2   Express the following functions *F*1 and *F*2 in terms of the inputs *A*, *B*, and *C*.
       What is the relationship between *F*1 and *F*2?

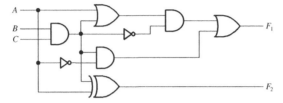

4.3   Given the following circuit:

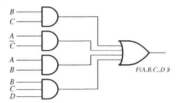

       (a)   Derive the Boolean expression for *F*(*A*, *B*, *C*, *D*).
       (b)   Derive the truth table.
       (c)   Determine the simplified expression for *F*(*A*, *B*, *C*, *D*) using
              a K-map.
       (d)   Draw the logic diagram for the simplified expression using
              NAND gates.

4.4   Determine the function *F* of the following logic diagram and then analyze the
       function using Boolean identities. Show that *F* can be represented by a single
       logic gate with *A* and *B* as inputs.

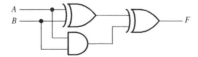

4.5   Design a combinational circuit that accepts a two-bit unsigned number, A and B
       such that the output, F = A plus B. Note that AB = 11_2 will never occur. Draw
       a schematic using logic gates. Show all steps clearly.

4.6   Design a combinational circuit with three inputs (*A*, *B*, *C*) and one output (*F*).
       The output is 1 when *A* + *C* = 0 or *AC* = 1; otherwise the output is 0. Draw a
       schematic using logic gates. Show all steps clearly.

4.7 Design a combinational circuit that accepts a three-bit unsigned number and generates an output binary number equal to the input number plus 1. Draw a schematic using logic gates. Show all steps clearly.

4.8 Design a combinational circuit with five input bits generating a 4-bit output that is the ones complement of four of the five input bits. Draw a logic diagram. Do not use NOT, NAND, or NOR gates.

4.9 Design a combinational circuit that converts a 4-bit BCD input to its nines complement output. Draw a schematic using logic gates. Show all steps clearly.

4.10 Design a BCD to seven-segment decoder that will accept a decimal digit in BCD and generate the appropriate outputs for the segments to display a decimal digit (0–9). Use a common anode display. Turn the seven segment OFF for non-BCD digits. Draw a logic circuit. What will happen if a common cathode display is used? Comment on the interface between the decoder and the display.

4.11 Design a combinational circuit for a BCD to seven-segment code converter that will input a BCD number and output it on a seven segment common-anode display. The code converter will only display the number 8. The converter will turn the display OFF for all other valid BCD digits except digit 9 which will never occur. Draw a schematic. Show all steps clearly.

4.12 Design a combinational circuit using a minimum number of full adders to decrement a 6-bit signed number by 2. Assume 6-bit result. Draw a logic diagram using the block diagram of a full adder as the building block.

4.13 Design a combinational circuit using full adders to multiply a 4-bit unsigned number by 2. Draw a logic diagram using the block diagram of a full adder as the building block.

4.14 Design a combinational circuit that adds two 4-bit signed numbers and generates an output of 1 if the 4-bit result is zero; the output is 0 if the 4-bit result is nonzero. Draw a logic circuit using the block diagram of a 4-bit binary adder as the building block and a minimum number of logic gates.

4.15 Design a 4 ×16 decoder using a minimum number of 74138 and logic gates.

4.16 Design a combinational circuit using a minimum number of 74138s (3 × 8 decoders) to generate the minterms $m1$, $m5$, and $m9$ based on four switch inputs $S3, S2, S1, S0$. Then display the selected minterm number (1 or 5 or 9) on a seven-segment display by generating a 4-bit input (W, X, Y, Z) for a BCD to seven-segment code converter. Ignore the display for all other minterms. Note that these four inputs (*W, X, Y, Z*) can be obtained from the selected output line (1 or 5 or 9) of the decoders that is generated by the four input switches (*S3, S2, S1, S0*). Use a minimum number of logic gates. Determine the truth table, and then draw a block diagram of your implementation by connecting the following building blocks (Figure P4.16):

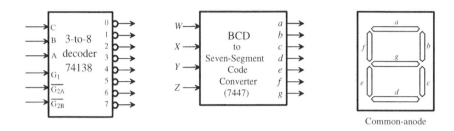

FIGURE P4.16.

4.17   A combinational circuit is specified by the following equations:

$$F_0(A, B, C) = \overline{A}\,\overline{B}\,\overline{C} + A\overline{B}C + \overline{A}BC \qquad F_2(A, B, C) = \overline{A}\,\overline{B}\,\overline{C} + ABC$$

$$F_1(A, B, C) = A\overline{B}\,\overline{C} + AB\overline{C} \qquad F_3(A, B, C) = \overline{A}BC + \overline{A}B\overline{C} + AB\overline{C}$$

Draw a logic diagram using a decoder and external gates. Assume that the decoder outputs a HIGH on the selected line.

4.18   Draw a logic diagram using a 74138 decoder and a minimum number of external gates to implement the following:

$F_0(A, B, C) = \Sigma m(1,3,4)$, $F_1(A, B, C) = \Sigma m(0,2,4,7)$, $F_2(A, B, C) = \Sigma m(0,1,3,5,6)$, $F_3(A, B, C) = \Sigma m(2,6)$

4.19   Draw a schematic of a full adder using a 74138 and a minimum number of external gates. Use minterms of the full adder.

4.20   Determine the truth table for a hexadecimal-to-binary priority encoder with line 0 having the highest priority and line 15 with the lowest.

4.21   Implement a digital circuit to increment (for $Cin = 1$) or decrement (for $Cin = 0$) a 4-bit signed number by 1 generating outputs in twos complement form. Note that Cin is the input carry to the full adder for the least significant bit. Draw a schematic:
        (a)   Using only a minimum number of full adders and multiplexers.
        (b)   Using only a minimum number of full adders and inverters. Do not use any multiplexers.

4.22 Using a minimum number of multiplexers, and by manipulating the inputs, draw a schematic that will perform the following operations:

S_1	S_0	Y
0	0	0
0	1	A
1	0	B
1	1	15_{10}

Assume that A is a 4-bit unsigned number where $A = a3a2a1a0$ and $B = \overline{a3}\,\overline{a2}\,\overline{a1}\,\overline{a0}$.

4.23 Implement each of the following using an 8-to-1 multiplexer:
(a) $F(A, B, C, D) = \overline{A}BC + \overline{A}BD + \overline{A}\,\overline{B}\,C + ACD$
(b) $F(W, X, Y, Z) = \Sigma m(2, 3, 6, 7, 8, 9, 12, 13, 15)$

4.24 What are the main logic elements/gates in a ROM chip?

4.25 Design a combinational circuit using a 16×16 ROM that will increment a 4-bit unsigned number by 1. Determine the truth table and then draw a block diagram of your implementation showing the addresses and their contents in binary along with one Output Enable (OE) input.

4.26 What are the basic differences among PROM, PLA, PAL, PEEL and FPGA?

4.27 What is the technology used to fabricate EPROMs and EEPROMs?

4.28 Design an 4K X 8 EPROM (with two enable lines, $\overline{CE}$ and $\overline{OE}$) based system to display the squares of BCD digits on seven segment displays using a minimum number of 74LS47 BCD to seven segment converters. Each BCD digit will be input to the EPROM via switches. The square of a particular BCD number will be displayed in BCD each time the 4-bit number is input to the EPROM via the switches. Draw a block diagram of your implementation showing the contents of memory along with addresses in hex.

4.29 Design a 4-bit adder/subtractor (Example 4.7) using only full adders and EXCLUSIVE-OR gates. Do not use any multiplexers.

4.30 Design a combinational circuit using a minimum number of full adders, and logic gates with one BCD to seven-segment converter (74LS47) and one common-anode seven-segment display, and which will perform A plus B or A minus B (A and B are signed numbers), depending on a mode select input, M. If M = 0,addition is carried out; if M = 1, subtraction is carried out. Assume A = A4 A3 A2 A1 A0 and B = B4 B3 B2 B1 B0 (Two 5-bit numbers). The circuit will be able to carry out the subtraction even if A < B. Use an LED to indicate the sign of the result (LED ON for negative result and LED OFF for positive result). The result of the operation should always appear in BCD form on the single seven-segment display. Assume that the result will be in the range of 0 through +9 in decimal and -1 through -9 in decimal. For example, if five-bit addition or subtraction provides a result of 10111 in binary, the circuit will take the two's complement of the number, and will display minus (Sign LED ON) and 9 on the single seven-segment display. The Overflow bit (V) should be indicated by another LED (LED ON for V=1 and LED OFF for V=0). Do not use any multiplexers.

4.31   Write a Verilog description for a 3-to-8 decoder with a LOW enable using:
        (a) Structural modeling
        (b) Behavioral modeling

4.32   Write a Verilog description for a four-bit unsigned comparator using dataflow modeling using the "logical equality", "less than", and "greater than" operators.

4.33   Write a Verilog description for a 4-to-1 multiplexer using dataflow modeling using the conditional operator.

4.34   Write a Verilog description for an 8-bit binary adder using hierarchical modeling.

4.35   Write a Verilog description for a 4-to-1 multiplexer using behavioral modeling with `case-endcase` construct.

4.36   Write a Verilog description of the 4-to-2 priority encoder of Figure 4.16a(iii) using modeling of your choice.

SEQUENTIAL LOGIC

5.1 Basic Concepts

So far, we have considered the design of combinational circuits. The main characteristic of these circuits is that the outputs at a particular time t are determined by the inputs at the same time t. This means that combinational circuits require no memory. However, in practice, most digital systems contain combinational circuits along with memory. These circuits are called *sequential*.

In sequential circuits, the present outputs depend on the present inputs and the previous states stored in the memory elements. These states must be fed back to the inputs in order to generate the present outputs. There are two types of sequential circuits: synchronous and asynchronous.

In a synchronous sequential circuit, a clock signal is used at discrete instants of time to ensure that all desired operations are initiated only by a train of synchronizing clock pulses. A timing device called the *clock generator* produces these clock pulses. The desired outputs of the memory elements are obtained upon application of the clock pulses and some other signal at their inputs. This type of sequential circuit is also called a *clocked sequential circuit*. The memory elements used in clocked sequential circuits are called *flip-flops*. The flip-flop stores only one bit. A clocked sequential circuit usually utilizes several flip-flops to store a number of bits as required. Synchronous sequential circuits are also called *state machines*.

In an asynchronous sequential circuit, completion of one operation starts the operation that is next in sequence. The asynchronous sequential circuit does not need a clock. Hence, synchronizing clock pulses are not required. Instead, time-delay devices are used in asynchronous sequential circuits as memory elements. Logic gates are typically used as time delay devices, because the propagation delay time associated with a logic gate is adequate to provide the required delay. A combinational circuit with feedback among logic gates can be considered as an asynchronous sequential circuit. One must be careful while designing asynchronous systems because feedback among logic gates may result in undesirable system operation. The logic designer is normally faced with many problems related to the instability of asynchronous system, so they are not commonly used. Most of the sequential circuits encountered in practice are synchronous because it is easy to design and analyze such circuits.

5.2 Latches and Flip-Flops

A flip-flop is a one-bit memory. As long as power is available, the flip-flop retains the bit. However, its output (stored bit) can be changed by a clock input. The basic type of flip-flop is called a *latch* which uses a level-sensitive clock. Flip-flops are designed using the latches. The most common latch is the SR (Set-Reset) latch. A flip-flop is a latch with an edge-trigerred clock input. Synchronous sequential circuits are designed using flip-flops whereas asynchronous sequential circuits are designed using latches.

5.2.1   SR Latch

Figure 5.1 shows a basic latch circuit using NOR gates along with its truth table. The SR latch has two inputs, S (Set) and R (Reset), and two outputs Q (true output) and $\overline{Q}$ (complement of Q). To analyze the SR latch of Figure 5.1(a), note that a NOR gate generates an output 1 when all inputs are 0; on the other hand, the output of a NOR gate is 0 if any input is 1. Now assume that $S = 1$ and $R = 0$; the $\overline{Q}$ output of NOR gate #2 will be 0. This places 0 at both inputs of NOR gate #1. Therefore, output Q of NOR gate #1 will be 1. Thus, Q stays at 1. This means that one of the inputs to NOR gate #2 will be 1, producing 0 at the $\overline{Q}$ output regardless of the value of S. Thus, when the pulse at S becomes 0, the output $\overline{Q}$ will still be 0. This will apply 0 at the input of NOR #1. Thus, Q will continue to remain at 1. This means that when the set input $S = 1$ and the reset (clear) input $R = 0$, the SR latch stores a 1 ($Q = 1$, $\overline{Q} = 0$). This means that the SR latch is set to 1.

   Consider $S = 0$, $R = 1$; the Q output of NOR gate #1 will be 0. This will apply 0 at both inputs of NOR gate #2. Thus, output $\overline{Q}$ will be 1. When the reset pulse input R returns to zero, the outputs continues to remain at $Q = 0$, and $\overline{Q} = 1$. This means that with set input $S = 0$ and reset input $R = 1$, the SR latch is cleared to 0 ($Q = 0$, $\overline{Q} = 1$).

   Next, consider $Q = 1$, $\overline{Q} = 0$. With $S = 0$ and $R = 0$, the NOR gate #1 will have both inputs at 0. This will generate 1 at the Q output. The output $\overline{Q}$ of NOR gate #2 will be zero. Thus, the outputs Q and $\overline{Q}$ are unchanged when $S = 0$ and $R = 0$.

   When $S = 1$ and $R = 1$, both Q and $\overline{Q}$ outputs are 0. This is an invalid condition because for the SR latch Q and $\overline{Q}$ must be complements of each other. Therefore, one must ensure that the condition $S = 1$ and $R = 1$ does not occur for the SR latch. This

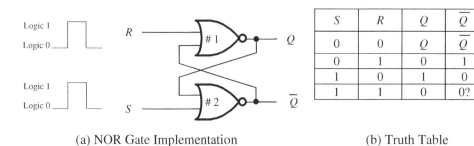

S	R	Q	$\overline{Q}$
0	0	Q	$\overline{Q}$
0	1	0	1
1	0	1	0
1	1	0	0?

(a) NOR Gate Implementation   (b) Truth Table

FIGURE 5.1   SR Latch using NOR gates.

S	R	Q	$\overline{Q}$
0	0	1	1?
0	1	1	0
1	0	0	1
1	1	Q	$\overline{Q}$

(a) NAND gate implementation   (b) Truth table

FIGURE 5.2   NAND implementation of an SR latch.

undesirable situation is indicated by a question mark (?) in the truth table shown in Figure 5.1(b). An SR latch can be built from NAND gates with active-low set and reset inputs. Figure 5.2(a) shows the NAND gate implementation of an SR latch.

The SR latch with S and R inputs will store a 1 ($Q = 1$ and $\overline{Q} = 0$) when the S input is activated by a low input (logic 0) and $R = 1$. On the other hand, the latch will be cleared or reset to 0 ($Q = 0$, $\overline{Q} = 1$) when the R input is activated by a low input (logic 0) and $S = 1$.

Note that an active low signal can be defined as a signal that performs the desired function when it is low or 0. In Figure 5.2(a), the SR latch stores a 1 when $S = 0 =$ active low and $R = 1$; on the other hand, the latch stores a 0 when $R = 0 =$ active low and $S = 1$.

Note that the NAND gate produces a 0 if all inputs are 1; on the other hand, the NAND gate generates a 1 if at least one input is 0. Now, suppose that $S = 0$ and $R = 1$. This implies that the output of NAND gate #2 is 1. Thus, $Q = 1$. This will apply 1 to both inputs of NAND gate #1. Thus, $\overline{Q} = 0$. Therefore, a 1 is stored in the latch. Similarly, with inputs $S = 1$ and $R = 0$, it can be shown that $Q = 0$ and $\overline{Q} = 1$. The latch stores a 0.

With $S = 1$ and $R = 1$, both outputs of the latch will remain at the previous values. There will be no change in the latch outputs. Finally, $S = 0$ and $R = 0$ will produce a invalid condition ($Q = 1$ and $\overline{Q} = 1$). This is indicated by a question mark (?) in the truth table of Figure 5.2(b).

An SR latch can be used for designing a switch debouncing circuit. Mechanical switches are typically used in digital systems for inputting binary data manually. These mechanical ON-OFF switches (e.g., the keys in a computer keyboard) vibrate or bounce several times such that instead of changing state once when activated, a key opens and closes several times before settling at its final values. These bounces last for several milliseconds before settling down.

A debouncer circuit, shown in Figure 5.3, can be used with each key to get rid of the bounces. The circuit consists of an SR latch (using NOR gates) and a pair of resistors. In the figure, a single-pole double-throw switch is connected to an SR latch. The center contact (Z) is tied to +5 V and outputs logic 1. On the other hand, contacts X or Y provide logic 0 when not connected to contact Z. The values of the resistors are selected in such a way that X is HIGH when connected to Z or Y is HIGH when connected to Z.

When the switch is connected to X, a HIGH is applied at the R input, and $S = 0$, then $Q = 0$, $\overline{Q} = 1$. Now, suppose that the switch is moved from X to Y. The switch is disconnected from R and $R = 0$ because the ground at the R input pulls R to 0. The outputs Q and $\overline{Q}$ of the SR latch are unchanged because both R and S inputs are at 0 during the switch transition from X to Y. When the switch touches Y, the S input of the latch goes to HIGH and thus $Q = 1$ and $\overline{Q} = 0$. If the switch vibrates, temporarily breaking the connection, the S input of the SR latch becomes 0, leaving the latch outputs unchanged. If the switch bounces back connecting Z to Y, the S input becomes 1, the latch is set again, and the outputs of the SR latch do not change. Similarly, the switch transition from Y to X will get rid of switch bounces and will provide smooth transition.

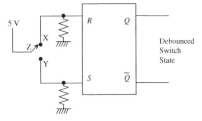

FIGURE 5.3   A debouncing circuit for a mechanical switch.

5.2.2 Gated SR Latch

Figure 5.4 shows the NAND gate implementation, the truth table, and the logic symbol of a gated SR latch. This latch includes an additional level-sensitive clock input. When a change in the level of the clock pulse occurs, the outputs of the latch change.

The gated SR latch contains the basic SR latch with two more NAND gates. It has three inputs (S, Clk, R) and two outputs (Q and $\overline{Q}$). When $S = 0$ and $R = 0$ and $Clk = 1$, the outputs of both NAND gates #1 and #2 are 1. This means that the output of NAND gate #3 is 0 if $\overline{Q} = 1$ and is 1 if $\overline{Q} = 0$. Hence, Q is unchanged as long as $S = 0$ and $R = 0$. On the other hand, the output of NAND gate #4 is 0 if $Q = 1$ and is 1 if $Q = 0$. Thus, $\overline{Q}$ is also unchanged. Suppose $S = 1$, $R = 0$, and $Clk = 1$. This will produce 0 and 1 at the outputs of NAND gates #1 and #2 respectively. This in turn will generate 1 and 0 at the outputs of NAND gates #3 and #4 respectively. Thus, the flip-flop is set to 1. When the clock is zero, the outputs of both NAND gates #1 and #2 are 1. This in turn will make the outputs of NAND gates #3 and #4 unchanged.

The other conditions in the function table can similarly be verified. Note that $S = 1$, $R = 1$, and $Clk = 1$ is combination of invalid inputs because this will make both outputs, Q and $\overline{Q}$ equal to 1. Also, Q and $\overline{Q}$ must be complements of each other in the gated SR latch. Note that Q^+ and $\overline{Q}^+$ are outputs of the gated SR latch after the clock (Clk) is applied.

5.2.3 Gated D Latch

Figure 5.5 shows the logic diagram, the truth table, and the logic symbol of a gated D latch. This type of latch ensures that the invalid input combinations $S = 1$ and $R = 1$ for the gated SR latch can never occur. The gated D latch has two inputs (D and Clk) and two outputs (Q and $\overline{Q}$). The D input is same as the S input and the complement of D is applied to the R input. Thus, R and S can never be equal to 1 simultaneously. Note that the Clk input is level sensitive.

The gated D latch transfers the D input to output Q when $Clk = 1$. Note that if $Clk = 0$, one of the inputs to each of the last two NAND gates will be 1; thus, outputs of the D flip-flop remain unchanged regardless of the values of the D input.

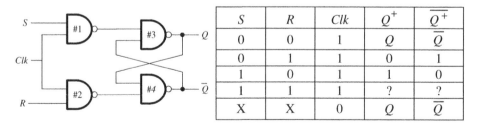

S	R	Clk	Q^+	$\overline{Q}^+$
0	0	1	Q	$\overline{Q}$
0	1	1	0	1
1	0	1	1	0
1	1	1	?	?
X	X	0	Q	$\overline{Q}$

(a) NAND gate implementation   (b) Truth Table

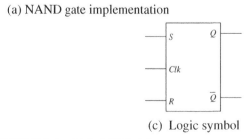

(c) Logic symbol

FIGURE 5.4   Gated SR latch.

The gated D latch is also called a "transparent latch." The term "transparent" is based on the fact that the output Q follows the D input when $Clk = 1$. Therefore, transfer of input to outputs is transparent, as if the gated D latch were not present.

5.2.4 Edge-Triggered D Flip-Flop

As mentioned before, sequential circuits contain combinational circuits and flip-flops. These flip-flops generate outputs at the clock based on the inputs from the combinational circuits. This can create an undesirable situation if the outputs from the combinational circuits that are connected to the flip-flop inputs change values at the clock pulse simultaneously when flip-flops change outputs. For example, assume that a positive pulse is used as the clock input of a gated D latch. With the D input $= 1$, the output of the latch will become 1 when the clock pulse reaches logic 1. Now, suppose that the D input changes to zero but the clock pulse is still at 1. This means that the latch will have a new output, 0. In this situation, the output of one latch cannot be connected to the input of another when both latches are enabled simultaneously by the same clock input. This situation can be avoided if the latch outputs do not change until the clock pulse goes back to 0. One way of accomplishing this is to ensure that the outputs of the latches are affected by the pulse transition rather than

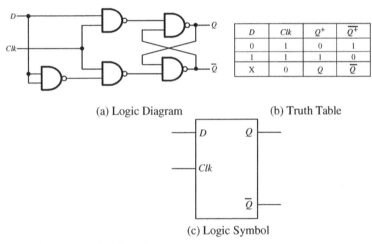

D	Clk	Q^+	$\overline{Q^+}$
0	1	0	1
1	1	1	0
X	0	Q	$\overline{Q}$

(a) Logic Diagram   (b) Truth Table

(c) Logic Symbol

FIGURE 5.5   Gated D Latch.

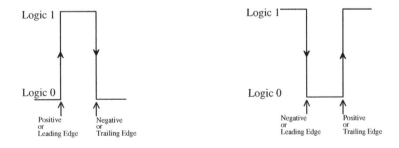

(a) Positive Pulse   (b) Negative Pulse

FIGURE 5.6   Clock Pulses.

pulse duration (level) of the clock input. To understand this concept, consider the clock pulses shown in Figure 5.6. There are two types of clock pulses: positive and negative. A positive pulse includes two transitions: logic 0 to logic 1 and logic 1 to logic 0. A negative pulse also goes through two transitions: logic 1 to logic 0 and logic 0 to logic 1.

Since flip-flops are constructed using the latches, the multiple-transition problem can be avoided in the flip-flops if they are clocked by either the positive (leading) edge or the negative (trailing) edge transition only rather than the signal level of the pulse as with the latches. A master–slave flip-flop is used to accomplish this. Figure 5.7 shows the logic diagram, truth table, and the logic symbol of a typical master-slave D flip-flop (or simply D flip-flop). A master-slave flip-flop contains two independent gated D latches. Gated D-latch #1 (GDL #1) works as a master flip-flop, whereas the gated D-latch #2 (GDL #2) is a slave. An inverter is used to invert the clock input to the slave gated D-latch.

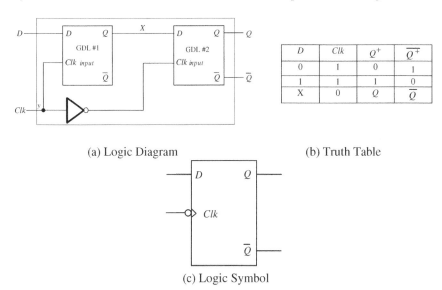

D	Clk	Q^+	$\overline{Q^+}$
0	1	0	1
1	1	1	0
X	0	Q	$\overline{Q}$

(a) Logic Diagram   (b) Truth Table

(c) Logic Symbol

FIGURE 5.7   Typical edge-triggered D Flip-Flop.

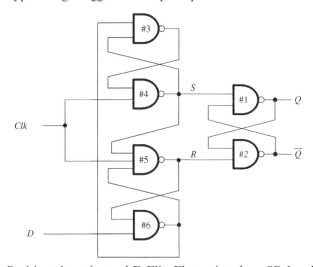

FIGURE 5.8   Positive edge-triggered D Flip-Flop using three SR Latches.

Assume that the *Clk* is a positive or leading edge. Suppose that the *D* input of the GDL #1 is 1 and the *Clk* input = 1 (leading edge). The output of the inverter will apply a 0 at the *Clk* input of the GDL #2. Thus, GDL #2 is disabled. The GDL #1 will transfer a 1 to its *Q* output. Thus, *X* will be 1.

At the trailing edge of the *Clk* input, the *Clk* input of the master GDL #1 is 0. Thus, GDL #1 is disabled. The inverter will apply a 1 at the *Clk* input of the GDL #2. Thus, 1 at the *X* input (*D* input of GDL #2) will be transferred to the *Q* output of GDL #2. When the *Clk* goes back to 0, the master gated D-latch is separated. This avoids any change in the other inputs to affect the master gated D-latch. The slave gated D-latch will have the same output as the master.

The D flip-flop of Figure 5.7(a) is negative edge-triggered since the outputs change at the negative edge of the clock. The triggering at the negative edge is indicated by a bubble with a triangular symbol at the *Clk* input of the logic symbol of the D-flip-flop of Figure 5.7(c). The logic symbol for the positive edge-triggered flip-flop, on the other hand, includes only the triangular symbol without any circle for the clock. Finally, note that the logic diagram of Figure 5.7(a) can be converted to a positive edge-triggered flip-flop by placing an additional inverter between *Clk* and *y*.

Master-slave flip-flops provide a simple way of understanding the meaning of edge-triggered flip-flops. A more efficient edge-triggered D-flip-flop can be obtained by using three SR latches as shown in Figure 5.8. This implementation requires six NAND gates while the master-slave D flip-flop needs eleven gates. In Figure 5.8, when *Clk* = 0, then $S = R = 1$, the outputs of NAND gate #1 and NAND gate #2 are maintained at the present outputs regardless of input D [(Figure 5.2(b)]. Now, if $D = 1$, when *Clk* goes to 1, then $R = 1$, $S = 0$ which means that the NAND-gate based SR latch (comprised of NAND #1 and NAND #2) is in the set state ($Q = 1$) regardless of changes in D. Now, when *Clk* = 1, then the *R* input of NAND gate #2 stays at 1 regardless of any changes in the *D* input. When *Clk* becomes 0, both *S* and *R* change to 1 without changing the outputs (Q and $\bar{Q}$).

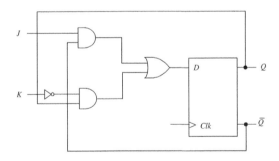

(a) Logic Diagram

J	*K*	*Clk*	*Q+*	$\bar{Q}+$
0	0	ʄ	*Q*	$\bar{Q}$
1	0	ʄ	1	0
0	1	ʄ	0	1
1	1	ʄ	$\bar{Q}$	*Q*
X	X	0	*Q*	$\bar{Q}$

(b) Truth Table

(c) Logic Symbol

FIGURE 5.9   Logic Diagram, Truth Table, and Logic Symbol of a JK Flip-Flop.

Similarly if D = 0 and *Clk* changes from 0 to 1, the input, *S* of NAND gate #1 becomes 0 resetting (*Q* = 0) the latch comprised of NAND gate #1 and NAND gate #2. Note that as long as *Clk* = 1, the output *Q* stays the same regardless of any changes in the *D* input. Hence, the output *Q* is the same as the *D* input for a positive-edge or a leading-edge clock. Therefore, this is a positive-edge or leading-edge triggered *D* flip-flop. Similarly, a negative-edge (trailing edge-triggered) *D* flip-flop can be obtained using NOR gates.

5.2.5   JK Flip-Flop

The JK flip-flop can be obtained from a *D* flip-flop and four logic gates (two AND, one OR, and one inverter) as shown in Figure 5.9 (a). From the figure, $D = J\,\overline{Q} + \overline{K}\,Q$. Let us now verify the truth table of Figure 5.9 (b) by determining the outputs (*Q* and $\overline{Q}$) by substituting the various combinations of the inputs (*J*, *K*) in the preceding equation as follows:

i)   **For *J* = *K* = 0**
   $D = 0.\overline{Q} + \overline{0}.Q = Q$
      Hence, the *D* input will be *Q*. Therefore, the outputs (*Q* and $\overline{Q}$) of the flip-flop will remain the same at the leading edge of *Clk*.

ii)   **For *J* = 1 and *K* = 0**
   $D = 1.\overline{Q} + \overline{0}.Q = 1$

      Hence, the *D* input will be one which will be transferred to the *Q* output at the leading edge of *Clk*. Therefore, *Q* = 1 and $\overline{Q}$ = 0. Thus the flip-flop is set to one.

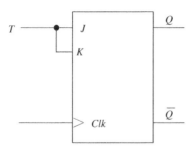

(a) T flip-flop by connecting JK inputs together.

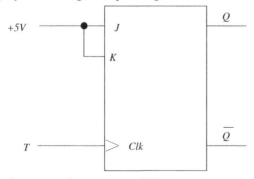

(b) T flip-flop by connecting JK to HIGH.

FIGURE 5.10   T flip-flop from JK flip-flop.

iii) **For $J = 0$ and $K = 1$**

$D = 0.\overline{Q} + \overline{1}.Q = 0$

Hence, the D input will be '0' which will be transferred to the Q output at the leading edge of *Clk*. Therefore, $Q = 0$ and $\overline{Q} = 1$. Thus the flip-flop is reset to zero.

iv) **For $J = 1$ and $K = 1$**

$D = 1.\overline{Q} + \overline{1}.Q = \overline{Q}$

Hence, the D input will be $.\overline{Q}$ which will be transferred to the output at the leading edge of *Clk*. Therefore, the flip-flop outputs will be complemented.

5.2.6 T Flip-Flop

The T (Toggle) flip-flop complements its output when the clock input is applied with $T = 1$; the output remains unchanged when $T = 0$. The name "toggle" is based on the fact that the T flip-flop toggles or complements its output when the clock input is 1 with $T = 1$.

The T flip-flop is not available commercially. That is you cannot buy T flip-flop. One needs to build it using other flip-flops such as the JK flip-flop. For example, T flip-flop can be obtained from JK flip-flop in two ways as shown in Figure 5.10 as follows:

i) In the first approach (Figure 5.10(a)), the J and K inputs of the JK flip-flop can be tied together to provide the T input; the output is complemented when $T = 1$ at the clock while the output remains unchanged when $T = 0$ at the clock.

ii) In the second approach, the J and K inputs can be tied to high; in this case, T is the clock input.

5.3 Flip-flop timing parameters for edge-triggered flip-flops

Certain timing parameters of an edge-triggered flip-flop must be considered when using the flip-flop. These are setup time (t_{su}), hold time (t_h), and propagation delay time of the flip-flop (t_{PHL}). These parameters are shown in the timing diagram of Figure 5.11 for a typical D flip-flop. The setup time is the minimum amount of time for which the input(s) of the flip-flop must remain stable before arrival of the clock edge. The hold time, on the other hand, is the minimum amount of time for which the input(s) of the flip-flop must remain stable after arrival of the clock edge. The propagation delay time of the flip-flop is the amount of time between the clock edge and the flip-flop achieving a new stable output. Note that the propagation delay is the time for a signal to travel from the flip-flop's input to output when the output changes its value; this delay is mainly due to the time associated with the internal transistors of the flip-flop to turn them ON and OFF.

As an example, consider 74LS74 D flip-flop. The timing parameters ((t_{su}, t_h, and t_{PHL}) for the 74LS74 according to the manufacturers' data sheet are provided in the following: $t_{su} = 20$ns, $t_h = 5$ ns, and $t_{PHL} = 40$ns. The designer should ensure that the flip-flop input (D in this case) is stable (either logic 0 or logic 1) at the clock edge so that Q will have a stable value for the system being designed. If the minimum setup and hold times specified by the manufacturers are not satisfied in a circuit, the flip-flop enters an undefined state (between 0 and 1) called *metastable state*. Metastability problem can be rectified by synchronizing any asynchronous input of a cirucit with the circuit's clock using a D flip-flop.

5.4   Preset and Clear Inputs

Commercially available flip-flops include separate inputs for setting the flip-flop to 1 or clearing the flip-flop to 0. These inputs are called *preset* and *clear* inputs respectively. These inputs are useful for initializing the flip-flops without the clock pulse. When the power is turned ON, the output of the flip-flop is in undefined state. The preset and clear inputs can directly set or clear the flip-flop as desired prior to its clocked operation.

Figure 5.12 shows a D flip-flop with a clear input. As mentioned before, the triangular symbol indicates that the flip-flop is clocked at the positive (leading) edge of the clock pulse. In Figure 5.12, a circle (inverter) is used with the triangular symbol. This means that the flip-flop is enabled at the negative (trailing) edge of the clock pulse.

The circle at the clear input means that the clear input must be 1 for normal operation. If the clear input is tied to ground (logic 0), the flip-flop is cleared to 0 ($Q = 0$, $\overline{Q} = 1$) irrespective of the clock pulse and the D input. The CLR input should be connected to 1 for normal operation. Some flip-flops may have a preset input that sets Q to 1 and $\overline{Q}$ to 0 when the preset input is tied to ground. The preset input is connected to 1 for normal operation.

5.5   Summary of the gated SR latch and the Flip-Flops

Figures 5.13 through 5.16 summarize operations of the gated SR latch and the three flip-flops along with the symbolic representations, characteristic and excitation tables respectively. In the figures, X represents don't care whereas $Q+$ indicates output Q after the clock pulse is applied.

The characteristic table of a flip-flop is similar to its truth table. It contains the input combinations along with the output after the clock pulse. The characteristic table is useful for analyzing the SR Latch and the flip-flops.

The present state (present output), the next state (next output) after the clock pulse, and the required inputs for the transition are included in the excitation table. This is useful for designing a sequential circuit, in which one normally knows the

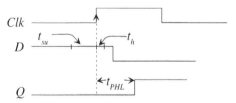

FIGURE 5.11   D Flip-Flop timing parameters.

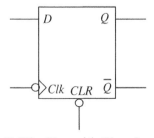

FIGURE 5.12   D Flip-Flop with Clear Input.

transition from the present to next state and wants to determine the required flip-flop inputs for the transition.

The D flip-flop is widely used in digital systems for transferring data. Several D flip-flops can be combined to form a register in the CPU of a computer. The 74HC374 is a 20-pin chip containing eight independent D flip-flops. It is designed using CMOS. The flip-flops are enabled at the leading edge of the clock. The 74LS374 is the same as the 74HC374 except that it is designed using TTL.

The JK flip-flop is a universal flip-flop and is typically used for general applications. Typical commercially available JK flip-flop includes the 74HC73 (or 74LS73A). The 74HC73 is a 14-pin chip. It contains two independent JK flip-flops in the same chip, designed using CMOS. Each flip-flop is enabled at the trailing edge of the clock pulse. Also, each flip-flop also contains a direct clear input. The 74HC73 is cleared to zero when the clear input is LOW. The 74LS73A is same as the 74HC73 except that

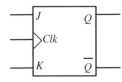

S	R	Q +	
0	0	Q	Unchanged
0	1	0	Reset
1	0	1	Set
1	1	?	Invalid

Q	Q+	S	R
0	0	0	X
0	1	1	0
1	0	0	1
1	1	X	0

(a) Symbolic representation   (b) Characteristic table   (c) Excitation table

FIGURE 5.13   Gated SR Latch.

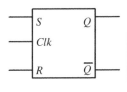

J	K	Q+	
0	0	Q	Unchanged
0	1	0	Reset
1	0	1	Set
1	1	$\overline{Q}$	Complement

Q	Q+	J	K
0	0	0	X
0	1	1	X
1	0	X	1
1	1	X	0

(a) Symbolic representation   (b) Characteristic table   (c) Excitation table

FIGURE 5.14   JK flip-flop.

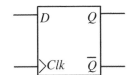

D	Q+	
0	0	Reset
1	1	Set

Q	Q+	D
0	0	0
0	1	1
1	0	0
1	1	1

(a) Symbolic representation   (b) Characteristic table   (c) Excitation table

FIGURE 5.15   D flip-flop.

T	Q+	
0	Q	Unchanged
1	$\overline{Q}$	Complement

Q	Q+	T
0	0	0
0	1	1
1	0	1
1	1	0

(a) Symbolic representation   (b) Characteristic table   (c) Excitation table

FIGURE 5.16   T flip-flop.

it is designed using TTL. The T flip-flop is normally used for designing binary counters because binary counters require complementation. The T flip-flop is not commercially available. One way of obtaining a T Flip-flop is by connecting the J and K inputs of a JK flip-flop together (Section 5.2.6).

An example of a commercially available level-triggered latch is the 74HC373 (or 74LS373). The 373 (20-pin chip) contains eight independent gated D latches with one enable input.

Sometimes the characteristic equations of the latches and flip-flops are useful in analyzing their operations. The characteristic equations for the gated SR latch and the flip-flops can be obtained from their respective truth tables. Figures 5.17 through 5.19

Q	S	R	$Q+$
0	0	0	0
0	0	1	0
0	1	0	1
0	1	1	Invalid
1	0	0	1
1	0	1	0
1	1	0	1
1	1	1	Invalid

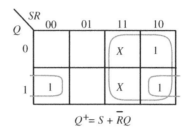

$$Q^+ = S + \bar{R}Q$$

(a) Truth Table for the Gated SR latch

(b) K-map for the characteristic equation of the SR latch

FIGURE 5.17   Truth table and K-map for the characteristic equation of the gated SR latch.

Q	J	K	$Q+$
0	0	0	0
0	0	1	0
0	1	0	1
0	1	1	1
1	0	0	1
1	0	1	0
1	1	0	1
1	1	1	0

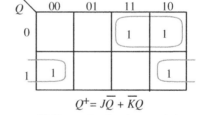

$$Q^+ = J\bar{Q} + \bar{K}Q$$

(a) Truth Table for the JK-FF

(b) K-map for the characteristic equation of the JK-FF

Q	T	$Q+$
0	0	0
0	1	1
1	0	1
1	1	0

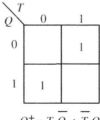

$$Q^+ = T\bar{Q} + \bar{T}Q$$

(c) Truth Table for the T-FF

(d) K-map for the characteristic equation of the T-FF.

FIGURE 5.18   Truth table and K-map for the characteristic equation of JK and T flip-flops.

Q	D	Q+
0	0	0
0	1	1
1	0	0
1	1	1

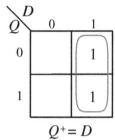

$$Q^+ = D$$

(a) Truth Table for D-FF | (b) K-map for the characteristic equation of D-FF.

FIGURE 5.19   Truth table and K-map for the characteristic equation of D flip-flop.

show how these equations are obtained using K-maps for the gated SR latch, JK, T, and D flip-flops.

Example 5.1

Given the following clock and the D inputs for a negative-edge-triggered D flip-flop of Figure 5.20, draw the timing diagram for the Q output for the first five cycles shown. Assume Q is preset to 1 initially.

Solution:

See Figure 5.20.

5.6   Analysis of Synchronous Sequential Circuits

A synchronous sequential circuit can be analyzed by determining the relationships between inputs, outputs, and flip-flop states. A state table or a state diagram illustrates how the inputs and the states of the flip-flops affect the circuit outputs. Boolean expressions can be obtained for the inputs of the flip-flops in terms of present states of the flip-flops and the circuit inputs. As an example consider analyzing the synchronous sequential circuit of Figure 5.21.

The logic circuit contains two D flip-flops (outputs X, Y), one input A and one output B. The equations for the next states of the flip-flops can be written as

$$X^+ = \overline{(X + Y) \cdot A}$$
$$Y^+ = A + \overline{X}$$

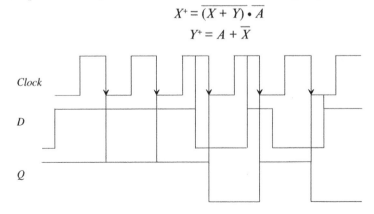

FIGURE 5.20   Figure for Example 5.1.

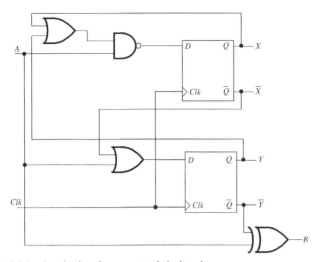

FIGURE 5.21   Analysis of a sequential circuit.

Here $X+$ and $Y+$ represent the next states of the flip-flops after the clock pulse. The right side of each equation denotes the present states of the flip-flops (X, Y) and the input (A) that will produce the next state of each flip-flop. The Boolean expressions for the next state are obtained from the combinational circuit portion of the sequential circuit. The outputs of the combinational circuit are connected to the D inputs of the flip-flops. These D inputs provide the next states of the flip-flops after the clock pulse. The present state of the output B can be derived from the figure as follows:

$$B = A \oplus \overline{Y}$$

A state table listing the inputs, the outputs, and the states of the flip-flops along with the required flip-flop inputs can be obtained for Figure 5.21. Table 5.1 depicts a typical state table. The state table is formed by using the following equations (shown earlier):

$$X^+ = \overline{(X + Y) \cdot A}$$
$$Y^+ = A + \overline{X}$$

To derive the state table, all combinations of the present states of the flip-flops and input A are tabulated. There are eight combinations for three variables from 000 to 111. The values for the flip-flop inputs (next states of the flip-flops) are determined using the equations. For example, consider the top row with $X = 0$, $Y = 0$, and $A = 0$. Substituting in the equations for next states.

$$X+ = \overline{(X + Y) \cdot A} = \overline{(0 + 0) \cdot 0} = 1$$
$$Y+ = A + \overline{X} = 0 + \overline{0} = 1$$

TABLE 5.1   State Table for Figure 5.21

| Present State | | Input | Next State | | Flip Flop Inputs | | Output |
X	Y	A	X+	Y+	D_X	D_Y	B
0	0	0	1	1	1	1	1
0	0	1	1	1	1	1	0
0	1	0	1	1	1	1	0
0	1	1	0	1	0	1	1
1	0	0	1	0	1	0	1
1	0	1	0	1	0	1	0
1	1	0	1	0	1	0	0
1	1	1	0	1	0	1	1

TABLE 5.2   Another Form of the State Table

Present State		Next State				Flip Flop Inputs				Outputs	
		A=0		A=1		A=0		A=1		A=0	A=1
X	Y	X+	Y+	X+	Y+	D_X	D_Y	D_X	D_Y	B	B
0	0	1	1	1	1	1	1	1	1	1	0
0	1	1	1	0	1	1	1	0	1	0	1
1	0	1	0	0	1	1	0	0	1	1	0
1	1	1	0	0	1	1	0	0	1	0	1

Now, to find the flip-flop inputs, one should consider each flip-flop separately. Two D flip-flops are used. Note that for a D flip-flop, the input at D is same as the next state. The D input is transferred to the output Q at the clock pulse. Therefore, $X+ = D_x$ and $Y+ = D_y$.

The characteristic table of a D flip-flop (Figure 5.15) is used to determine the flip-flop inputs that will change present states of the flip-flops to next state. The characteristic table of D flip-flop is provided here for reference:

D	Q^+
0	0
1	1

Therefore, for D flip-flops, the next states and the flip-flop inputs will be same in the state table. By inspecting the top row of the state table, it can be concluded that $D_x = 1$ and $D_y = 1$ because the next states $X+ = 1$ and $Y+ = 1$.

Finally, the output B can be obtained from the equation,

$$B = A \oplus \overline{Y}$$

For example, consider the top row of the state table. $A = 0$ and $Y = 0$. Thus,

$$B = 0 \oplus \overline{0} = 0 \oplus 1 = 1$$

All other rows of the state table can similarly be verified. The state table of Table 5.1 can be shown in a slightly different manner. Table 5.2 depicts another form of the state table of Table 5.1.

A state table can be depicted in a graphical form called the *state diagram*. All information in the state table can be represented in the state diagram. A circle is used to represent a state in the state diagram. A straight line with an arrow indicator is used to show direction of transition from one state to another. Figure 5.22 shows the state diagram for Table 5.1.

Because there are two flip-flops (X, Y) in Figure 5.21, there are four states: 00, 01, 10 and 11. These are shown in the circle of the state diagram. Also, transition from one state to another is represented by a line with an arrow. Each line is assigned with a/b where a is input and b is output. From the state diagram of Figure 5.22, with

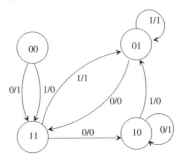

FIGURE 5.22   State diagram for Table 5.1.

present state 10 and an input of 1, the output is 0 and the next state is 01. If the input (and/or output) is not defined in a problem, the input (and/or output) will be deleted in the state table and the state diagram.

The inputs of the flip-flops (D_x and D_y) in the state table are not necessary to derive the state diagram. In analyzing a synchronous sequential circuit, the logic diagram is given. The state equation, the state table, and the state diagram are obtained from the logic diagram. However, in order to design a sequential circuit, the designer has to derive the state table and the state diagram from the problem definition. The flip-flop inputs will be useful in the design. One must express the flip-flop inputs and outputs in terms of the present states of the flip-flops and the inputs. The minimum forms of these expressions can be obtained using a K-map. From these expressions, the logic diagram can be drawn.

5.7   Types of Synchronous Sequential Circuits

There are two types of Synchronous sequential circuits: the Mealy circuit and the Moore circuit. A synchronous sequential circuit typically contains inputs, outputs, and flip-flops. In the Mealy circuit, the outputs depend on both the inputs and the present states of the flip-flops. In the Moore circuit, on the other hand, the outputs are obtained from the flip-flops, and depend only on the present states of the flip-flops . Therefore, the only difference between the two types of circuits is in how the outputs are produced.

The state table of a Mealy circuit must contain an output column. The state table of a Moore circuit may contain an output column, which is dependent only on the present states of the flip-flops. A Moore machine normally requires more states to generate identical output sequence compared to a Mealy machine. This is because the transitions are associated with the outputs in a Mealy machine.

5.8   Minimization of States

A simplified form of a synchronous sequential circuit can be obtained by minimizing the number of states. This will reduce the number of flip-flops and simplify the complexity of the circuit implementations. However, logic designers rarely use the minimization procedures. Also, there are sometimes instances in which design of a synchronous sequential circuit is simplified if the number of states is increased. The techniques for reducing the number of states presented in this section are merely for illustrative purpose.

The number of states can be reduced by using the concept of equivalent states. Two states are equivalent if both states provide the same outputs for identical inputs.

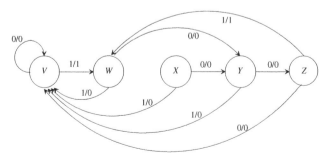

FIGURE 5.23   State diagram for minimization.

TABLE 5.3   State table for minimization of states

Present State	Next State		Output	
	d=0	*d=1*	*d=0*	*d=1*
V	*V*	*W*	*0*	*1*
W	*Y*	*V*	*0*	*0*
X	*Y*	*V*	*0*	*0*
Y	*Z*	*V*	*0*	*0*
Z	*V*	*W*	*0*	*1*

TABLE 5.4   Replacing states by their equivalents

Present State	Next State		Output	
	d=0	*d=1*	*d=0*	*d=1*
V	*V*	*W*	*0*	*1*
W	*Y*	*V*	*0*	*0*
X̶	*Y*	*V*	*0*	*0*
Y	*Z̶V*	*V*	*0*	*0*
Z̶	*V*	*W*	*0*	*1*

One of the states can be eliminated if two states are equivalent. Thus, the number of states can be reduced.

For example, consider the state diagram of Figure 5.23. Each state is represented by a circle with transition to the next state based on either an input of 0 or 1 generating an output.

Next, consider that a string of input data bits (*d*) in the sequence 0100111101 is applied at state *V* of a synchronous sequential circuit. For the given input sequence, the output and the state sequence can be obtained as follows:

```
State    V   V   W   Y   Z   W   V   W   V   V   W
Input    0   1   0   0   1   1   1   1   0   1
Output   0   1   0   0   1   0   1   0   0   1
```

With the sequential circuit in initial state *V*, a 0 input generates a 0 output and the circuit stays in state *V*, whereas in state *V*, an input of 1 produces an output 1 and the circuit will move to the next state *W*. In state *W* and input = 0, the output is 0 and the next state is *Y*. The process thus continues.

The state table shown in Table 5.3 for the state diagram in Figure 5.23 can be obtained. Next, the equivalent states will be determined to reduce the number of states. *V* and *Z* are equivalent because they have same next states of *V* and *W* with identical inputs *d* = 0 and *d* = 1. Similarly, *W* and *X* are equivalent states. Table 5.4 shows the process of replacing of a state by its equivalent.

Because *V* and *Z* are equivalent, one of the states can be eliminated; *Z* is removed. Also, *W* and *X* are equivalent, so one of the states can be removed; *X* is thus eliminated in the state table. The row with present states *X* and *Z* is also eliminated. If they appear in the next state columns, they must be replaced by their equivalent states. In our case, the row for state *Y* contains *Z* in the next column. This is replaced its equivalent state *V*. By inspecting the modified state table further, no more equivalent states are found. The state table after elimination of equivalent states is shown in Table 5.5.

Note that the original state diagram in Figure 5.23 requires five states. Figure 5.24 shows the reduced form of the state diagram with only three states. Three flip-flops are required to represent five states whereas two flip-flops will represent three states. Thus, one flip-flop is eliminated and the complexity of implementation may be reduced. Note that a synchronous sequential circuit can be minimized by determining the equivalent states, provided the designer is only concerned with the output sequences due to input sequences.

TABLE 5.5   State table after the elimination of equivalent states

| Present State | Next State | | Output | |
	d=0	*d=1*	*d=0*	*d=1*
V	*V*	*W*	0	1
W	*Y*	*V*	0	0
Y	*V*	*V*	0	0

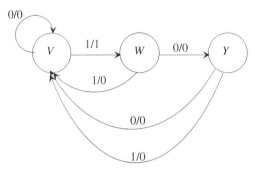

FIGURE 5.24   Reduced form of the state diagram.

5.9   Design of Synchronous Sequential Circuits

The procedure for designing a synchronous sequential circuit is a three-step process as follows:

1. Derive the state table and state diagram from the problem definition. If the state diagram is given, determine the state table.
2. Obtain the minimum form of the Boolean equations for flip-flop inputs and outputs, if any, using K-maps.
3. Draw the logic diagram. Note that a combinational circuit is designed using a truth table whereas the synchronous sequential circuit design is based on the state table.

Example 5.2

Design a synchronous sequential circuit for the state diagram of Figure 5.25 using D flip-flops.

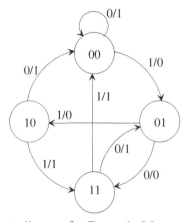

FIGURE 5.25   State diagram for Example 5.2.

Solution

Step 1: Derive the state table. The state table is derived from the state diagram (Figure 5.25) and the excitation table (Figure 5.15(c)) of the D flip-flop. Table 5.6 shows the state table. The state table is obtained directly from the state diagram. In the state table, the next states are the same as the flip-flop inputs because D flip-flops are used. This is evident from the excitation table of Figure 5.15 (c).

Step 2: Obtain the minimum forms of the equations for the flip-flop inputs and the output. Using K-maps and the output, the equations for flip-flop inputs are simplified as shown in Figure 5.26.

Step 3: Draw the logic diagram. The logic diagram is shown in Figure 5.27.

TABLE 5.6   State Table for Example 5.2

| Present State | | Input | Next State | | Flip Flop Inputs | | Output |
X	Y	A	$X+$	$Y+$	D_X	D_Y	Z
0	0	0	0	0	0	0	1
0	0	1	0	1	0	1	0
0	1	0	1	1	1	1	0
0	1	1	1	0	1	0	0
1	0	0	0	0	0	0	1
1	0	1	1	1	1	1	1
1	1	0	0	1	0	1	1
1	1	1	0	0	0	0	1

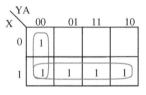

(a) K-map for D_X (b) K-map for D_Y (c) K-map for Z

$D_X = X\bar{Y}A + \bar{X}Y$ $D_Y = \bar{Y}A + Y\bar{A} = Y \oplus A$ $Z = \bar{Y}\bar{A} + X$

FIGURE 5.26   K-maps for Example 5.2.

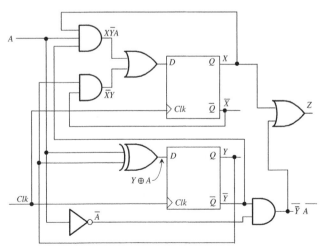

FIGURE 5.27   Logic diagram for Example 5.2.

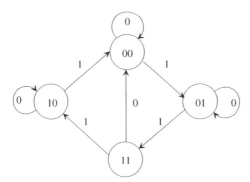

FIGURE 5.28   State diagram for Example 5.3.

TABLE 5.7   State and Excitation Tables for Example 5.3
         (a) Excitation Table of JK flip-flop from Figure 5.14 (c)

Q	Q+	J	K
0	0	0	X
0	1	1	X
1	0	X	1
1	1	X	0

(b) State Table for Example 5.2

Present State		Input	Next State		Flip Flop Inputs			
X	Y	A	X+	Y+	J_X	K_X	J_Y	K_Y
0	0	0	0	0	0	X	0	X
0	0	1	0	1	0	X	1	X
0	1	0	0	1	0	X	X	0
0	1	1	1	1	1	X	X	0
1	0	0	1	0	X	0	0	X
1	0	1	0	0	X	1	0	X
1	1	0	0	0	X	1	X	1
1	1	1	1	0	X	0	X	1

Example 5.3

Design a synchronous sequential circuit for the state diagram of Figure 5.28 using JK flip-flops.

Solution

Step 1: Derive the state table. The state table can be directly obtained from the state diagram (Figure 5.28) and the excitation table [Table 5.7(a)]. Table 5.7(b) shows the state table. For convenience, the excitation table of JK flip-flop of Figure 5.14 (c) is re-drawn in Table 5.7(a).

Let us explain how the state table is obtained. The input A is 0 or 1 at each state, so the left three columns show all eight combinations for X, Y, and A. The next state column is obtained from the state diagram. The flip-flop inputs are then obtained using the excitation table for the JK flip-flop. For example, consider the top row. From the state diagram, the present state (00) remains in the same state (00) when input $A = 0$ and the clock pulse is applied. The output of flip-flop X goes from 0 to 0 and the output of flip-flop Y goes from 0 to 0. From the excitation table of the JK flip-flop, $J_x = 0$, $K_x = X$, $J_x = 0$, and $K_x = X$. The other rows are obtained similarly.

Step 2: Obtain the minimum forms of the equations for the flip-flop inputs. Using K-maps, the equations for flip-flop inputs are simplified as shown in Figure 5.29.

Step 3: Draw the logic diagram as shown in Figure 5.30.

Example 5.4

Design a synchronous sequential circuit with one input X and an output Z. The input X is a serial message and the system reads X one bit at a time. The output $Z = 1$ whenever the pattern 101 is encountered in the serial message. For example,

$$\text{If input:} \quad 0\ 0\ 1\ 0\ 1\ 0\ 1\ 1\ 1\ 0\ 1\ 0\ 0\ 0\ 1\ 0\ 1$$
$$\text{then output:} \quad 0\ 0\ 0\ 0\ 1\ 0\ 1\ 0\ 0\ 0\ 1\ 0\ 0\ 0\ 0\ 0\ 1$$

Use T flip-flops.

Solution

Step 1: Derive the state diagram and the state table.

Figure 5.31 shows the state diagram. In this diagram, each node represents a state. The labeled arcs (lines joining two nodes) represent state transitions. For example, when the system is in state C, if it receives an input 1, it produces an output 1 and makes a

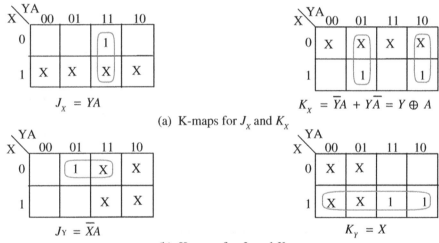

(a) K-maps for J_X and K_X

(b) K-maps for J_Y and K_Y

FIGURE 5.29 K-maps for Example 5.3.

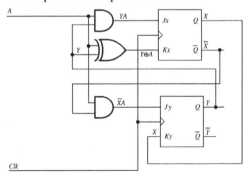

FIGURE 5.30 Logic Diagram for Example 5.3.

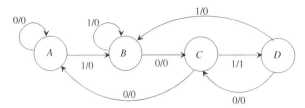

FIGURE 5.31   State Diagram for Example 5.4.

TABLE 5.8   State Table for Example 5.4

Present State	Next State		Output Z	
	$X=0$	$X=1$	$X=0$	$X=1$
A	A	B	0	0
B	C	B	0	0
C	A	D	0	1
D	C	B	0	0

TABLE 5.9   Modified State Table for Example 5.4

Present State		Next State		Output Z	
y_1	y_0	$y_1^+ y_0^+$	$y_1^+ y_0^+$	Input	Input
		$X=0$	$X=1$	$X=0$	$X=1$
0	0	0 0	0 1	0	0
0	1	1 1	0 1	0	0
1	1	0 0	1 0	0	1
1	0	1 1	0 1	0	0

transition to the state D after the clock. Similarly, when the system is in state C and receives a 0 input, it generates a 0 output and moves to state A after the clock. This type of sequential circuit is called a *Mealy machine* because the output generated depends on both the input X and the present state of the system. It should be emphasized that each state in the state diagram actually performs a bookkeeping operation; these operations are summarized as follows:

State	Interpretation
A	Looking for a new pattern
B	Received the first 1
C	Received a 1 followed by a 0
D	Recognized the pattern 101

The state diagram can be translated into a *state table,* as shown in Table 5.8. Each state can be represented by the binary assignment as follows:

Symbolic State	Binary State	
	y_1	y_0
A	0	0
B	0	1
C	1	1
D	1	0

The state table in Table 5.8 can be modified to reflect this state assignment, as illustrated in Table 5.9(a). Note that the excitation table actually describes the required excitation for a particular state transition to occur. For example, with respect to a T flip-flop, for the transition $0 \rightarrow 1$ or $1 \rightarrow 0$, a 1 must be applied to the T input. Similarly, for transitions $0 \rightarrow 0$ or $1 \rightarrow 1$ (that is, no change of state), the T input must be made 0. Using this excitation table, the flip-flop input equations can be derived as illustrated in Table 5.9(b).

In this figure, the entries corresponding to the flip-flop inputs T_{y1} and T_{y0} are directly derived using the T flip-flop excitation table. For example, consider the present state $y_1y_0 = 00$. When the input $X = 1$, the next state is 01. This means that flip-flop y_1 should not change its states and flip-flop y_0 must change its state to 1. It follows that $T_{y1} = 0$ (because a $0 \rightarrow 0$ transition is required) and $T_{y0} = 1$ (because a $0 \rightarrow 1$ transition is required). The other entries for T_{y1} and T_{y0} may be obtained in a similar manner.

Step 2: Derive the minimum forms of the equations for the flip-flop inputs and the output.

Using K-maps, the simplified equations for the flip-flops inputs and the output can be obtained as shown in Figure 5.32.

Step 3: Draw the logic diagram as shown in Figure 5.33.

TABLE 5.9(b)   State Table of Figure 5.9(a) with ff inputs

Present State		Input	Next State		Flip Flop Inputs		Ouput
y_1	y_0	X	y_1^+	y_0^+	T_{y1}	T_{y0}	Z
0	0	0	0	0	0	0	0
0	0	1	0	1	0	1	0
0	1	0	1	1	1	0	0
0	1	1	0	1	0	0	0
1	0	0	1	1	0	1	0
1	0	1	0	1	1	1	0
1	1	0	0	0	1	1	0
1	1	1	1	0	0	1	1

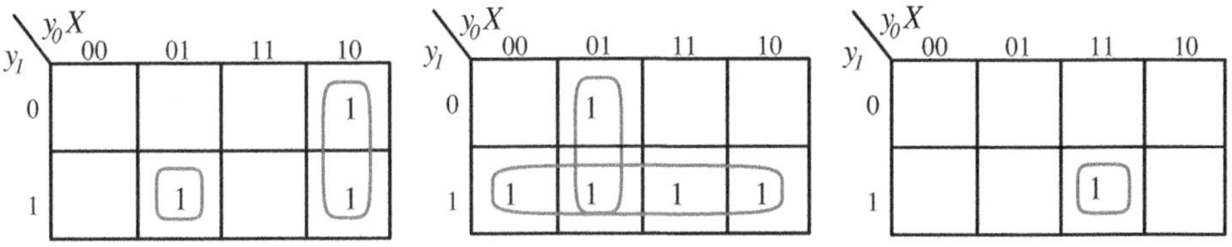

(a) K-map for T_{y1}

$$T_{y1} = y_0\overline{X} + y_1\overline{y_0}X$$

(b) K-map for T_{y0}

$$T_{y0} = y_1 + \overline{y_0}X$$

(c) K-map for Z

$$Z = y_1 y_0 X$$

FIGURE 5.32   K-maps for Example 5.4.

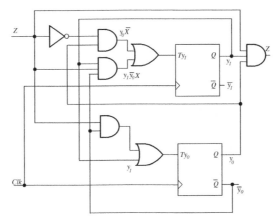

FIGURE 5.33   Logic Diagram for Example 5.4.

5.10 Design of Counters

A counter is a synchronous sequential circuit that moves through a predefined sequence of states upon application of clock pulses. A binary counter, which counts binary numbers in sequence at each clock pulse, is the simplest example of a counter. An n-bit binary counter contains n flip-flops and can count binary numbers from 0 to 2^{n-1}. Other binary counters may count in an arbitrary manner in a nonbinary sequence. The following examples will illustrate the straight binary sequence and nonbinary sequence counters.

Example 5.5

Design a two-bit counter to count in the sequence 00, 01, 10, 11, and repeat. Use T flip-flops.

Solution

Step 1: Derive the state diagram and the state table.

Figure 5.34 shows the state diagram. Note that state transition occurs at the clock pulse. No state transitions occurs if there is no clock pulse. Therefore, the clock pulse does not appear as an input. Table 5.10 shows the state table.

The excitation table of the T flip-flop is used for deriving the state table. For example, consider the top row. The state remains unchanged ($a_1 = 0$ and $a_{1+} = 0$) requiring a T input of 0 and thus $T_{A1} = 0$. a_0 (top row) is complemented from the present state to the next state, and thus $T_{A0} = 1$.

Step 2: Derive the minimum forms of the equations for the flip-flop inputs.

Using K-maps, the simplified equations for the flip-flop inputs can be obtained as shown in Figure 5.35.

Step 3: Draw the logic diagram as shown in Figure 5.36.

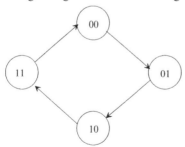

FIGURE 5.34   State Diagram for Example 5.5.

TABLE 5.10   Stable table for Example 5.5

Present State		Next State		Flip Flop inputs	
a_1	a_0	a_1+	a_0+	T_{A1}	T_{A0}
0	0	0	1	0	1
0	1	1	0	1	1
1	0	1	1	0	1
1	1	0	0	1	1

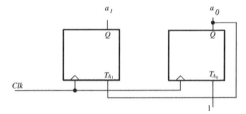

(a) K-map for T_{A_1}

$T_{A_1} = a_0$

(b) K-map for T_{A_0}

$T_{A_0} = 1$

FIGURE 5.35   K-maps for Example 5.5.

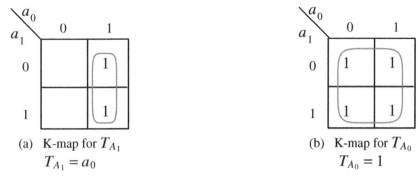

FIGURE 5.36   Logic Diagram for two-bit Counter of Example 5.5.

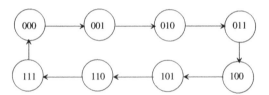

FIGURE 5.37   State Diagram for Example 5.6.

Example 5.6

Design a three-bit counter to count in the sequence 000 through 111, return to 000 after 111, and then repeat the count. Use JK flip-flops.

Solution

Step 1: Derive the state diagram and the state table.

Figure 5.37 shows the state diagram. Table 5.11 shows the JK ff excitation table, and the state table. Consider the top row. The present state of a_2 changes from 0 to 0 at the clock, a_1 changes from 0 to 0, and a_0 changes from 0 to 1. From the JK flip-flop excitation table, for these transitions, $J_{a2} = 0$, $K_{a2} = X$, $J_{a1} = 0$, $K_{a1} = X$, and $J_{a0} = 1$, $K_{a0} = X$.

Step 2: Derive the minimum forms of the equations for the flip-flop inputs. Using K-maps, the simplified equations for the flip-flop inputs can be obtained as shown in Figure 5.38.

Step 3: Draw the logic diagram as shown in Figure 5.39.

TABLE 5.11   JK ff excitation table and State Table for Example 5.6
(a) Excitation Table of JK Flip-flop

Q	$Q+$	J	K
0	0	0	X
0	1	1	X
1	0	X	1
1	1	X	0

(b) State Table for Example 5.6

Present State			Next State			Flip-Flop Inputs					
a_2	a_1	a_0	a_2+	a_1+	a_0+	Ja_2	Ka_2	Ja_1	Ka_1	Ja_0	Ka_0
0	0	0	0	0	1	0	X	0	X	1	X
0	0	1	0	1	0	0	X	1	X	X	1
0	1	0	0	1	1	0	X	X	0	1	X
0	1	1	1	0	0	1	X	X	1	X	1
1	0	0	1	0	1	X	0	0	X	1	X
1	0	1	1	1	0	X	0	1	X	X	1
1	1	0	1	1	1	X	0	X	0	1	X
1	1	1	0	0	0	X	1	X	1	X	1

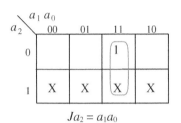

$Ja_2 = a_1a_0$

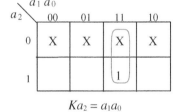

$Ka_2 = a_1a_0$

(a)   K-maps for Ja_2 and Ka_2

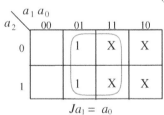

$Ja_1 = a_0$

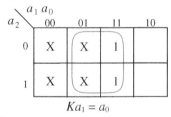

$Ka_1 = a_0$

(b) K-maps for Ja_1 and Ka_1

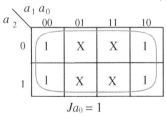

$Ja_0 = 1$

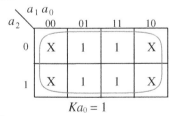

$Ka_0 = 1$

(c) K-maps for Ja_0 and Ka_0

FIGURE 5.38   K-Maps for Example 5.6.

Example 5.7

Design a three-bit counter that will count in the sequence 000, 010, 011, 101, 110, 111, and repeat the sequence. The counter has two unused states. These are 001 and 100.

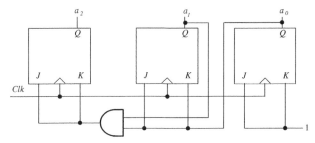

FIGURE 5.39   Logic Diagram for Example 5.6.

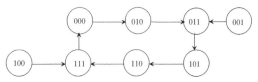

FIGURE 5.40   State Diagram for Example 5.7.

Implement the counter as a self-correcting such that if the counter happens to be in one of the unused states (001 or 100) upon power-up or due to error, the next clock pulse puts it in one of the valid states and the counter provides the correct count. Use T Flip-flops. Note that the initial states of the flip-flops are unpredictable when power is turned ON. Therefore, all the unused (don't care) states of the counter should be checked to ensure that the counter eventually goes into the desirable counting sequence. This is called a self-correcting counter.

Solution

Step 1: Derive the state diagram and the state table. Figure 5.40 shows the state diagram. Note that in the state diagram it is shown that if the counter goes to an invalid state such as 001 upon power-up, the counter will then go to the valid state 011 and will count correctly. Similarly, for the invalid state 100, the counter will be in state 111 and the correct count will continue. This self-correcting feature will be verified from the counter's state table using T flip-flops as shown in Table 5.12.

Step 2: Derive the minimum forms of the equations for the flip-flop inputs.

Using K-maps, the simplified equations for the flip-flop inputs can be obtained, as shown in Figure 5.41. The unused states 001 and 100 are invalid and can never occur, so they are don't care conditions.

Now, let us verify the self-correcting feature of the counter. The flip-flop input equations are

$$Ta_2 = a_1 a_0$$

$$Ta_1 = \overline{a_1} + a_0$$

$$Ta_0 = a_2 + a_1 \overline{a_0}$$

Suppose that the counter is in the invalid state 001 upon power-up or due to error, therefore, in this state, $a_2 = 0$, $a_1 = 0$, and $a_0 = 1$. Substituting these values in the flip-flop input equations, the values for Ta_2, Ta_1, and Ta_0 are obtained as follows:

$$Ta_2 = 0 \cdot 1 = 0$$

$$Ta_1 = \bar{0} + 1 = 1$$

$$Ta_0 = 0 + 0 \cdot \bar{1} = 0$$

TABLE 5.12   T-ff excitation table and State Table for Example 5.7

(a) Excitation Table for T Flip-Flop

Q	Q+	T
0	0	0
0	1	1
1	0	1
1	1	0

(b) State Table for Example 5.7

Present State			Next State			Flip Flop Inputs		
a_2	a_1	a_0	a_2+	a_1+	a_0+	Ta_2	Ta_1	Ta_0
0	0	0	0	1	0	0	1	0
0	1	0	0	1	1	0	0	1
0	1	1	1	0	1	1	1	0
1	0	1	1	1	0	0	1	1
1	1	0	1	1	1	0	0	1
1	1	1	0	0	0	1	1	1

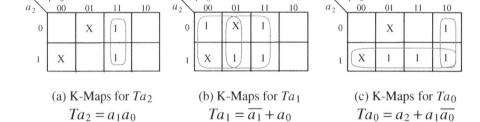

(a) K-Maps for Ta_2

$$Ta_2 = a_1 a_0$$

(b) K-Maps for Ta_1

$$Ta_1 = \overline{a_1} + a_0$$

(c) K-Maps for Ta_0

$$Ta_0 = a_2 + a_1 \overline{a_0}$$

FIGURE 5.41   K-maps for example 5.7.

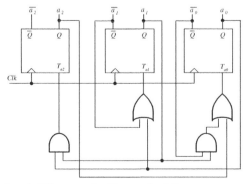

FIGURE 5.42   Logic Diagram for Example 5.7.

Note that with $a_2a_1a_0 = 001$ and $Ta_2Ta_1Ta_0 = 010$, the state changes from 001 to 011. Therefore, the next state will be 011. The correct count will resume. Next, if the flip-flop goes to the invalid state 100 due to error or when power is turned ON. Substituting $a_2 = 1$, $a_1 = 0$, and $a_0 = 0$ gives

$$Ta_2 = 0 \cdot 0 = 0$$

$$Ta_1 = \overline{0} + 0 = 1$$

$$Ta_0 = 1 + 0 \cdot \overline{0} = 1$$

Note that with $a_2a_1a_0 = 100$ and $Ta_2Ta_1Ta_0 = 011$, the state changes from 100 to 111. Hence, the next state for the counter will be 111. The correct count will continue. Therefore, the counter is self-correcting.

Step 3: Draw the logic diagram as shown in Figure 5.42.

5.11 Examples of Synchronous Sequential Circuits

Typical examples include registers, modulo-*n* counters and RAMs (Random Access Memories). They play an important role in the design of digital systems, especially computers.

5.11.1 Registers

A register contains a number of flip-flops for storing binary information in a computer. The register is an important part of any CPU. A CPU with many registers reduces the number of accesses to the main memory, therefore simplifying the programming task and shortening execution time. A general-purpose register (GPR) is designed in this section. The primary task of the GPR is to store address or data for an indefinite amount of time, then to be able to retrieve the data when needed. A GPR is also capable of manipulating the stored data by shift left or right operations. Figure 5.43 contains a summary of typical shift operations. In logical shift operation, a bit that is shifted out will be lost, and the vacant position will be filled with a 0. For example, if we have the number $(11)_{10}$, after right shift, the following occurs:

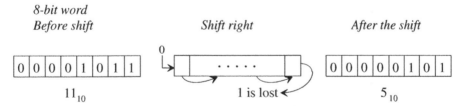

8-bit word
Before shift   *Shift right*   *After the shift*

11_{10} 1 is lost 5_{10}

Note that a logical left or right shift of an unsigned number by n positions implies multiplication or division of the number by 2^n, respectively, provided that a 1 is not shifted out during the operation.

In the case of true arithmetic left or right shift operations, the sign bit of the number to be shifted must be retained. However, in computers, this is true for right shift and not for left shift operation. For example, if a register is shifted right arithmetically, the most significant bit (MSB) of the register is preserved, thus ensuring that the sign of the number will remain unchanged. This is illustrated next.

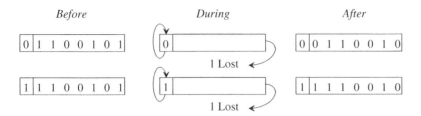

There is no difference between arithmetic and logical left shift operations. If the most significant bit changes from 0 to 1, or vice versa, in an arithmetic left shift, the result is incorrect and the computer sets the overflow flag to 1. For example, if the original value of the register is $(3)_{10}$, the results of two successive arithmetic left shift operations are interpreted as follows:

Original	After first shift	After second shift
$0011_2 = (3)_{10}$	$0110_2 = (6)_{10}$	$1100_2 = (-4)$
	$3 \times 2 = 6$, correct	$6 \times 2 = 12$ not -4, incorrect and the overflow bit is set to 1.

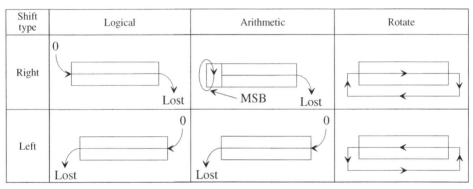

FIGURE 5.43   Summary of Typical Shift Operations.

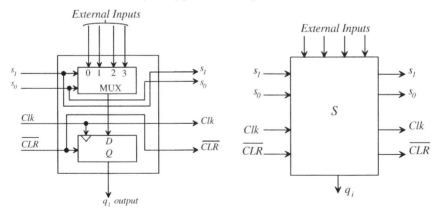

(a) Internal Organization of the   (b) Block Diagram of the
     Basic Cell S Basic Cell S

FIGURE 5.44   A Basic Cell for Designing a GPR.

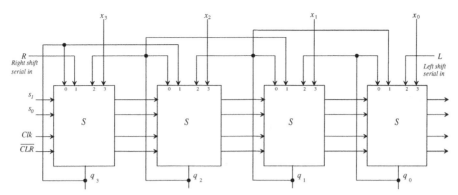

FIGURE 5.45 A 4-bit General Register.

TABLE 5.13 Truth Table for the General Register

Selection Input		Clock input	Clear Input	Operation
S_1	S_0	*Clk*	$\overline{CLR}$	
X	X	X	0	Clear
0	0	⌐	1	No Operation
0	1	⌐	1	Shift Right
1	0	⌐	1	Shift Left
1	1	⌐	1	Parallel Load

X means "don't care"

To design a GPR, first let us propose a basic cell S. The internal organization of the S cell is shown in Figure 5.44. A 4-input multiplexer selects one of the external inputs as the D flip-flop input, and the selected input appears as the flip-flop output Q after the clock pulse. The $\overline{CLR}$ input is an asynchronous clear input, and whenever this input is asserted (held low), the flip-flop is cleared to zero. Using the basic cell S as the building block, a 4-bit GPR can be designed. Its schematic representation is shown in Figure 5.45.

The truth table illustrating the operation of this register is shown in Table 5.13. This table shows that manipulation of the selection inputs S_1 and $S_0 = 11$, the external inputs x_3 through x_0 are selected as the D inputs for the flip-flop, the output q_i will follow the input x_i after the clock. By choosing the correct values for the serial shift inputs R and L, logical, arithmetic, or rotating shifts can be achieved.

This register can be loaded with any desired data in a serial fashion. For example, after four successive right shift operations, data $a_3\, a_2\, a_1\, a_0$ will be loaded into the register if the register is set in the right shift mode and the required data $a_3\, a_2\, a_1\, a_0$ is applied serially to input R.

5.11.2 Modulo-*n* Counters

The modulo-*n* counter counts in a sequence and then repeats the count. Modulo-*n* counters can be used to generate timing signals in a computer. The control unit inside the CPU of a computer translates instructions. The control unit utilizes timing signals that determines the time sequences in which the operations required by an instruction are executed. These timing signals shown in Figure 5.46 can be generated by a special modulo-*n* counter called the *ring counter*. For proper operation, a ring counter must be initialized with one flip-flop in the high state (Q=1) and all other flip-flops in the zero state (Q=0).

An *n*-bit ring counter transfers a single bit among the flip-flops to provide *n* unique states. Figure 5.47 shows a 4-bit ring counter. Note that the ring counter requires no decoding but contains n flip-flops for an *n*-bit ring counter. The circuit will count in the sequence 1000, 0100, 0010, 0001, and repeat. Although the circuit does not count in the usual binary counting sequence, it is still called a *counter* because each count corresponds to a unique set of flip-flop states. The state table for the 4-bit ring counter is provided below:

Present State				Next State				FF Inputs			
W	X	Y	Z	W+	X+	Y+	Z+	D_W	D_X	D_Y	D_Z
1	0	0	0	0	1	0	0	0	1	0	0
0	1	0	0	0	0	1	0	0	0	1	0
0	0	1	0	0	0	0	1	0	0	0	1
0	0	0	1	1	0	0	0	1	0	0	0

From the above, using the present states along with the unused present states (not shown above) as don't cares, the following equations can be obtained using four K-maps (one for each FF input): $D_w=Z$, $D_x=W$, $D_y=X$, $D_z=Y$. This circuit is also known as a *circular shift register,* because the least significant bit shifted is not lost. This is the simplest shift-register counter. Thus, the schematic of Figure 5.47 can be obtained.

The main advantages of this circuit are design simplicity and the ability to generate timing signals without a decoder. Nevertheless, *n* flip-flops are required to

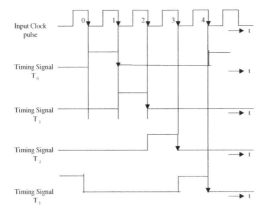

FIGURE 5.46   Timing Signals.

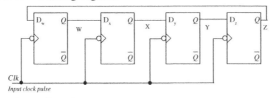

FIGURE 5.47   Ring Counter.

generate n timing signals. This approach is not economically feasible for large values of n. To generate timing signals economically, a new approach is used. A modulo-2^n counter is first designed using n flip-flops. The n outputs from this counter are then connected to a n-to-2^n decoder as inputs to generate 2^n timing signals. The circuit depicted in Figure 5.48 shows how to generate four timing signals using a modulo-4 counter and a 2-to-4 decoder.

In the circuit of Figure 5.48, the Boolean equation for each timing signal can be derived as

$$T_0 = \overline{A}\,\overline{B}$$
$$T_1 = \overline{A}B$$
$$T_2 = A\overline{B}$$
$$T_3 = AB$$

These equations show that four two-input AND gates are needed to derive the timing signals (assuming single-level decoding). The main advantage of this approach is that 2^n timing signals using only n flip-flops are generated. In this method, though, 2^n (n-input) AND gates are required to decode the n-bit output from the flip-flops into 2^n different timing signals. Yet the ring counter approach requires 2^n flip-flops to accomplish the same task.

Typical modulo-n counters provide trade-offs between the number of flip-flops and the amount of decoding logic needed. The binary counter uses the minimum number of flip-flops but requires a decoder. On the other hand, the ring counter uses the maximum number of flip-flops but requires no decoding logic. The Johnson counter (also called the *Switch-tail counter* or the *Mobius counter*) is very similar to a ring counter. Figure 5.49 shows a four-bit Johnson counter using JK flip-flops. Note that the Q output of the right-hand flip-flop is connected to the J input of the left-most flip-flop while the Q output of the right-most flip-flop is connected to the K input of the left-most flip-flop.

A Johnson counter requires the same hardware as a ring counter of the same size but can represent twice as many states. Assume that the flip-flops are initialized at 1000.

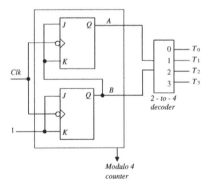

FIGURE 5.48   Modulo-4 Counter with a Decoder.

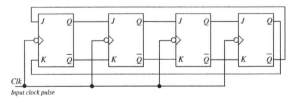

FIGURE 5.49   Four-bit Johnson Counter.

The counter will count in the sequence 1000, 1100, 1110, 1111, 0111, 0011, 0001, 0000 and repeat. The Johnson counter can be designed using the steps described earlier.

5.11.3   Random-Access Memory (RAM)

As mentioned before, a RAM is read/write volatile memory. RAM can be classified into two types: static RAM (SRAM) and dynamic RAM (DRAM). A static RAM stores each bit in a flip-flop whereas the dynamic RAM stores each bit as charge in a capacitor. As long as power is available, the static RAM retains information. Because the capacitor can hold charge for a few milliseconds, the dynamic RAM must be refreshed every few milliseconds. This means that a circuit must rewrite that stored bit in a dynamic RAM every few milliseconds. Let us now discuss a typical SRAM implementation using a gated D-latch. Figure 5.50 shows a typical SRAM cell.

In Figure 5.50(a), $R/\overline{W} = 1$ means READ whereas $R/\overline{W} = 0$ indicates a WRITE operation. Select = 1 indicates that the one-bit RAM is selected. In order to read the cell, $R/\overline{W} = 1$ and select = 1. A '1' appears at the input of AND gate 3. This will transfer Q to the output. This is a READ operation. Note that the inverted $R/\overline{W}$ to the input of AND gate 2 is 0. This will apply a 0 at the input of the CLK input of the gated D-latch. The output of the gated D-latch is unchanged. In order to write into the one-bit RAM, $R/\overline{W}$ must be zero. This will apply a 1 at the input of AND gate 2. The output of AND gate 2 (CLK input) is 1. The D input is connected to the value of the bit (1 or 0) to be written into the one-bit RAM. With CLK = 1, the input bit is transferred at the output. The one-bit RAM is, therefore, written into with the input bit. Figure 5.51 shows a 4×2 RAM. It includes 8 RAM cells providing 2-bit output and 4 locations.

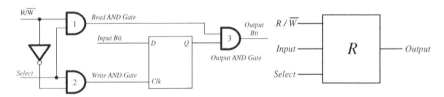

    (a)   A one-bit RAM (R)   (b)   Block diagram of the one-bit RAM

FIGURE 5.50   A typical SRAM cell.

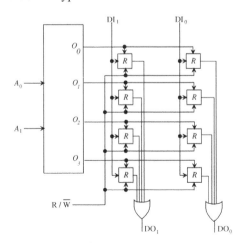

FIGURE 5.51 4×2 RAM.

The RAM contains a 2×4 decoder and 8 RAM cells implemented with gated D-latches and gates. In contrast, a ROM consists of a decoder and OR gates. The four locations (00, 01, 10, 11) in the RAM are addressed by 2 bits (A_1, A_0). In order to read from location 00, the address $A_1A_0 = 00$ and $R/\overline{W} = 1$. The decoder selects O_0 high. $R/\overline{W} = 1$ will apply 0 at the clock inputs of the two RAM cells of the top row and will apply 1 at the inputs of the output AND gates, thus transferring the outputs of the two gated D-latches to the inputs of the two OR gates. The other inputs of the OR gate will be 0. Thus, the outputs of the two RAM cells of the top row will be transferred to DO_1 and DO_0, performing a READ operation. On the other hand, consider a WRITE operation: The 2-bit data to be written is presented at DI_1 DI_0. Suppose $A_1A_0 = 00$. The top row is selected ($O_0 = 1$). Input bits at DI_1 and DI_0 will respectively be applied at the inputs of the gated D-latches of the top row. Because $R/\overline{W} = 0$, the clock inputs of both the gated D-latches of the top row are 1; thus, the D inputs are transferred to the outputs of the latches. Therefore, data at DI_1 DI_0 will be written into the RAM.

5.12   Algorithmic State Machines (ASM) Chart

The performance of a synchronous sequential circuit (also referred to as a *state machine*) can be represented in a systematic way by using a flowchart called the *Algorithmic State Machines* (ASM) chart. This was an alternative approach to the state diagram. In the previous sections, it was shown how state diagrams could be used to design synchronous sequential circuit. An ASM chart was widely used in the past along with the state diagram for designing a synchronous sequential circuit. However, with the popularity of hardware description languages (HDLs), use of ASM chart has been reduced significantly. Note that HDLs provide a lot more structured features including C-constructs. Hence, a brief coverage of ASM is provided in this section.

An ASM chart is similar to a flowchart for a computer program. The main difference is that the flowchart for a computer program is translated into software whereas an ASM chart is used to implement hardware. An ASM chart specifies the sequence of operations of the state machine along with the conditions required for their execution. Three symbols are utilized to develop the ASM chart: the state symbol, the decision symbol, and the conditional output symbol (see Figure 5.52).

The ASM chart utilizes one state symbol for each state. The state symbol includes the state name, binary code assignment, and outputs (if any) that are asserted during the specified state. The decision symbol indicates testing of an input and then going to an exit if the condition is true and to another exit if the condition is false. The entry of the conditional output symbol is connected to the exit of the decision symbol.

The ASM chart and the state diagram are very similar. Each state in a state diagram is basically similar to the state symbol. The decision symbol is similar to the binary information written on the lines connecting two states in a state diagram. Figure 5.53 shows an example of an ASM chart for a modulo-7 counter (counting the sequence 000, 001, ..., 111 and repeat) with an enable input. Q_2, Q_1, and Q_0 at the top of the ASM chart represent the three flip-flop states for the three-bit counter.

Each state symbol is given a symbolic name at the upper left corner along with a binary code assignment of the state at the upper right corner. For example, the state 'a' is assigned with a binary value of 000. The enable input E can only be checked at state 'a', and the counter can be stopped if $E = 0$; the counter continues if

$E = 1$. This is illustrated by the decision symbol. Figure 5.54 shows the equivalent state diagram of the ASM chart for the three-bit counter.

The ASM chart describes the sequence of events and the timing relationship between the states of a synchronous sequential circuit and the operations that occur for

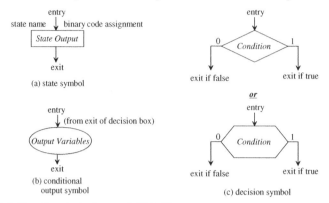

FIGURE 5.52   Symbols for an ASM Chart.

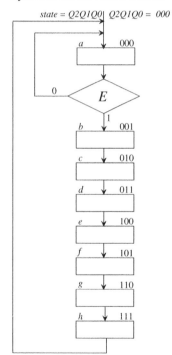

FIGURE 5.53   An ASM Chart for a 3-bit Counter with Enable Input.

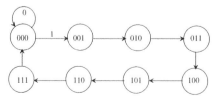

FIGURE 5.54   State Diagram for the 3-bit Counter.

transition from one state to the next. An arbitrary ASM chart depicted in Figure 5.55 illustrates this. The chart contains three ASM blocks. Note that an ASM block must contain one state symbol and may include any number of decisions and conditional output symbols connected to the exit. The three ASM blocks are the ASM block for T_0 surrounded by the dashed lines and the simple ASM block defined by T_1 and T_2. Figure 5.56 shows the state diagram.

From the ASM chart of Figure 5.55, there are three states: T_0, T_1, and T_2. A ring counter can be used to generate these timing signals. During T_0, register X is cleared and flip-flop A is checked. If $A = 0$, the next state will be T_1. On the other hand, if $A = 1$, the circuit increments register X by 1 and then moves to the next state, T_2. Note that the following operations are performed by the circuit during state T_0:

1. Clear register X.
2. Check flip-flop A for 1 or 0.
3. If $A = 1$, increment X by 1.

On the other hand, state machines do not perform any operations during T_1 and T_2.

Note that in contrast, state diagrams do not provide any timing relationship between states. ASM charts are utilized in designing the controller of digital systems such as the control unit of a CPU. It is sometimes useful to convert an ASM chart to a state diagram and then utilize the procedures of synchronous sequential circuits to design the control logic.

State Machine Design using ASM chart

As mentioned before, an ASM chart is used to define digital hardware algorithms which can be utilized to design and implement state machines. This section describes a procedure for designing state machines using the ASM chart. This is a three step process as follows:

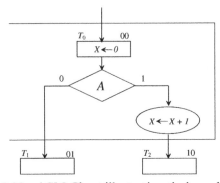

FIGURE 5.55 ASM Chart illustrating timing relationships between states.

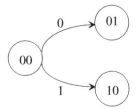

FIGURE 5.56 State Diagram for the ASM Chart of Figure 5.51.

1. Draw the ASM chart from problem definition.
2. Derive the state transition table representing the sequence of operations to be performed.
3. Derive the logic equations and draw the hardware schematic. The hardware can be designed using either classical sequential design or PLAs as illustrated by the examples provided below.

In the following, a digital system is designed using an ASM chart that will operate as follows:

The system will contain a 2-bit binary counter. The binary counter will count in the sequence 00, 01, 10, and 11. The most significant bit of the binary count XY is X while Y is the least significant bit. The system starts with an initial count of 3. A start signal I (represented by a switch) initiates a sequence of operations. If $I = 0$, the system stays in the initial state T_0 with count of 3. On the other hand, $I = 1$ starts the sequence.

When $I = 1$, counter Z (represented by XY) is first cleared to zero. The system then moves to state T_1. In this state, counter Z is incremented by 1 at the leading edge of each clock pulse. When the counter reaches 3, the system goes back to the initial state T_0, and the process continues depending on the status of the start switch I. The counter output will be displayed on a seven-segment display. An LED will be connected at the output of flip-flop W. The system will turn the LED ON for the count sequence 1, 2 by clearing flip-flop W to 0.

The flip-flop W will be preset to 1 in the initial state to turn the LED OFF. This can be accomplished by using input I as the PRESET input of flip-flop W. Use D flip-flops for the system.

Step 1: **Draw the ASM chart**. Figure 5.57 shows the ASM chart. The symbol T_n is used without its binary value for the state boxes in all ASM charts in this section.

In the ASM chart of Figure 5.57, when the system is in initial state T_0, it waits for the start signal (I) to become HIGH. When $I=1$, Counter Z is cleared to zero and the system goes to state T_1. The counter is incremented at the leading edge of each clock pulse. In state T_1, one of the following possible operations occurs after the next clock pulse transition:

Either, if counter Z is 1 or 2, flip-flop W is cleared to zero and control stays in state T_1;
or
If the Counter Z counts to 3, the system goes back to initial state T_0.

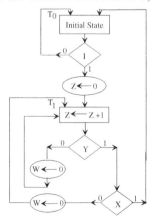

FIGURE 5.57   ASM Chart showing the sequence of operations for the binary counter.

TABLE 5.14   State Transition Table

COUNTER		FLIP-FLOP W	CONDITIONS	STATE
X	Y	(Q)		
0	0	1	X = 0, Y = 0	T_0
0	1	0	X = 0, Y = 1	T_1
1	0	0	X = 1, Y = 0	T_1
1	1	1	X = 1, Y = 1	T_0

The ASM chart consists of two states and two blocks. The block associated with T_0 includes one state box, one decision box, and one conditional box. The block in T_1 consists of one state box, two decision boxes and two conditional boxes.

Step 2:   Derive the state transition table representing the sequence of operations.

One common clock pulse specifies the operations to be performed in every block of an ASM chart. Table 5.14 shows the State Transition Table.

The binary values of the counter along with the corresponding outputs of flip-flop W is shown in the transition table. In state T_0, if I = 1, Counter Z is cleared to zero (XY = 00) and the system moves from state T_0 to T_1. In state T_1, Counter Z is first incremented to XY = 01 at the leading edge of the clock pulse; Counter Z then counts to XY = 10 at the leading edge of the following clock pulse. Finally, when XY = 11, the system moves to state T_0. The system stays in the initial state T_0 as long as I = 0; otherwise the process continues.

The operations that are performed in the digital hardware as specified by a block in the ASM chart occur during the same clock period and not in a sequence of operations following each other in time, as is usually interpreted in a conventional flowchart. For example, consider state T_1. The value of Y to be considered in the decision box is taken from the value of the counter in the present state T_1. This is because the decision boxes for Flip-flop W belong to the same block as state T_1. The digital hardware generates the signals for all operations specified in the present block before arrival of the next clock pulse.

Step 3:   Derive the logic equations and draw the hardware.

The system can be divided into two sections. These are data processor and controller. The requirements for the design of the data processor are defined inside the state and conditional boxes. The logic for the controller, on the other hand, is determined from the decision boxes and the necessary state transitions.

The design of the data processor is typically implemented by using digital components such as registers, counters, multiplexers, and adders. The system can be designed using the theory of sequential logic already discussed. Figure 5.58 shows the hardware block diagram. The Controller is shown with the required inputs and outputs. The data processor includes a two -bit counter, one flip-flop, and one AND gate. The counter is incremented by one at the positive edge of every clock pulse when control is in state T_1. The counter is assumed to be in count 3 initially. It is cleared to zero only when control is in state T_0 and I=1. Therefore, T_0 and I are logically Andes. The D-input of Flip-flop W is connected to output X of the counter to clear Flip-flop W during state T_0. This is because if present count is 00 (X=0), the counter will be 01 after the next clock. On the other hand, if the present count is 01 (X=0), the count will be 10 after the next clock. Hence, X is connected to the D-input of Flip-flop W to turn the LED ON for count sequence 1, 2. A common clock is used for all flip-flops in the system including the flip-flops in the counter and Flip-flop W.

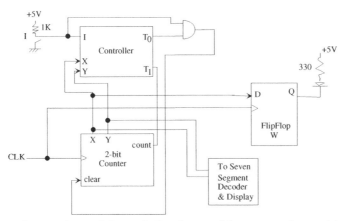

FIGURE 5.58   Hardware Schematic for the two-bit counter along with associated
               blocks.

This example illustrates a technique of designing digital systems using the
ASM chart. The two-bit counter can be designed using the concepts already described.
In order to design the Controller, a state table for the controller must be derived. Table
5.15 shows the state table for the Controller. There is a row in the table for each possible
transition between states. Initial state T_0 stays in T_0 or goes from T_0 to T_1 depending
on the status of the switch input (I). The same procedure for designing a sequential
circuit described in Chapter 5 can be utilized. Since there are two controller outputs
(T_1,T_0) and three inputs (I, X, Y), a three-variable K-map is required. The design of
the final hardware schematic is left as an exercise to the reader. The system will contain
D flip-flops with the same common clock and a combinational circuit. The design of
the system using classical sequential design method may be cumbersome. Hence, other
simplified methods using PLAs can be used as illustrated below.

A second example is provided below for designing a digital system using an ASM
chart. The system has three inputs (X, Y, Z) and a 2-bit MOD-4 counter (W) to count
from 0 to 3. The four counter states are T_0, T_1, T_2, and T_3. The operation of the system is
initiated by the counter clear input, C. When $C = 0$, the system stays in initial state T_0. On
the other hand, when $C = 1$, state transitions to be handled by the system are as follows:

INPUTS	STATE TRANSITIONS
$X = 0$	The system moves from T_0 to T_1
$X = 1$	The system stays in T_0
$Y = 0$	The system moves back from T_1 to T_0
$Y = 1$	The system goes from T_1 to T_2
$Z = 0$	The system stays in T_2
$Z = 1$	The system moves from T_2 to T_3 and then stays in T_3 indefinitely (for counter clear input C=1) until counter W is reset to zero (state T_0) by activating the counter clear input C to 0 to start a new sequence.

Use counter, decoder, and a PLA. Figure 5.59 shows the block diagram of the
MOD-4 counter to be used in the design.

Step 1:   **Draw an ASM chart**.
The ASM chart is shown in Figure 5.60

TABLE 5.15 State Table for the Controller

Present State (Controller)	Present States (counter)		Inputs (Controller)			Next States (counter)		Next Output States (controller)	
	X	Y	I	X	Y	X+	Y+	T_1	T_0
T_0	1	1	0	1	1	1	1	0	1
T_0	1	1	1	1	1	0	0	0	1
T_0	0	0	1	0	0	0	1	1	0
T_1	0	1	1	0	1	1	0	1	0
T_1	1	0	1	1	0	1	1	0	1

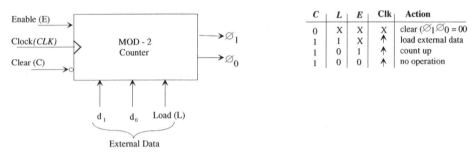

FIGURE 5.59   Block diagram and truth table of the 2-bit counter.

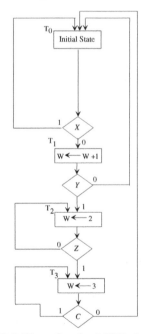

FIGURE 5.60   ASM Chart for the MOD-4 counter along with transitions.

Step 2:   **Derive the inputs, outputs, and a sequence of operations.**

The system will be designed using a PLA, a MOD-4 counter, and a 2 to 4 decoder. The MOD-4 counter is loaded or initialized with the external data if the counter control inputs C and L are both ones. The counter load control input L overrides the counter enable control input E.

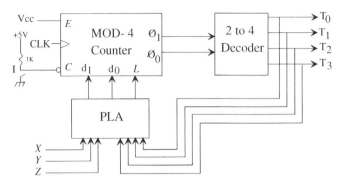

FIGURE 5.61   Hardware Schematic of the MOD-4 counter with PLA and decoder.

The counter counts up automatically in response to the next clock pulse when the counter load control input $L = 0$ and the enable input E is tied to HIGH. Such normal activity is desirable for the situation (obtained from the ASM chart) when the counter goes through the sequence T_0, T_1, T_2, T_3 for the specified inputs.

However, if the following situations occur, the counter needs to be loaded with data out of its normal sequence: If the counter is in initial state T_0 (Counter W=0 with $C= 0$) , it stays in T_0 for $X = 1$. This means that if the counter output is 00 and if $X = 1$, the counter must be loaded with external data $d_1d_0 = 00$. Similarly, the other out of normal sequence count includes transitions ($C = 1$) from T_1 to T_0 ($X = 0, Y = 0$), T_2 to T_2 ($X = 0$, $Y = 1$, $Z = 0$) with count 2, and T_3 to T_3 ($X = 0$, $Y = 1$, $Z = 1$); C is assumed to be HIGH during these transitions. Finally, if $C = 0$, transition from T_3 to T_0 occurs regardless of the values of X, Y, Z and the process continues. The appropriate external data must be loaded into the counter for out of normal count sequence by the PLA using the L input of the counter.

Step 3: **Derive the logic equations and draw a hardware schematic**.

Figure 5.61 depicts the logic diagram. Figure 5.62 shows the truth table and hardware schematic for PLA-based implementation.
The equations for the product terms are: $P_0 = X\,T_0\,C$, $P_1 = \overline{X}\,\overline{Y}\,T_1 C$, $P_2 = \overline{X}\,Y\,ZT_2 C$, $P_3 = \overline{X}Y\,ZT_3 C$, $P_4 = T_3\overline{C}$, $L = P_0 + P_1 + P_2 + P_3 + P_4$, $d_1 = P_2 + P_3$, $d_0 = P_3$

5.13   Asynchronous Sequential Circuits

Asynchronous sequential circuits do not require any synchronizing clocks. As mentioned before, a sequential circuit basically consists of a combinational circuit with memory. In synchronous sequential circuits, memory elements are clocked flip-flops. In contrast, memory in asynchronous sequential circuits includes either unclocked flip-flop latches or time-delay devices. The propagation delay time of a logic gate (finite time for a signal to propagate through a gate) provides its memory capability. Note that a sequential circuit contains inputs, outputs, and states. In synchronous sequential circuits, changes in states take place due to clock pulses. On the other hand, asynchronous sequential circuits typically contain a combinational circuit with feedback. The timing problems in the feedback may cause instability. Asynchronous sequential circuits are, therefore, more difficult to design than synchronous sequential circuits.

Asynchronous sequential circuits are used in applications in which the system must take appropriate actions to input changes rather than waiting for a clock to

Inputs								Outputs		
C	X	Y	Z	T_0	T_1	T_2	T_3	L	d_1	d_0
1	1	X	X	1	X	X	X	1	0	0
1	0	0	X	X	1	X	X	1	0	0
1	0	1	0	X	X	1	X	1	1	0
1	0	1	1	X	X	X	1	1	1	1
0	X	X	X	X	X	X	1	1	0	0
X = don't cares										

(a)   Truth Table for out of normal Count sequence

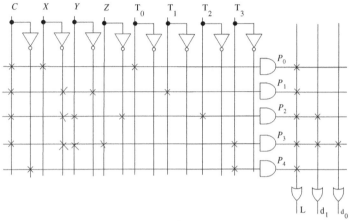

(b)   PLA implementation

FIGURE 5.62   PLA-based System.

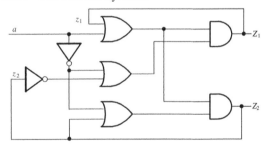

FIGURE 5.63   Asynchronous Sequential Circuit.

initiate actions. For proper operation of an asynchronous sequential circuit, the inputs must change one at a time when the circuit is in a stable condition (called the *fundamental mode of operation*). The inputs to the asynchronous sequential circuits are called *primary variables* whereas outputs are called *secondary variables*.

Figure 5.63 shows an asynchronous sequential circuit. In the feedback loops, the uppercase letters are used to indicate next values of the secondary variables and the lowercase letters indicate present values of the secondary variables. For example, Z_1, and Z_2 are next values whereas z_1 and z_2 are present values. The output equations can be derived as follows:

$$Z_1 = (a + z_1)(\bar{a} + \bar{z_2})$$
$$Z_2 = (a + z_1)(\bar{a} + \bar{z_2})$$

The delays in the feedback loops can be obtained from the propagation delays between z_1 and Z_1 or z_2 and Z_2. Let us now plot the functions Z_1 and Z_2 in a map, and a transition table as shown in Figure 5.64.

The map for Z_1 in Figure 5.63 is obtained by substituting the values z_1, z_2, and a for each square into the equation for $Z1$. For example, consider $z_1 z_2 = 11$ and $a = 0$.

$$Z_1 = (a + z_1)(\overline{a} + \overline{z_2})$$
$$= (0 + 1)(\overline{0} + \overline{1})$$
$$= 1$$

$$Z_2 = (a + z_1)(\overline{a} + z_2)$$
$$= (0 + 1)(\overline{0} + 1)$$
$$= 1$$

Similarly, values for all other sequences can be obtained similarly. The transition table of Figure 5.60(c) can be obtained by combining the binary values of two squares in the same position and placing them in the corresponding square in the transition table. Thus, the variable $Z = Z_1 Z_2$ is placed in each square of the transition table. For example, from the first square of Figure 5.60(a) and (b), $Z = 00$. This is shown in the first square of Figure 5.64 (c). The squares in the transition table in which $z_1 z_2 = Z_1 Z_2$ are circled to show that they are stable. The uncircled squares are unstable states.

Let us now analyze the behavior of the circuit due to change in the input variable. Suppose $a = 0$, $z_1 z_2 = 00$, then the output is 00. Thus, 00 is circled and shown in the first square of Figure 5.64 (c). Z is the next value of $z_1 z_2$ and is a stable state. Next suppose that a goes from 0 to 1 and the value of Z changes from 00 to 01. Note that this causes an interim unstable situation because $Z_1 Z_2$ is initially equal to $z_1 z_2$. This is because as soon as the input changes from 0 to 1, this change in input travels through the circuit to change $Z_1 Z_2$ from 00 to 01. The feedback loop in the circuit eventually makes $z_1 z_2$ equal to $Z_1 Z_2$; that is, $z_1 z_2 = Z_1 Z_2 = 01$. Because $z_1 z_2 = Z_1 Z_2$, the circuit attains a stable state. The state 01 is circled in the figure to indicate this. Similarly, it can be shown that as the input to an asynchronous sequential circuit changes, the circuit goes to a temporary unstable condition until it reaches a stable state when $Z_1 Z_2$ = present state, $z_1 z_2$. Therefore, as the input moves between 0 and 1, the circuit goes through the states 00, 01, 11, 10, and repeats the sequence depending on the input changes. A state table can be derived from the transition table. This is shown in Table 5.16, which is the state table for Figure 5.64 (c).

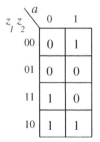

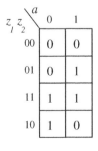

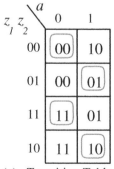

(a) Map for Z_1 (b) Map for Z_2 (c) Transition Table

FIGURE 5.64   Map and Transition Table.

TABLE 5.16   Transition Table

Present State		Next State			
		$a=0$		$a=1$	
0	0	0	0	1	0
0	1	0	0	0	1
1	0	1	1	1	0
1	1	1	1	0	1

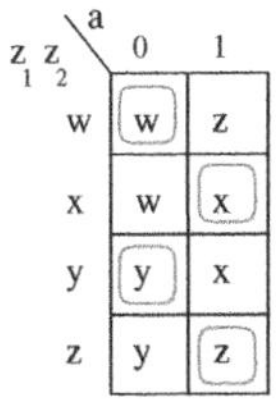

FIGURE 5.65   Flow Table.

A flow table obtained from the transition table is normally used in designing an asynchronous sequential circuit. A flow table resembles a transition table except that the states are represented by letters instead of binary numbers. The transition table of Figure 5.64(c) can be translated into a flow table as shown in Figure 5.65. Note that the states are represented by binary numbers as follows: $w = 00$, $x = 01$, $y = 11$, $z = 10$. The flow table in Figure 5.65 is called a *primitive flow table* because it has only one stable state in each row.

An asynchronous sequential circuit can be designed using the primitive flow table from the problem definition. The flow table is then simplified by combining squares to a minimum number of states. The transition table is then obtained by assigning binary numbers to the states. Finally, a logic diagram is obtained from the transition table. The logic diagram includes a combinational circuit with feedback.

The design of an asynchronous sequential circuit is more difficult than the synchronous sequential circuit because of the timing problems associated with the feedback loop. This topic is beyond the scope of this book.

5.14   Verilog description of typical synchronous sequential circuits

An introduction to Verilog is provided in Chapter 4. Several examples for writing Verilog description of typical combinational circuits are also covered in Section 4.12. In this section, basics of Verilog description of typical synchronous sequential circuits will be provided.

Synchronous sequential circuits are typically described in Verilog using behavioral modeling. The behavior is defined in a module using the keywords `initial` or `always` followed by one or more statements enclosed by keywords `begin` and `end`. The `initial` block is typically used to provide initializations for a simulation and produce stimulus waveforms for a simulation test bench. An `always` block, on the other hand, is defined using an `always` statement. A typical structure for the `always` block is provided below:

```
always @(sensivity list)
Procedural statements
```

The `always` statement specifies certain events in the sensitivity list. Any changes in these events will initiate execution of the procedural statements that follow.

The events can be either level-sensitive (combinational circuits and latches) or edge-sensitive (flip-flops). The sensitivity list can be delay- based timing control or event control. In delay-based timing control, the `always` statement is controlled by delays that wait for a specified amount time. The event control, on the other hand, defines a condition based on the change in value typically in a register that to trigger execution of a statement or a block of statements. An event control is defined by the symbol @ along with the keyword `always`. The event control condition will be covered in the following.

In synchronous sequential circuits, level-sensitive latches and edge-triggered flip-flops are encountered. The level-sensitive latch can be accomplished by the following statement:

```
always @ (x,enable)
Procedural statements
```

As soon as a change in 'x' or 'enable' occurs, the procedural statements in the `always` block will be executed. Note that Verilog also allows separation of events 'x' or 'enable' by the keyword 'or' rather than by commas. This implies that `always @ (x,enable)` and `always @ (x or enable)` mean the same thing. Also, note that "@ (x,enable)" specifies the procedural control. Verilog provides the keywords `posedge` and `negedge` to implement positive-edge triggered (positive clock) or negative-edge triggered (negative clock) clock respectively. For example, the statements

```
always @ (posedge clock)
Procedural statements
```

will initiate execution of the procedural statements in the `always` block for positive clock. The following statements, on the other hand,

```
always @ (negedge clock)
Procedural statements
```

will initiate execution of the procedural statements in the `always` block for negative clock.

Since a sequential circuit is comprised of flip-flops and combinational circuits, it can be represented using behavioral and data flow modeling. Flip-flops can be described with behavioral modeling using `always` keyword while the combinational circuit part can be assigned with data flow modeling using `assign` keyword and Boolean equations.

Note that a behavioral model in Verilog is defined using the keyword `always` followed by one or more procedural statements. The final output of these statements must be of the `reg` data type rather than `wire` (normally used for structural) data type. As mentioned before, `wire` continuously updates the output while the `reg` stores the value until a new value is provided.

Next, the meaning of "procedural statement" will be discussed. A procedural statement is an assignment in an `always` statement. Also, procedural statement assigns value to a register (data objects of type `reg`). There are three types of procedural assignments. These are procedural assignment (uses = as the operator), continuous procedural assignment (uses keyword `assign` with = as the operator), and non-blocking procedural assignment (uses <= as the operator). The right hand side of a procedural assignment is an expression which must evaluate to a value while the left hand side is typically a `reg`. The procedural continuous assignment retains the last output (when a digital circuit is disabled) until it is enabled again. This is useful in modeling latches and flip-flops. The first two procedural assignments that use the = operator execute the statements sequentially. These statements are called *blocking assignments*. This means that in blocking assignment, the next procedural assignment must wait until the present one is completed. This means that these types of procedural statements are executed sequentially in the order they are listed in the source code. Note that the continuous assignment statements are only used for level-sensitive behavior such as combinational design and latches. They cannot be used for modeling edge-trigerred behavior such as flip -flops. Next, an always block with an * inside the bracket will be dicussed.

Note that always@(*) blocks are used to describe combinational logic. Only blocking assignment (=) should be used in an always@(*) block. One should never use non-blocking assignments (<=) in an always@(*) block. The always@(*) block should only be used when one wants to infer an element(s) that changes its value as soon as one or more of its inputs change. However, it is not recommentded to use the always@(*) block since it does not list the actual input(s) in the block.

Next, consider the following two procedural blocking assignments (uses '=' as the operator) included in the always block :

```
b = 0;
x = y;
```

In the above, the statement x = y is executed only after b = 0 is executed. The statements in the always block are executed in sequence since blocking statements are used.

In non-blocking procedural assignment, the right hand side of each expression is evaluated first using the values at the beginning of the always block and the new values are then updated concurrently at the end of the always block as soon as changes in the condition in the sensitivity list occurs. Next, consider an example of the following blocking assignments included in the always block:

```
z <= y;
a <= b;
```

In the above, variables 'z' and 'a' have certain values when the always block is entered. The new values of 'z' and 'a' are then calculated and assigned simultaneously at the end of the always block as soon as one or more conditions in the sensitivity list change.

Non-blocking assignments are used in digital design where multiple concurrent data transfers such as in a register transfer, take place after a common event (positive or negative edge triggered clock).

For state machines, the inputs including clock, and outputs can be declared at the beginning of a Verilog program. The states can be defined using parameter keyword in Verilog which defines constants in a module. Statement using always along with posedge or negedge can be used for the clock. Statements using case and if-else can then be used to implement various state transitions.

EXAMPLE 5.8

Write a Verilog description using behavioral modeling for the gated D latch of Figure 5.66 (re-drawn from Figure 5.13 (a)). Use conditional operator.

Solution

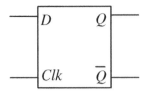

FIGURE 5.66 Block diagram of gated D-latch of Figure 5.15 (a).

```
// Gated D latch
module dlatch(q,nq, d, clk);
   input clk, d;
   output q, nq;
   reg q, nq;
   assign q = clk ? d:q;
   assign nq = !q;
endmodule
```

EXAMPLE 5.9

Write a Verilog description using behavioral modeling for a D flip-flop
(a) with a positive edge reset and a negative edge triggered clock. Use `if-else`.
(b) with a positive edge triggered clock and a negative edge clear input. Use `if-else`.

Solution

5.9 (a)
```
// D Flip-Flop
// Module DFF with synchronous reset
module dfflop(q, d, clk, reset);
   input d, clk, reset;
   output reg q;
                  //always do the following when the reset
                  //is positive edge
                  //or clock is negative edge
   always @(posedge reset , negedge clk)
   begin
                  // if it's reset q will be equal to zero
   if (reset)
      q = 1'b0;
                  // if it's clock ,q will equal to d
   else
      q = d;
   end
endmodule
```

5.9 (b)

```
//D flipflop

module D_ff(Q, Q_bar, CLR, CLK, D);
```

```
 output reg Q, Q_bar;
 input CLR, CLK, D;

always @(posedge CLK , negedge CLR)
begin
//When CLR == 0 (neg logic) Q is always 0
//else @ rising edge of clock, Q <-- D
 if(!CLR)
  begin
   Q <= 1'b0;
   Q_bar <= 1'b1;
  end
 else
 begin
   Q <= D;
   Q_bar <= !D; // Q_bar = not (D);
  end
  end
endmodule
```

EXAMPLE 5.10

Write a Verilog description using behavioral modeling for a JK flip-flop with negative edge triggered clock. Use `case` statements.

Solution

```
// JK ff using case statements

// J=A and K=B as inputs

// Q and nQ are outputs

module jk_ff(A,B,clock,Q,nQ);

        input A,B,clock;
        output Q,nQ;
        reg Q;
        assign nQ=~Q;
        always @ (negedge clock)
              case ({A,B})
                2'b00:Q=Q;
                2'b01:Q=1'b0;
                2'b10:Q=1'b1;
                2'b11:Q=~Q;
              endcase
endmodule
```

Note: In the above, the statement "`assign nQ = ~Q;`" is placed outside the "`always`" block. This is because the "`assign`" statement works in an asynchronous manner. That is, whenever "`always @`" changes anywhere in a behavioral block the, "`assign`" statement will automatically be executed and produce the correct value for "~q".

EXAMPLE 5.11

Write a Verilog description using behavioral modeling for the state diagram of Figure 5.67 (re-drawn from Figure 5.25). Use a reset input so that the hardware can be initialized.

Solution

```
//Description:state machine of Example 5.11
//implementation of state machine of figure 5.67
//APROACH : behavioral

module fig5-67( Z , state , A , clk , reset);
output  Z ;
output [1:0] state;
reg    [1:0] currentstate , state;
reg   Z ;
input  A , clk , reset;
always @ (posedge clk)
begin
if (reset == 1)  //need to asynchronously reset to start
currentstate = 0 ; //from a known state at some point
case (currentstate) //step thru all states per state table
     0:
     if(A == 1)
     begin
      state=1;
      Z =  0;
     end
else
  begin
    state=0;
    Z=1;
  end

1:
 if ( A==1)
   begin
     state=2;
     Z = 0;
   end
 else
   begin
     state=3;
     Z = 0;
   end
2:
```

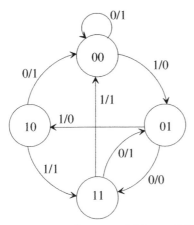

FIGURE 5.67   State diagram of Figure 5.25.

```
if (A == 1)
 begin
 state 3:
 Z = 1;
 end
 else

 begin
 state=0;
 Z=1;
 end

3:
 if (A==1)
   begin
     state = 0;
     Z=1;
   end
else
   begin
   state=1;
   Z=1;
   end

default:
   if (A == 1)
   begin
   state = 2'bxx;
   Z = 1'bx;
   end
else
     begin
       state = 2'bxx ;
```

```
      Z =1'bx;
   end
 endcase
currentstate = state ;  //update state for next time pass
end
endmodule


module fig5-67-0 test;
reg A , clk, reset;
wire [1:0] state;
wire Z ;
fig5-67          fig5-67-0 ( Z,state,A,clk , reset);

initial
   $monitor( "Time %0d, state=%b, A= %b, Z= %b, reset= %b",
       $time,   state,  A,   Z,   reset );
       initial
begin
   #0
 A=   1'b0;        //reset to state 0
 reset=1'b1;
 clk =1'b0;
 #20
 clk =1'b1;
 #20
 A=   1'b0;        //Input 1 to go to state 1
 reset=1'b0;
 clk =1'b0;
 #20
 clk =1'b1;
 #20
 A=   1'b0;        //Input 0 to go to state 3
 reset=1'b0;
 clk =1'b0;
 #20
 clk =1'b1;
 #20
 A=   1'b1;        //Input 1 to go to state 0
 reset=1'b0;
 clk =1'b0;
 #20
 clk =1'b1;
 #20
 A=   1'b0;        //Input 0 to stay at state 0
 reset=1'b0;
 clk =1'b0;
 #20
 clk =1'b1;
```

```
   #20
   A=    1'b0;        //Input 1 to go to state 1
   reset=1'b0;
   clk =1'b0;
   #20
   clk =1'b1;
   #20
   A=    1'b1;        //Input 1 to go to state 2
   reset=1'b0;
   clk =1'b0;
   #20
   clk =1'b1;
    #20
   A=    1'b1;        //Input 1 to go to state 3
   reset=1'b0;
   clk =1'b0;
   #20
   clk =1'b1;
   #20
   A=    1'b1;        //Input 1 to go to state 0
   reset=1'b0;

   clk =1'b0;
   #20
   clk =1'b1;
   #20
   A=    1'b1;        //done
   reset=1'b0;
   clk =1'b0;
   #20
   clk =1'b1;
   end
 endmodule
```

EXAMPLE 5.12

Write a Verilog description using behavioral modeling for the two-bit counter of
Example 5.5. Figure 5.68 shows the state diagram (re-drawn from Figure 5.34).

Solution

```
// example 5.5
  module counter2bit(clock, reset, state);
  input clock, reset;
  output [1:0] state;
  reg [1:0] state, next_state;
  parameter s00 = 2'b00,
            s01 = 2'b01,
            s10 = 2'b10,
            s11 = 2'b11;
```

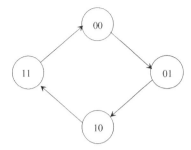

FIGURE 5.68   State diagram for the two-bit counter.

```
  always @ (posedge clock or posedge reset)
   begin
     if (reset == 1)
       state <= s00;
     else
       state <= next_state;
   end

  always @ (state)
   begin
    case(state)
     s00 : next_state <= s01;
     s01 : next_state <= s10;
     s10 : next_state <= s11;
     s11 : next_state <= s00;
    endcase
   end

 endmodule

 module test;
  reg clock, reset;
  wire [1:0] state;

  counter2bit c2bit(clock, reset, state);

 initial
  begin
  $display(" clock  reset\tstate binary \tstate decimal");
   $monitor ( "   %b\t    %b\t  %b\t  %d ",
                 clock, reset,state,state);
  #0 reset = 0;
  #1 reset = 1;
  #1 reset = 0;
   end
   initial
     begin
```

```
   #0 clock = 0;
   #40 $finish;
   end
  always #1 clock = ~clock;

 endmodule
```

Note: In the above, inclusion of \t with statements for $display and $monitor
provides horizontal tab.

clock	reset	state binary	state decimal
0	0	xx	x
1	1	00	0
0	0	00	0
1	0	01	1
0	0	01	1
1	0	10	2
0	0	10	2
1	0	11	3
0	0	11	3
1	0	00	0
0	0	00	0
1	0	01	1
0	0	01	1
1	0	10	2
0	0	10	2
1	0	11	3
0	0	11	3
1	0	00	0
0	0	00	0
1	0	01	1
0	0	01	1
1	0	10	2
0	0	10	2
1	0	11	3
0	0	11	3
1	0	00	0
0	0	00	0
1	0	01	1
0	0	01	1
1	0	10	2
0	0	10	2
1	0	11	3
0	0	11	3
1	0	00	0
0	0	00	0
1	0	01	1
0	0	01	1
1	0	10	2
0	0	10	2
1	0	11	3

EXAMPLE 5.13

Write a Verilog description using behavioral modeling for the three-bit counter of
Example 5.7. Figure 5.69 shows the state diagram (re-drawn from Figure 5.37).

Solution

```
// example 5.7
 module nonbinarycounter(clock, reset, state);
 input clock, reset;
 output [2:0] state;
 reg [2:0] state, next_state;
 parameter s0 = 3'b000, s1 = 3'b001,
       s2 = 3'b010, s3 = 3'b011,
       s4 = 3'b100, s5 = 3'b101,
       s6 = 3'b110, s7 = 3'b111;

  always @ (posedge clock or posedge reset)
   begin
     if (reset == 1)
       state <= s0;
     else
       state <= next_state;
   end

  always @ (state)
   begin
    case(state)
     s0 : next_state <= s2;
     s1 : next_state <= s3;
     s2 : next_state <= s3;
     s3 : next_state <= s5;
     s4 : next_state <= s7;
     s5 : next_state <= s6;
     s6 : next_state <= s7;
     s7 : next_state <= s0;
    endcase
   end
 endmodule

 module test;
  reg clock, reset;
  wire [2:0] state;
  nonbinarycounter nbc(clock, reset, state);
 initial
 begin
 $display(" clock reset\tstate binary \tstate decimal");
  $monitor ( "  %b\t  %b\t  %b\t  %d ",
             clock, reset, state, state);
```

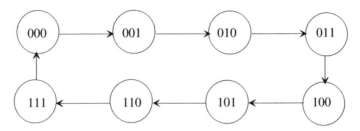

FIGURE 5.69   State Diagram for Example 5.7.

```
#0 reset = 0;
#1 reset = 1;
#1 reset = 0;
end
  initial
    begin
    #0 clock = 0;
    #40 $finish;
    end
  always #1 clock = ~clock;
endmodule
```

Note: In the above, inclusion of `\t` with statements for `$display` and
`$monitor` provides horizontal tab.

clock	reset	state binary	state decimal
0	0	xxx	x
1	1	000	0
0	0	000	0
1	0	010	2
0	0	010	2
1	0	011	3
0	0	011	3
1	0	101	5
0	0	101	5
1	0	110	6
0	0	110	6
1	0	111	7
0	0	111	7
1	0	000	0
0	0	000	0
1	0	010	2
0	0	010	2
1	0	011	3
0	0	011	3
1	0	101	5
0	0	101	5
1	0	110	6
0	0	110	6

1	0	111	7
0	0	111	7
1	0	000	0
0	0	000	0
1	0	010	2
0	0	010	2
1	0	011	3
0	0	011	3
1	0	101	5
0	0	101	5
1	0	110	6
0	0	110	6
1	0	111	7
0	0	111	7
1	0	000	0
0	0	000	0
1	0	010	2

EXAMPLE 5.14

Write a Verilog description using behavioral modeling for the General Purpose
Register specified by the truth table of Table 5.13.
Figure 5.70 (re-drawn from Figure 5.44) and Figure 5.71 (re-drawn from Figure 5.45)
show relevant figures.

Solution

```
/**********************************************************
Description: Basic Cell
File Name: BasicCell.v
**********************************************************/
module BasicCell( q, CLR, CLK, s, A );
output q;
input CLK, CLR;
input [1:0] s;
input [3:0] A;
wire data, q_bar;
mux4to1 M1( data, s, A );
D_ff D0( q, q_bar, CLR, CLK, data );
endmodule


/**********************************************************
Description: D Flip Flop
File Name: D.v
**********************************************************/
module D_ff( Q, Q_bar, CLR, CLK, D );
output Q, Q_bar;
```

```
input CLR, CLK, D;
reg Q, Q_bar;
always @(posedge CLK or negedge CLR)
begin                   //When CLR == 0 (neg logic) Q is always 0
                        //else @ rising edge of clock, Q <-- D
        if(!CLR)
                begin
                        Q <= 1'b0;
                        Q_bar <= 1'b1;
                end
        else
                begin
                        Q <= D;
                        Q_bar <= !D;
                end
end


endmodule
// The code for the 4 to 1 multiplexer used in the Basic cell is:
// Filename : mux4to1.v
//description: 4 to 1 multiplexer

module mux4to1(X, s, A);
        output X;
        input [1:0] s;
        input [3:0] A;
        assign X =  (s == 2'b00)? A[0]:
                    (s == 2'b01)? A[1]:
                    (s == 2'b10)? A[1]: A[3];

endmodule


//description: General purpose register

module GPR (Q, CLR, CLK, S, X, r_in, l_in) ;
output [3:0] Q;
input CLR, CLK, r_in, l_in;
input [ 1: 0] S;
input [3:0] X;
wire [3:0] A;
BasicCell Cell3 (A[3] , CLR, CLK, S, {X[3] , A[2] , r_in , A[3]} );
BasicCell Cell2 (A[2] , CLR, CLK, S, {X[2] , A[1] , A[3] , A[2]} );
BasicCell Cell1 (A[1] , CLR, CLK, S, {X[1] , A[0] , A[2] , A[1]} );
BasicCell Cell0 (A[0] , CLR, CLK, S, {X[0] , l_in, A[1] , A[0] } );
assign Q = A;
endmodule
```

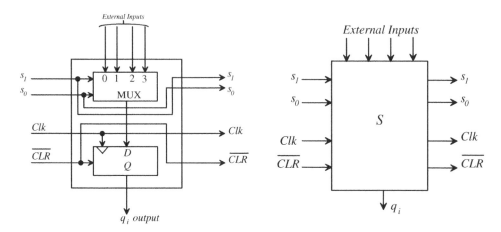

(a) Internal Organization of   (b) Block Diagram of
     the Basic Cell S the Basic Cell S

FIGURE 5. 70   A Basic Cell for Designing a GPR.

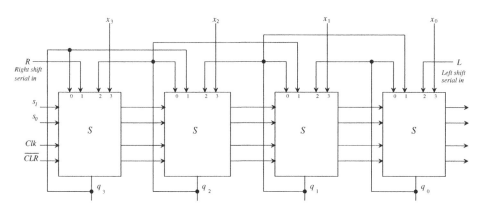

FIGURE 5.71   A 4-bit General Register.

Example 5.15

All microcontrollers contain a register usually called the *Status register* which contains typical flags such as carry (C), sign (S), zero (Z), overflow (V), and parity (P) flags. Hence, in this example,

(a) determine the carry (C), sign (S), zero (Z), overflow (V), and parity (P) flags for the following operation: 0110_2 plus 1010_2 based on the concepts covered in Chapter 2. Assume the parity bit = 1 for ODD parity in the result; otherwise the parity bit = 0. Also, assume that the numbers are signed.

(b) Draw a logic diagram for implementing the flags in a five-bit register (called status register) using D flip-flops; use P = bit 0, V = bit 1, Z = bit 2, S = bit 3, and C = bit 4.

(c) Write a Verilog description using behavioral modeling of the status register.

Solution

(a)

$$
\begin{array}{r}
1\ 1\ 0 \longleftarrow \text{Intermediate Carries} \\
0\ 1\ 1\ 0 \\
+\ 1\ 0\ 1\ 0 \\
\hline
\text{Result} = 0\ 0\ 0\ 0
\end{array}
$$

$C_f = C = 1 \longleftarrow$ $Z = 1$ since result $= 0$
$S = 0 \longleftarrow$ $P = 0$ since even parity
$C_p = 1 \longleftarrow$ $V = C_f \oplus C_p = 1 \oplus 1 = 0$

The equations for the flags are provided below:
Let us name the status register as "stat". Also, Result = r[3] r[2] r[1] r[0]

Bit 0 of the stat register, stat[0] =Parity Flag = $r[3] \oplus r[2] \oplus r[1] \oplus r[0]$
Bit 1 of the stat register, stat[1] = Overflow Flag $= \overline{C_f \oplus Cp}$
Bit 2 of the stat register, stat[2] = Zero Flag = $\overline{r[3] + r[2] + r[1] + r[0]}$
Bit 3 of the stat register, stat[3] = Sign Flag = r[3]
Bit 4 of the stat register, stat[4] = Carry Flag = Carry final = C_f

(b) Figure 5.72 shows the logic diagram of the status register. Note that a typical microcontroller's ALU (Arithmetic Logic Unit) performs this addition using appropriate instruction, and will then provide the result usually in a register. In this case, suppose the 4-bit result register is r[3] r[2] r[1] r[0]. The bits in the result register can be connected to the hardware of Figure 5.68 to affect the flags accordingly.
The flag register can be implemented from the 4-bit result in Figure 5.72.

(c)Verilog description of the status register is provided below:

```
// Status Register
module statsreg(stat,cfinal,cprev,clk,r);
input [3:0] r;
input cfinal,cprev,clk;
output [4:0] stat;
reg [4:0] stat;
/* The status register is 5-bit. They will be latched and
   the output is shown at a positive edge of the clock.*/
 always@(posedge clk)
    begin
      stat[0] <= r[3]^r[2]^r[1]^r[0];   //Parity flag
      stat[1] <= cfinal^cprev;          //Overflow flag
      stat[2] <= ~(r[3]|r[2]|r[1]|r[0]); //Zero flag
      stat[3] <= r[3];                   //Sign flag
      stat[4] <= cfinal;               //Final carry
    end
endmodule
```

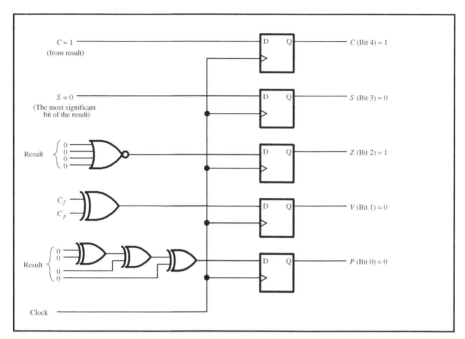

FIGURE 5.72   Logic diagram of the status register.

Note: In the above, `stat[0]  <=  r[3]^r[2]^r[1]^r[0];` can be replaced by `stat[0]  <=  ^(r);`

Also, `stat[2]  <=  ~(r[3]|r[2]|r[1]|r[0]);` can be replaced using the NOR reduction operator as `stat[2]  <=  ~|(r);`

Questions and Problems

5.1   What is the basic difference between a combinational circuit and a sequential circuit?

5.2   Identify the main characteristics of a synchronous sequential circuit and an asynchronous sequential circuit.

5.3   What is the basic difference between a latch and a flip-flop?

5.4   Draw the logic diagram of a D flip-flop using OR gates and inverters.

5.5   Assume that initially $x = 1$, $A = 0$, and $B = 1$ in Figure P5.5. Determine the values of A and B after the positive edge of *Clk*.

5.6   Draw the logic diagram of a JK flip-flop using AND gates and inverters.

5.7   Assume that initially $X = 0$, $A = 0$, and $B = 1$ in Figure P5.7. Determine the values of A and B after one *Clk* pulse. Note that the latches are gated SR.

5.8   Given Figure P5.8, draw the timing diagram for Q and $\overline{Q}$ assuming a negative-edge triggered JK flip-flop. Assume Q is preset to 1 initially.

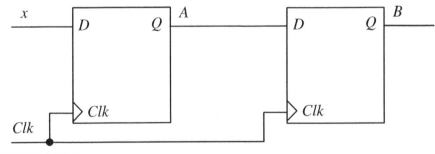

FIGURE P5.5.

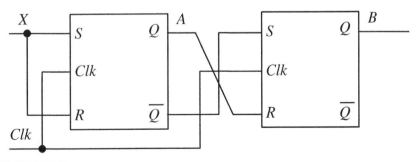

FIGURE P5.7.

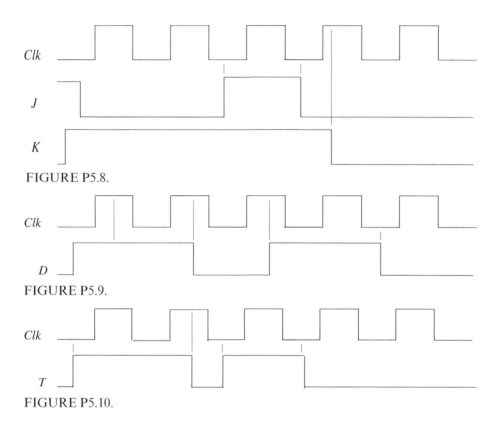

FIGURE P5.8.

FIGURE P5.9.

FIGURE P5.10.

5.9   Given the timing diagram for a positive-edge triggered D flip-flop in Figure
        P5.9, draw the timing diagrams for Q and $\overline{Q}$. Assume Q is cleared to zero
        initially.

5.10   Given the timing diagram for a negative-edge triggered T flip-flop in Figure
        P5.10, draw the timing diagram for Q. Assume Q is preset to 1 initially.

5.11   What is the basic difference between an edge-triggered flip-flop and a latch?

5.12   What are the advantages of a master–slave flip-flop?

5.13   Draw the block diagram of a T flip-flop using (a) JK ff (b) D ff.

5.14   Draw a logic circuit of the switch debouncer circuit using NAND gates.

5.15   Analyze the clocked synchronous circuit shown in Figure P5.15. Express the
        next state in terms of the present state and inputs, derive the state table, and
        draw the state diagram.

5.16   A synchronous sequential circuit with two D flip-flops *(a,b as outputs)*, one
        input *(x)*, and an output *(y)* is expressed by the following equations:

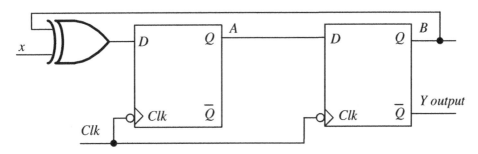

FIGURE P5.15.

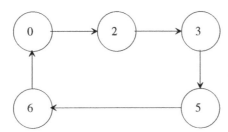

FIGURE P5.17.

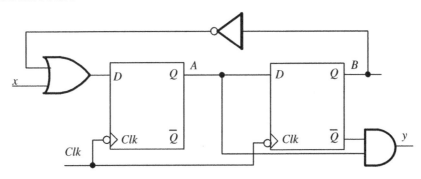

FIGURE P5.18.

$$D_a = a \bar{b} x + \bar{a} b, \; D_b = \bar{x} b + \bar{b} x$$
$$y = \bar{b} \bar{x} + a$$

(a) Derive the state table and state diagram for the circuit.
(b) Draw a logic diagram.

5.17 A synchronous sequential circuit is represented by the state diagram shown in Figure P5.17. Using JK flip-flops and undefined states as don't-cares:
(a) Derive the state table.
(b) Minimize the equation for flip-flop inputs using K-maps.
(c) Draw a logic diagram.

5.18 A sequential circuit contains two D flip-flops (*A*, *B*), one input (*x*), and one output (*y*), as shown in Figure P5.18.
Derive the state table and the state diagram of the sequential circuit.

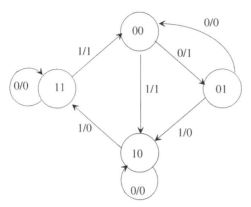

FIGURE P5.19

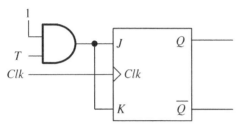

FIGURE P5.22.

5.19   Design a synchronous sequential circuit using D flip-flops for the state
        diagram shown in Figure P5.19.

5.20   Design a two-bit counter that will count in the following sequence: 00, 11, 10,
        01, and repeat. Using T flip-flops:
        (a) Draw a state diagram.
        (b) Derive a state table.
        (c) Implement the circuit.

5.21   Design a synchronous sequential circuit with one input x and one output
        y. The input x is a serial message, and the system reads x one bit at a time.
        The output y is 1 whenever the binary pattern 000 is encountered in the
        serial message. For example: If the input is 01000000, then the output will be
        00001010. Use T flip-flops.

5.22   Analyze the circuit shown in Figure P5.22 and show that it is equivalent to a
        T flip-flop.

5.23   Design a BCD counter to count in the sequence 0000, 0001, 0010, 0011, 0100,
        0101, 0110, 0111, 1000, 1001, and repeat. Use T flip-flops.

5.24   Design the following nonbinary sequence counters using the type of flip-
        flop specified. Assume the unused states as don't cares. Is the counter self-
        correcting? Justify your answer.

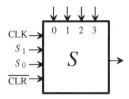

FIGURE P5.25.

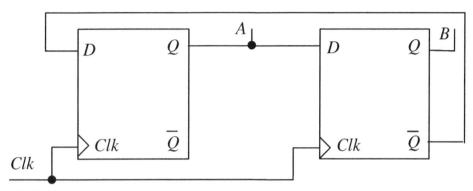

FIGURE P5.27.

(a) Counting sequence 0, 1, 3, 4, 5, 6, 7, and repeat. Use JK flip-flops.
(b) Counting sequence 0, 2, 3, 4, 6, 7, and repeat. Use D flip-flops.
(c) Counting sequence 0, 1, 2, 4, 5, 6, 7, and repeat. Use T flip-flops.

5.25 Design a 4-bit general-purpose register as follows:

S_1	S_0	Function
0	0	Load external data
0	1	Rotate left; $(A_0 \leftarrow A_3$, $A_i \leftarrow A_{i-1}$ for $i = 1,2,3)$
1	0	Rotate right; $(A_3 \leftarrow A_0$, $A_i \leftarrow A_{i+1}$ for $i = 0,1,2)$
1	1	Increment

Use Figure P5.25 as the building block:

5.26 Design a logic diagram that will generate 19 timing signals. Use a ring counter with JK flip-flops.

5.27 Consider the 2-bit Johnson counter shown in Figure P5.27. Derive the state diagram. Assume the D flip-flops are initialized to $A = 0$ and $B = 0$.

5.28 Assuming $AB = 10$, verify that the 2-bit counter shown in Figure P5.28 is a ring counter. Derive the state diagram.

5.29 What is the basic difference between SRAM and DRAM?

5.30 Given a memory with a 24-bit address and 8-bit word size,
(a) How many bytes can be stored in this memory?
(b) If this memory were constructed from 1K × 1-bit RAM chips, how many memory chips would be required?

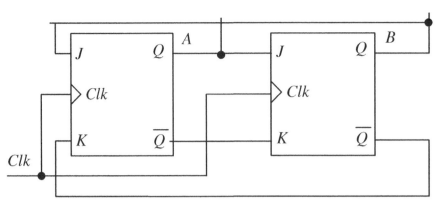

FIGURE P5.28.

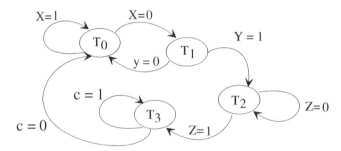

FIGURE P5.33.

5.31   Draw an ASM chart for the following: Assume three states (a, b, c) in the system
        with one input x and two registers R_1 and R_2. The circuit is initially in state a. If
        $x = 0$, the control goes from state a to state b and, clears registers R_1 to 0 and sets
        R_2 to 1, and then moves to state c. On the other hand if $x = 1$, the control goes
        to state c. In state c, R_1 is subtracted from R_2 and the result is stored in R_1. The
        control then moves back to state a and the process continues.

5.32   Draw an ASM chart for each of the following sequence of operations:
        (a)   The ASM chart will define a conditional operation to perform the
               operation $R_2 \leftarrow R_2 - R_1$ during State T_0 and will transfer control to State
               T_1 if the control input c is 1; if c=0, the system will stay in T_0. Assume
               that R_1 and R_2 are 8-bit registers.
        (b)   The ASM chart in which the system is initially in State T_0 and then
               checks a control input c. If c=1, control will move from State T_0 to State
               T_1; if c=0, the system will increment an 8-bit register R by 1 and control
               will return to the initial state.

5.33   Draw an ASM chart for the following state diagram of Figure P5.33:
        Assume that the system stays in initial state T_0 when control input c = 0 and
        input X = 1. The sequence of operations is started from T_0 when X = 0. When
        the system reaches state T_3, it stays in T_3 indefinitely as long as c = 1; the
        system returns to state T_0 when c = 0.

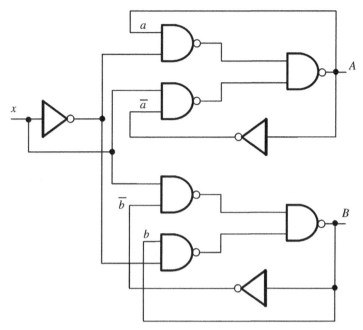

FIGURE P5.34.

5.34 Derive the output equations for the asynchronous sequential circuit shown in Figure P5.34. Also, determine the state table and flow table.

5.35 Write a Verilog description using behavioral modeling for:
(a) the positive edge-triggered JK Flip-Flop of Figure 5.9.
(b) a D flip-flop with a synchronous reset input and a positive edge triggered clock. Use synchronous reset such that if reset ==0, the flip-flop is cleared to 0; on the other hand, if reset==1, the output of the flip-flop is unchanged until the procedural statements are evaluated at the positive edge of the clock.
(c) the T flip-flop of Figure 5.10(b).
(d) the T flip-flop of Problem 5.13(b) with a HIGH reset input.
(e) the state machine of Problem 5.19.
(f) a 4-bit binary ripple counter. Note that in a binary ripple counter, the clock inputs of high order flip-flops are not triggered by the common clock, but by the transition outputs of the low order flip-flops. The 4-bit binary ripple counter contains four T flip-flops (obtained from D-ffs), with the output of each ff connected to the clock input of the next higher-order ff. The clock input is connected to the least significant T-ff. The 4-bit ripple counter can be designed using four T flip- flops (tff0 through tff3). Each T-ff can be obtained from a D-ff by connecting its output q to the input of an inverter, and then connecting the inverter output to the D input; the T-ff has one input (T input is the same as the clock input). This T-ff toggles every clock. The 4-bit ripple counter can be obtained by connecting the clock to the tff0 clock input, q0 of tff0 to

clock input of tff1, q1 output of tff1 to clock input of tff2, and q2 output of tff2 to the clock input of tff3. Use negative edge-triggered D-ffs. Each D-ff will have a reset input to clear the ff.

(g)   a 4-bit serial shift (right) register with a positive edge triggered reset and a positive edge triggered clock. The 4-bit serial shift register can be obtained by connecting four D-ff's to a common clock and a common reset. The four D-ff's are cleared to 0 at the positive edge triggered clock and positive edge triggered reset. Assume, v as the serial input bit connected to the D input of the left-most D-ff with z as its output; z is connected to the D input of the next right D-ff with y as its output; y is connected to the D input of the next right D-ff with x as its output; finally, x is connected to the D input of the right-most D-ff with w as its output.

(h)   a 4-bit register with a reset input, a parallel load input and a positive edge-triggered clock. The 4-bit register is cleared to 0 at the positive edge of the reset. On the other hand, if the load input is high, 4-bit data is transferred to the register at the positive edge of the clock. Use behavioral modeling.

(i)   the counters of Problems 5.24 (a) through 5.24 (c).

(j)   the general purpose register of Problem 5.25.

6

CPU, MEMORY, AND I/O

This chapter first describes the design of a simple Central Processing Unit (CPU). Since the basic CPU contains a set of registers, an Arithmetic Logic Unit (ALU), and a Control unit, design of these components are covered. Also, in this chapter, basic concepts of memory organization and Input/Output (I/O) techniques associated with typical microcontrollers are described.

6.1 Design of the CPU

As mentioned before, the CPU contains three elements: registers, ALU (Arithmetic Logic Unit), and control unit.

Since programs contain instructions and data, the register section of the CPU contains four basic registers to hold instructions, data, and their corresponding addresses. These registers are: Instruction Register (IR), Program Counter (PC), Memory Address Register (MAR), and General Purpose Register (GPR).

The Instruction Register (IR) stores instructions. The contents of the instruction register are always translated by the CPU as an instruction. After reading (fetching) an instruction from memory, the CPU stores it in the instruction register. The instruction is decoded (translated) internally by the CPU, which then performs the required operation.

The program counter contains the address of the instruction. The program counter normally contains the address of the next instruction to be executed.

The Memory Address Register (MAR) contains the address of data. The CPU uses the address, which is stored in the memory address register, as a direct pointer to memory. The contents of the address is the actual data that is being transferred.

The General Purpose Register (GPR) stores data addressed by MAR. The CPU typically has instructions to shift the contents of the GPR one bit to the right or left. The CPU can execute an instruction to retrieve the contents of this register when needed.

A basic ALU performs arithmetic and logic operations. The Control unit, on the other hand, translates instructions.

To execute a program, a conventional CPU repeats the following three steps for completing each instruction:

1. *Fetch.* The CPU fetches (instruction read) the instruction from the program memory (external to the CPU) into the instruction register.
2. *Decode.* The CPU decodes or translates the instruction using the control unit. The control unit inputs the contents of the instruction register, and then decodes (translates) the instruction to determine the instruction type.
3. *Execute.* The CPU executes the instruction using the control unit. To accomplish the task, the control unit generates a number of enable signals required by the instruction.

For example, suppose that it is desired to add the contents of two general-purpose registers, X and Y, and store the result in another general purpose register , Z. To accomplish this, the conventional CPU performs the following steps:

1. The CPU fetches the instruction into the instruction register.
2. The control unit (CU) decodes the contents of the instruction register.
3. The CU executes the instruction by generating enable signals for the register and ALU sections to perform the following:
    a. The CU transfers the contents of registers X and Y from the Register section into the ALU.
    b. The CU commands the ALU to ADD.
    c. The CU transfers the result from the ALU into register Z of the register section.
    We now describe the design of registers, ALU, and control unit.

6.1.1 Register Design

The design of general-purpose and flag registers is provided in Chapters 5. This section includes combinational shifter design and the concepts associated with barrel shifters.

A high-speed shifter can be designed using combinational circuit components such as a multiplexer. The block diagram, internal organization, and truth table of a typical combinational shifter are shown in Figure 6.1. From the truth table, the following equations can be obtained:

$$y_3 = \overline{s_1}\,\overline{s_0}i_3 + \overline{s_1}s_0i_2 + s_1\overline{s_0}i_1 + s_1s_0i_0$$

$$y_2 = \overline{s_1}\,\overline{s_0}i_2 + \overline{s_1}s_0i_1 + s_1\overline{s_0}i_0 + s_1s_0i_{-1}$$

$$y_1 = \overline{s_1}\,\overline{s_0}i_1 + \overline{s_1}s_0i_0 + s_1\overline{s_0}i_{-1} + s_1s_0i_{-2}$$

$$y_0 = \overline{s_1}\,\overline{s_0}i_0 + \overline{s_1}s_0i_{-1} + s_1\overline{s_0}i_{-2} + s_1s_0i_{-3}$$

The 4×4 shifter of Figure 6.1 can be expanded to obtain a system capable of rotating 16-bit data to the left by 0, 1, 2, or 3 positions. Table 6.1 shows the truth table. Figure 6.2 shows the logic diagram of the combinational shifter capable of rotating 16-bit data to the left by 0, 1, 2, or 3 positions.

This design can be extended to obtain a more powerful shifter called the *barrel shifter*. The shift is a cycle rotation, which means that the input binary information is shifted in one direction; the most significant bit is moved to the least significant position.

Table 6.2 shows the truth table representing the operation of a 16×16 barrel shifter. The block-diagram representation of the barrel shifter is shown in Figure 6.3. This shifter is capable of rotating the given 16-bit data to the left by n positions, where $0 \le n \le 15$. The barrel shifter is an on-chip component for typical 32-bit microprocessors such as the Pentium.

6.1.2 Arithmetic Logic Unit (ALU)

Functionally, an ALU can be divided into two segments: the arithmetic unit and the logic unit. The arithmetic unit performs typical arithmetic operations such as addition, subtraction, and increment or decrement by 1. Usually, the data involved may be signed or unsigned integers. In some cases, however, an arithmetic unit must handle

4-bit binary-coded decimal (BCD) numbers and floating-point numbers. Therefore, this unit must include the circuitry necessary to manipulate these data types. As the name implies, the logic unit contains hardware elements that perform typical operations such as Boolean NOT and OR. In this section, typical arithmetic operations performed by an ALU will be covered first. Design of a simple ALU will then be described.

Adders Addition is the basic arithmetic operation performed by an ALU. Other operations such as subtraction and multiplication can be obtained via addition. Thus,

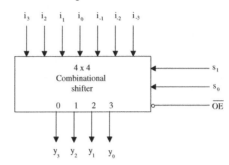

(a)   Block Diagram

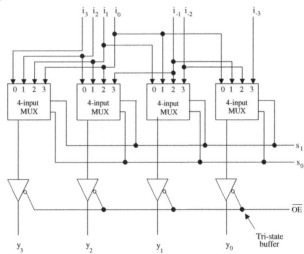

(b)   Internal Schematic

	Shift Count		Output				Comment
$\overline{OE}$	s_1	s_0	y_3	y_2	y_1	y_0	
1	X	X	Z	Z	Z	Z	Outputs are tristated
0	0	0	i_3	i_2	i_1	i_0	Pass (no shift)
0	0	1	i_2	i_1	i_0	i_{-1}	Left Shift once
0	1	0	i_1	i_0	i_{-1}	i_{-2}	Left shift twice
0	1	1	i_0	i_{-1}	i_{-2}	i_{-3}	Left shift three times

(c)   Truth Table (X is don't care in the above)

FIGURE 6.1 4×4 combinational shifter.

TABLE 6.1   Combinational shifter capable of rotating 16-bit data to the left by 0, 1, 2, or 3 positions

Shift Count		Output																
S_1	S_0	y_{15}	y_{14}	y_{13}	y_{12}	y_{11}	y_{10}	y_9	y_8	y_7	y_6	y_5	y_4	y_3	y_2	y_1	y_0
0	0	x_{15}	x_{14}	x_{13}	x_{12}	x_{11}	x_{10}	x_9	x_8	x_7	x_6	x_5	x_4	x_3	x_2	x_1	x_0
0	1	x_{14}	x_{13}	x_{12}	x_{11}	x_{10}	x_9	x_8	x_7	x_6	x_5	x_4	x_3	x_2	x_1	x_0	x_{15}
1	0	x_{13}	x_{12}	x_{11}	x_{10}	x_9	x_8	x_7	x_6	x_5	x_4	x_3	x_2	x_1	x_0	x_{15}	x_{14}
1	1	x_{12}	x_{11}	x_{10}	x_9	x_8	x_7	x_6	x_5	x_4	x_3	x_2	x_1	x_0	x_{15}	x_{14}	x_{13}

the time required to add two numbers plays an important role in determining the speed of the ALU. The basic concepts of half-adder, full adder, and binary adder are discussed in Section 4.5.6. The following equations for the full-adder were obtained assuming $x_i = x$, $y_i = y$, $c_i = z$, and $C_{i+1} = c_i$:

$$\text{Sum, } S_i = \overline{x_i}\,\overline{y_i}c_i + \overline{x_i}y_i\overline{c_i} + x_i\,\overline{y_i}\,\overline{c_i} + x_i y_i c_i$$

$$= x_i \oplus y_i \oplus c_i$$

From Table 4.6,

$$\text{Carry, } C_{i+1} = \overline{x_i}\,y_i c_i + x_i\overline{y_i}c_i + x_i\,y_i\,\overline{c_i} + x_i y_i c_i$$

$$= (\overline{x_i}\,y_i c_i + x_i y_i c_i) + (x_i\overline{y_i}c_i + x_i y_i c_i) + (x_i\,y_i\,\overline{c_i} + x_i y_i c_i)$$

$$= y_i c_i + x_i c_i + x_i y_i$$

The logic diagrams for implementing these equations are given in Figure 6.4.

As has been made apparent by Figure 6.4, for generating C_{i+1} from c_i, two gate delays are required. To generate S_i from c_i, three gate delays are required because c_i

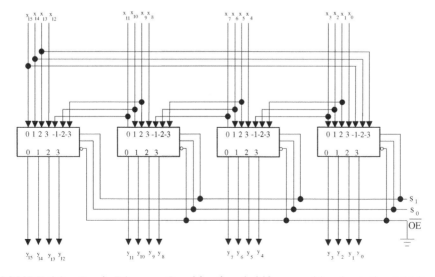

FIGURE 6.2   Logic Diagram Combinational shifter capable of rotating 16-bit data to the left by 0, 1, 2, or 3 positions.

TABLE 6.2   Truth table of a 16 x 16 Barrel Shifter

Shift Count				Output															
s_3	s_2	s_1	s_0	y_{15}	y_{14}	y_{13}	y_{12}	y_{11}	y_{10}	y_9	y_8	y_7	y_6	y_5	y_4	y_3	y_2	y_1	y_0
0	0	0	0	x_{15}	x_{14}	x_{13}	x_{12}	x_{11}	x_{10}	x_9	x_8	x_7	x_6	x_5	x_4	x_3	x_2	x_1	x_0
0	0	0	1	x_{14}	x_{13}	x_{12}	x_{11}	x_{10}	x_9	x_8	x_7	x_6	x_5	x_4	x_3	x_2	x_1	x_0	x_{15}
0	0	1	0	x_{13}	x_{12}	x_{11}	x_{10}	x_9	x_8	x_7	x_6	x_5	x_4	x_3	x_2	x_1	x_0	x_{15}	x_{14}
0	0	1	1	x_{12}	x_{11}	x_{10}	x_9	x_8	x_7	x_6	x_5	x_4	x_3	x_2	x_1	x_0	x_{15}	x_{14}	x_{13}
0	1	0	0	x_{11}	x_{10}	x_9	x_8	x_7	x_6	x_5	x_4	x_3	x_2	x_1	x_0	x_{15}	x_{14}	x_{13}	x_{12}
0	1	0	1	x_{10}	x_9	x_8	x_7	x_6	x_5	x_4	x_3	x_2	x_1	x_0	x_{15}	x_{14}	x_{13}	x_{12}	x_{11}
0	1	1	0	x_9	x_8	x_7	x_6	x_5	x_4	x_3	x_2	x_1	x_0	x_{15}	x_{14}	x_{13}	x_{12}	x_{11}	x_{10}
0	1	1	1	x_8	x_7	x_6	x_5	x_4	x_3	x_2	x_1	x_0	x_{15}	x_{14}	x_{13}	x_{12}	x_{11}	x_{10}	x_9
1	0	0	0	x_7	x_6	x_5	x_4	x_3	x_2	x_1	x_0	x_{15}	x_{14}	x_{13}	x_{12}	x_{11}	x_{10}	x_9	x_8
1	0	0	1	x_6	x_5	x_4	x_3	x_2	x_1	x_0	x_{15}	x_{14}	x_{13}	x_{12}	x_{11}	x_{10}	x_9	x_8	x_7
1	0	1	0	x_5	x_4	x_3	x_2	x_1	x_0	x_{15}	x_{14}	x_{13}	x_{12}	x_{11}	x_{10}	x_9	x_8	x_7	x_6
1	0	1	1	x_4	x_3	x_2	x_1	x_0	x_{15}	x_{14}	x_{13}	x_{12}	x_{11}	x_{10}	x_9	x_8	x_7	x_6	x_5
1	1	0	0	x_3	x_2	x_1	x_0	x_{15}	x_{14}	x_{13}	x_{12}	x_{11}	x_{10}	x_9	x_8	x_7	x_6	x_5	x_4
1	1	0	1	x_2	x_1	x_0	x_{15}	x_{14}	x_{13}	x_{12}	x_{11}	x_{10}	x_9	x_8	x_7	x_6	x_5	x_4	x_3
1	1	1	0	x_1	x_0	x_{15}	x_{14}	x_{13}	x_{12}	x_{11}	x_{10}	x_9	x_8	x_7	x_6	x_5	x_4	x_3	x_2
1	1	1	1	x_0	x_{15}	x_{14}	x_{13}	x_{12}	x_{11}	x_{10}	x_9	x_8	x_7	x_6	x_5	x_4	x_3	x_2	x_1

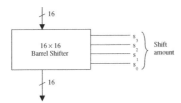

FIGURE 6.3   Block Diagram of a 16 × 16 Barrel Shifter.

must be inverted to obtain $\bar{c_i}$. Note that no inverters are required to get $\bar{x_i}$ or $\bar{y_i}$ from x_i or y_i, respectively, because the numbers to be added are usually stored in a register that is a collection of flip-flops. The flip-flop generates both normal and complemented outputs. For the purpose of discussion, assume that the gate delay is Δ time units, and the actual value of Δ is decided by the technology. For example, if transistor translator logic (TTL) circuits are used, the value of Δ will be 10 ns.

By cascading n full adders, an n-bit binary adder capable of handling two n-bit data (X and Y) can be designed. The implementation of a 4-bit ripple-carry or binary adder is shown in Figure 6.5. When two unsigned integers are added, the input carry, c_0, is always zero. The 4-bit adder is also called a *carry-propagate adder* (CPA), because the carry is propagated serially through each full adder. This hardware can be cascaded to obtain a 16-bit CPA, as shown in Figure 6.6.

Although the design of an n-bit CPA is straightforward, the carry propagation time limits the speed of operation. For example, in the 16-bit CPA (Figure 6.6), the addition operation is completed only when the sum bits s_0 through s_{15} are available.

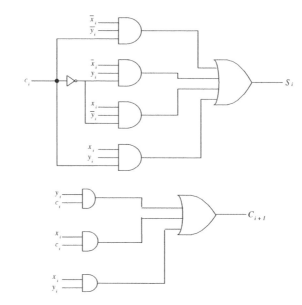

(a)   Full adder

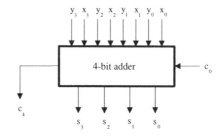

(b)   Carry

FIGURE 6.4   Logic circuit of the full adder.

To generate s_{15}, c_{15} must be available. The generation of c_{15} depends on the availability of c_{14}, which must wait for c_{13} to become available. In the worst case, the carry process propagates through 15 full adders. Therefore, the worst-case add-time of the 16-bit CPA can be estimated as follows:

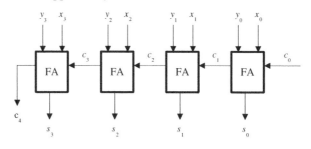

(a)   Block Diagram of a 4-bit Ripple-Carry Adder

(b)   Four 4-bit Full Adders are Cascaded to implement a 4-Bit Ripple-Carry Adder

FIGURE 6.5   Implementation of a 4-bit Ripple-Carry Adder.

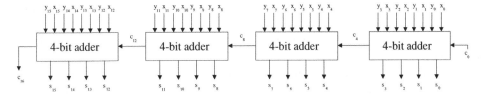

FIGURE 6.6 Implementation of a 16-bit adder using 4-Bit Adders as Building Blocks.

Time taken for carry to propagate
through 15 full adders (the delay
involved in the path from c_0 to c_{15}) $= 15 * 2\,\Delta$
Time taken to generate s_{15} from c_{15} $= 3\,\Delta$
Total $= 33\,\Delta$

If $\Delta = 10$ ns, then the worst-case add-time of a 16-bit CPA is 330 ns. This delay is prohibitive for high-speed systems, in which the expected add-time is typically less than 100 ns, which makes it necessary to devise a new technique to increase the speed of operation by a factor of 3. One such technique is known as the *carry look-ahead*. In this approach the extra hardware is used to generate each carry ($c_i,\ i > 0$) directly from c_0. To be more practical, consider the design of a 4-bit carry look-ahead adder (CLA). Let us see how this may be used to obtain a 16-bit adder that operates at a speed higher than the 16-bit CPA.

Recall that in a full adder for adding X_i, Y_i, and C_i, the output carry C_{i+1} is related to its carry input C_i, as follows:

$$C_{i+1} = X_iY_i + X_iC_i + Y_iC_i$$

The result can be rewritten as

$$C_{i+1} = G_i + P_iC_i$$

where $G_i = X_iY_i$ and $P_i = X_i + Y_i$

The function G_i is called the *carry-generate function*, because a carry is generated when $X_i = Y_i = 1$. If X_i or Y_i is a 1, then the input carry C_i is propagated to the next stage. For this reason, the function P_i is often referred to as the *carry-propagate* function. Using G_i and P_i, C_1, C_2, C_3, and C_4 can be expressed as follows:

$$C_1 = G_0 + P_0C_0$$

$$C_2 = G_1 + P_1C_1$$

$$C_3 = G_2 + P_2C_2$$

$$C_4 = G_3 + P_3C_3$$

All high-order carries can be generated in terms of C_0 as follows:

$$C_1 = G_0 + P_0C_0$$

$$C_2 = G_1 + P_1(G_0 + P_0C_0) = G_1 + P_1G_0 + P_1P_0C_0$$

$$C_3 = G_2 + P_2C_2 = G_2 + P_2(G_1 + P_1G_0 + P_1P_0C_0)$$

$$= G_2 + P_2G_1 + P_2P_1G_0 + P_2P_1P_0C_0$$

$$C_4 = G_3 + P_3C_3 = G_3 + P_3(G_2 + P_2G_1 + P_2P_1G_0 + P_2P_1P_0C_0)$$

$$= G_3 + P_3G_2 + P_3P_2G_1 + P_3P_2P_1G_0 + P_3P_2P_1P_0C_0$$

Therefore, C_1, C_2, C_3, and C_4 can generated directly from C_0. For this reason, these equations are called *carry look-ahead equations*, and the hardware that implements these equations is called a *4-stage look-ahead circuit* (4-CLC). The block diagram of such circuit is shown in Figure 6.7.

The following are some important points about this system:

- A 4-CLC can be implemented as a two-level AND-OR logic circuit (The first level consists of AND gates, whereas the second level includes OR gates).

- The outputs g_0 and p_0 are useful to obtain a higher-order look-ahead system.

To construct a 4-bit CLA, assume the existence of the basic adder cell shown in Figure 6.8. Using this basic cell and 4 - CLC, the design of a 4-bit CLA can be completed as shown in Figure 6.9. Using this cell as a building block, a 16-bit adder can be designed as shown in Figure 6.10.

The worst-case add-time of this adder can be calculated as follows:

		Delay
For P_i, G_i generation from X_i, Y_i ($0 \le i \le 15$)	...	Δ
To generate C_4 from C_0	...	2Δ
To generate C_8 from C_4	...	2Δ
To generate C_{12} from C_8	...	2Δ
To generate C_{15} from C_{12}	...	2Δ
To generate S_{15} from C_{15}	...	3Δ
Total delay	...	12Δ

A graphical illustration of this calculation can be shown as follows:

$$\text{Data available} \xrightarrow{\Delta} G_iP_i \xrightarrow{2\Delta} C_4 \xrightarrow{2\Delta} C_8 \xrightarrow{2\Delta} C_{12} \xrightarrow{2\Delta} C_{15} \xrightarrow{3\Delta} S_{15}$$

From this calculation, it is apparent that the new 16-bit adder is faster than the 16-bit CPA by a factor of 3 approximately. In fact, this system can be speeded up further by employing another 4-bit CLC and eliminating the carry propagation between the 4-bit CLA blocks. For this purpose, the g_i and p_i outputs generated by the 4-bit CLA are used. This design task is left as an exercise to the reader.

g0 and p0 shown in the figure are defined as follows:

$$g_0 = G_3 + P_3G_2 + P_3P_2G_1 + P_3P_2P_1G_0$$

$$p_0 = P_3P_2P_1P_0$$

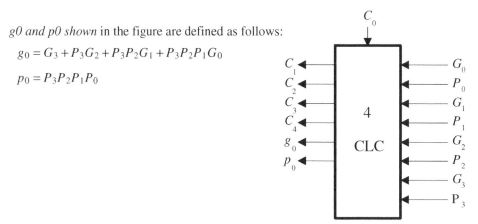

FIGURE 6.7   A Four-Stage Carry Look-ahead Circuit (4 - CLC).

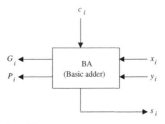

FIGURE 6.8   Basic CLA cell.

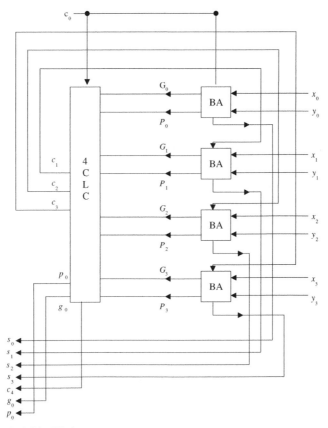

FIGURE 6.9   A 4-bit CLA.

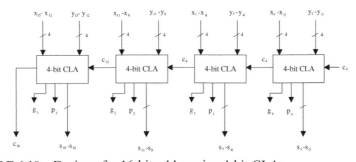

FIGURE 6.10   Design of a 16-bit adder using 4-bit CLAs.

If there is a need to add more than three data, a technique known as *carry-save addition* is used. To see its effectiveness, consider the following example:

$$
\begin{array}{r}
44 \\
28 \\
32 \\
\underline{79} \\
\underline{63} \leftarrow \text{Sum vector} \\
\underline{12} \leftarrow \text{Carry vector} \\
\underline{183} \leftarrow \text{Final answer}
\end{array}
$$

In this example, four decimal numbers are added. First, the unit digits are added, producing a sum of 3 and a carry digit of 2. Similarly, the tens digits are added, producing a sum digit of 6 and a carry digit of 1. Because there is no carry propagation from the unit digit to the tenth digit, these summations can be carried out in parallel to produce a sum vector of 63 and a carry vector of 12. When all data are added, the sum and the shifted carry vector are added in the conventional manner, which produces the final answer. Note that the carry is propagated only in the last step, which generates the final answer no matter how many operands are added. The concept is also referred to as *addition by deferred carry assimilation*.

Typical multiplication and division algorithms   Two unsigned integers can be multiplied using repeated addition as mentioned in Chapter 2. Also, they can be multiplied in the same way as two decimal numbers are multiplied by paper and pencil method. Consider the multiplication of two unsigned integers, where the multiplier Q = 15 and the multiplicand is M = 14, as illustrated:

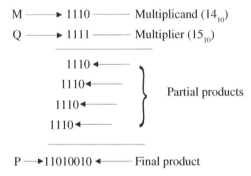

In the paper and pencil algorithm, shifted versions of multiplicands are added. This procedure can be implemented by using combinational circuit elements such as AND gates and full adders. Generally, a 4-bit unsigned multiplier Q and a 4-bit unsigned multiplicand M can be written as M: $m_3\, m_2\, m_1\, m_0$ and Q: $q_3\, q_2\, q_1\, q_0$. The process of generating the partial products and the final product can also be generalized as shown in Figure 6.11. Each cross-product term ($m_i\, q_j$) in this figure can be generated using an AND gate. This requires 16 AND gates to generate all cross-product terms that are summed by the full adders of Figure 6.12(b) drawn using the basic cell of Figure 6.12(a).

Consider the generation of p_2 in Figure 6.12. From Figure 6.11, p_2 is the sum of $m_2 q_0$, $m_1 q_1$ and $m_0 q_2$. The sum of these three elements is obtained by using two full adders. (See column for p_2 in Figure 6.12). The top full-adder in this column generates the sum $m_2q_0 + m_1q_1$. This sum is then added to m_0q_2 by the bottom full-adder along with any carry from the previous full-adder for p_1.

The time required to complete the multiplication can be estimated by considering the longest carry propagation path comprising of the rightmost diagonal (which includes the full-adder for p_1 and the bottom full-adders for p_2 and p_3), and the last row (which includes the full-adder for p_6 and the bottom full-adders for p_4 and p_5). The time taken to multiply two n-bit numbers can be expressed as follows:

$$T(n) = \Delta_{ANDgate} + (n - 1) \Delta_{\text{carry propagation}} + (n - 1) \Delta_{\text{carry propagation}}$$

In this equation, all cross-product terms m_iq_i can be generated simultaneously by an array of AND gates. Therefore, only one AND gate delay is included in the equation. Also, the rightmost diagonal and the bottom row contain (n - 1) full-adders each for the n × n multiplier.

m_3	m_2	m_1	m_0
q_3	q_2	q_1	q_0

FIGURE 6.11   Generalized Version of the Multiplication of Two 4-bit Numbers Using the Paper and Pencil Algorithm.

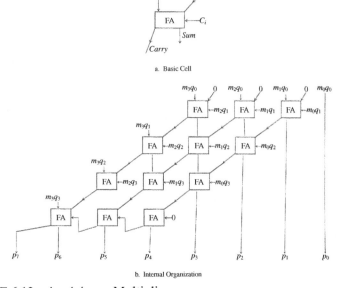

a. Basic Cell

b. Internal Organization

FIGURE 6.12   4 × 4 Array Multiplier.

Assuming that $\Delta_{\text{AND gate}} = \Delta_{\text{carry propogation}} = $ *Two gate delays* $= 2\Delta$, the preceding expression can be simplified as shown:

$T(n) = 2\Delta + (2n - 2)2\Delta = (4n - 2)\Delta$

The array multiplier that has been considered so far is known as Braun's multiplier. The hardware is often called a *nonadditive multiplier* (NM), since it does not include any additive inputs. An additive multiplier (AM) includes an extra input R; it computes products of the form P = M * Q + R. This type of multiplier is useful in computing the sum of products of the form XiYi.

As mentioned in Chapter 2, signed multiplication can be performed using various algorithms. A simple algorithm follows.

In the case of signed numbers, there are three possibilities:
1. *M* and *Q* are in sign-magnitude form.
2. *M* and *Q* are in ones complement form.
3. *M* and *Q* are in twos complement form.

For the first case, perform unsigned multiplication of the magnitudes without the sign bits. The sign bit of the product is determined as $M_n \oplus Q_n$, where M_n and Q_n are the most significant bits (sign bits) of the multiplicand (*M*) and the multiplier (*Q*), respectively. For the second case, proceed as follows:

Step 1: If $M_n = 1$, then compute the ones complement of *M*.
Step 2: If $Q_n = 1$, then compute the ones complement of *Q*.
Step 3: Multiply the $n - 1$ bits of the multiplier and the multiplicand.
Step 4: $S_n = M_n \oplus Q_n$
Step 5: If $S_n = 1$, then compute the ones complement of the result obtained in
       Step 3.

Whenever the ones complement of a negative number (sign bit = 1) is taken, the sign is reversed. Hence, with respect to the multiplier, the inputs are always a positive quantity. When the sign of the bit is negative, however ($M_n \oplus Q_n = 1$), the result must be presented in the ones complement form. This is why the ones complement of the product found by the unsigned multiplier is computed. When *M* and *Q* are in twos complement form, the same procedure is repeated, with the exception that the twos complement must be determined when $Q_n = 1$, $M_n = 1$, or $M_n \oplus Q_n = 1$. Consider *M* and *Q* as twos complement numbers. Suppose $M = 1100_2$ and $Q = 0111_2$. Because $M_n = 1$, take the twos complement of $M = 0100_2$; because $Q_n = 0$, do not change *Q*. Multiply 0111_2 and 0100_2 using the unsigned multiplication method discussed before. The product is 00011100_2. The sign of the product $S_n = M_n \oplus Q_n = 1 \oplus 0 = 1$. Hence, take the twos complement of the product 00011100_2 to obtain 11100100_2, which is the final answer: -28_{10}.

As mentioned in Chapter 2, unsigned division can be performed using repeated subtraction. However, the general equation for division can be used for signed division. Note that the general equation for division is *Dividend = Quotient* Divisor + Remainder*. For example, consider dividend $= -9$, divisor = 2. Three possible solutions are shown next.

(a) $-9 = -4 * 2 - 1$, Quotient $= -4$, Remainder $= -1$.
(b) $-9 = -5 * 2 + 1$, Quotient $= -5$, Remainder $= +1$.
(c) $-9 = -6 * 2 + 3$, Quotient $= -6$, Remainder $= +3$.

However, the correct answer is shown in (a) in which, Quotient $= -4$ and Remainder $= -1$. Hence, for signed division, the sign of the remainder is the same as the sign of the dividend, unless the remainder is zero. Typical 32-bit microprocessors such as the Pentium follow this convention.

6.1.3 ALU Design

In this section, the design of a simple ALU using typical combinational elements such as gates, multiplexers, and a 4-bit binary adder is discussed. First, an arithmetic unit and a logic unit are designed separately; then they are combined to obtain an ALU.

For the first step, a two-function arithmetic unit, as shown in Figure 6.13 is designed. The key element of this system is the 4-bit binary adder. The multiplexers select one of the inputs (Y or $\overline{Y}$) of the binary adder. For example, if $s_0 = 0$, then $B = Y$; otherwise $B = \overline{Y}$. Because the selection input (s_0) also controls the input carry (c_{in}), the following results are obtained:

if $s_0 = 0$ then $F = X$ plus Y
else $F = X$ plus $\overline{Y}$ plus 1
    $= X$ minus Y

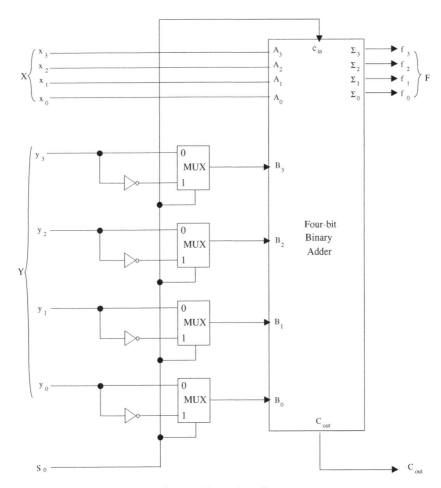

FIGURE 6.13   Organization of an arithmetic unit.

This arithmetic unit generates addition and subtraction operations. For the second step, let us design a two-function logic unit as shown in Figure 6.14. From the figure, it can be seen that when $s_0 = 0$, the output $G = X$ AND Y; otherwise the output $G = X \oplus Y$.

The outputs generated by the arithmetic and logic units can be combined by using a set of multiplexers, as shown in Figure 6.15. From the figure, when the select line $s_1 = 1$, the multiplexers select outputs generated by the logic unit; otherwise, the outputs of the arithmetic unit are selected.

More commonly, the select line, s_1, is referred to as the *mode input* because it selects the desired mode of operation (arithmetic or logic). A complete block diagram schematic of this ALU is shown in Figure 6.16. The truth table illustrating

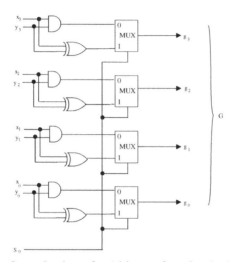

FIGURE 6.14   Organization of a 4-bit two-function logic unit.

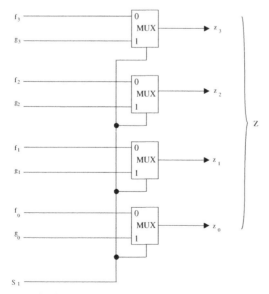

FIGURE 6.15   Combining the outputs generated by the arithmetic and logic units.

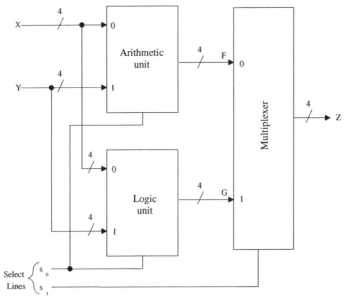

FIGURE 6.16   Schematic representation of the four functions.

the operation of this ALU is shown in Table 6.3. This table shows that this ALU is capable of performing two arithmetic and two logic operations on 4-bit data X and Y.

Note that the size of the ALU defines the size of a microcontroller. For example, the Microchip PIC18F4321 is an 8-bit microcontroller since it contains an 8-bit ALU.

6.1.4   Control Unit Design

The main purpose of the control unit is to translate or decode instructions and generate appropriate enable signals to accomplish the desired operation. Based on the contents of the instruction register, the control unit sends the selected data items to the appropriate processing hardware at the right time. The control unit drives the associated processing hardware by generating a set of signals that are synchronized with a master clock.

The control unit performs two basic operations: instruction interpretation and instruction sequencing. In the interpretation phase, the control unit reads (fetches) an instruction from the memory addressed by the contents of the program counter into the instruction register. The control unit inputs the contents of the instruction register. It recognizes the instruction type, obtains the necessary operands, and routes them to the appropriate functional units of the execution unit (registers and ALU). The control unit then issues necessary signals to the execution unit to perform the desired operation and routes the results to the specified destination.

TABLE 6.3   Truth table selecting the operations of the ALU of Figure 6.16

Select Lines		Output Z	Comment
s_1	s_0		
0	0	X plus Y	Addition
0	1	X plus $\overline{Y}$ Plus 1	2's Complement subtraction
1	0	X ∧ Y	Boolean AND
1	1	X ⊕ Y	Exclusive-OR

In the sequencing phase, the control unit generates the address of the next instruction to be executed and loads it into the program counter. To design a control unit, one must be familiar with some basic concepts such as register transfer operations, types of bus structures inside the control unit, and generation of timing signals. These are described in the next section.

There are two methods for designing a control unit: hardwired control and microprogrammed control. In the hardwired approach, synchronous sequential circuit design procedures are used in designing the control unit. Note that a control unit is a clocked sequential circuit. The name "hardwired control" evolved from the fact that the final circuit is built by physically connecting the components such as gates and flip-flops. In the microprogrammed approach, on the other hand, all control functions are stored in a ROM inside the control unit. This memory is called the *control memory*. RAMs and PALs are also used to implement the control memory. The words in this memory are called *control words*, and they specify the control functions to be performed by the control unit. The control words are fetched from the control memory and the bits are routed to appropriate functional units to enable various gates. An instruction is thus executed. Design of control units using microprogramming (sometimes called *firmware* to distinguish it from hardwired control) is more expensive than using hardwired controls. To execute an instruction, the contents of the control memory in microprogrammed control must be read, which reduces the overall speed of the control unit. The most important advantage of microprogramming is its flexibility; many changes can be made by simply altering the microprogram in the control memory. A small change in the hardwired approach may lead to redesigning the entire system.

There are two types of microcontroller/microprocessor architectures: CISC (Complex Instruction Set Computer) and RISC (Reduced Instruction Set Computer). CISC microcontrollers contain a large number of instructions and many addressing modes while RISC microcontrollers include a simple instruction set with a few addressing modes. Almost all computations can be obtained from a few simple operations. RISC basically supports a small set of commonly used instructions which are executed at a fast clock rate compared to CISC which contains a large instruction set (some of which are rarely used) executed at a slower clock rate. In order to implement fetch /execute cycle for supporting a large instruction set for CISC, the clock is typically slower. In CISC, most instructions can access memory while RISC contains mostly load/store instructions. The complex instruction set of CISC requires a complex control unit, thus requiring microprogrammed implementation. RISC utilizes hardwired control which is faster. CISC is more difficult to pipeline while it is easier to implement pipelining with RISC. An advantage of CISC over RISC is that complex programs require fewer instructions in CISC with a fewer fetch cycles while the RISC requires a large number of instructions to accomplish the same task with several fetch cycles. However, RISC can significantly improve its performance with a faster clock, more efficient pipelining and compiler optimization. Microchip Technology PIC18F and Motorola/Freescale HC11 utilize RISC and CISC architectures respectively. Intel Pentium microprocessor family, on the other hand, utilizes a combination of RISC and CISC architectures for providing high performance. The Pentium uses RISC (hardwired control) to implement efficient pipelining for simple instructions. CISC (microprogrammed control) for complex instructions is utilized by the Pentium to provide upward compatibility with the Intel 8086/80X86 family.

Basic Concepts Register transfer notation is the fundamental concept associated with the control unit design. For example, consider the register transfer operation of Figure 6.17. The contents of 16-bit register R_0 are transferred to 16-bit register R_1 as described by the following notation: $R_1 \leftarrow R_0$

The symbol $\leftarrow$ is called the *transfer operator*. However, this notation does not indicate the number of bits to be transferred. A declaration statement specifying the size of each register is used for the purpose:

```
Declare registers R0 [16], R1 [16]
```

The register transfer notation can also be used to move a specific bit from one register to a particular bit position in another. For example, the statement

$$R_1 [1] \leftarrow R_0 [14]$$

means that bit 14 of register R_0 is moved to bit 1 of register R_1.

An enable signal usually controls transfer of data from one register to another. For example, consider Figure 6.18. In the figure, the 16-bit contents of register R_0 are transferred to register R_1 if the enable input E is HIGH; otherwise the contents of R_0 and R_1 remain the same. Such a conditional transfer can be represented as

$$E: R_1 \leftarrow R_0$$

Figure 6.19 shows a hardware implementation of transfer of each bit of R_0 and R_1. The enable input may sometimes be a function of more than one variable. For example, consider the following statement involving three 16-bit registers: If $R_0 < R_1$ and $R_2 [1] = 1$ then $R_1 \leftarrow R_0$.

The condition $R_0 < R_1$ can be determined by an 8-bit comparator such that the output y of the comparator goes to 0 if $R_0 < R_1$. The conditional transfer can then be expressed as follows: $E: R_1 \leftarrow R_0$ where $E = y \cdot R_2 [1]$. Figure 6.20 depicts the hardware implementation.

FIGURE 6.17   16-Bit register transfer from $\boldsymbol{R_0}$ to $\boldsymbol{R_1}$.

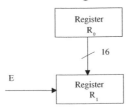

FIGURE 6.18   An enable input controlling register transfer.

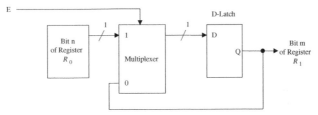

FIGURE 6.19   Hardware for each bit transfer from R_0 to R_1.

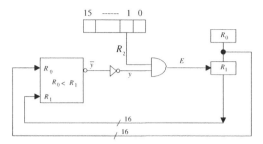

FIGURE 6.20   Hardware implementation E: R1 $\leftarrow R_0$ where E $= y\,R_2$ [1].

A number of wires called *buses* are normally used to transfer data in and out of a digital processing system. Typically, there will be a pair of buses ("inbuses" and "outbuses") inside the CPU to transfer data from the external devises into the processing section and vice versa. Like the registers, these buses are also represented using register transfer notations and declaration statements. For example, "Declare inbus [16] and outbus [16]" indicate that the digital system contains two 16-bit wide data buses (inbus and outbus). $R_0 \leftarrow$ inbus means that the data on the inbus is transferred into register R_0 when the next clock arrives. An equate (=) symbol can also be used in place of $\leftarrow$. For example, "outbus $= R_1$ [15:8]" means that the high-order 8 bits of the 16-bit register R_1 are made available on the outbus for one clock period. An algorithm implemented by a digital system can be described by using a set of register transfer notations and typical control structures such as if-then and go to. For example, consider the description shown in Figure 6.21 for multiplying two 8-bit unsigned numbers (Multiplication of an 8-bit unsigned multiplier by an 8-bit multiplicand) using repeated addition.

The hardware components for the preceding description include an 8-bit inbus, an 8-bit outbus, an 8-bit parallel adder, and three 8-bit registers, R, M, and Q. This hardware performs unsigned multiplication by repeated addition. This is equivalent to unsigned multiplication performed by assembly language instruction.

A distinguishing feature of this description is to describe concurrent operations. For example, the operations $R \leftarrow 0$ and $M \leftarrow$ inbus can be performed simultaneously. As a general rule, a comma is inserted between operations that can be executed concurrently. On the other hand, a semicolon between two transfer operations indicates that they must be performed serially. This restriction is primarily due to the data path provided in the hardware. For example, in the description, because there is only one input bus, the operations $M \leftarrow$ inbus and $Q \leftarrow$ inbus cannot be performed simultaneously. Rather, these two operations must be carried out serially. However, one of these operations may be overlapped with the operation $R \leftarrow 0$ because the

```
Declare registers R[8],M[8],Q[8];
Declare buses inbus[8],outbus[8];
  Start:    R ← 0, M ← inbus;          Clear register R to 0 and move multiplicand
            Q ← inbus;                  Transfer multiplier
  Loop:     R ← R + M, Q ← Q-1;         Add multiplicand
            If Q < > 0 then go to loop;  repeat if Q≠ 0
            Outbus ← R;
  Halt:     Go to Halt;
```

FIGURE 6.21   Register transfer description of 8 × 8 unsigned multiplication
(Assume 8-bit result).

operation does not use the inbus. The description also includes labels and comments to improve readability of the task description. Operations such as $R \leftarrow 0$ and $M \leftarrow$ inbus are called *micro-operations*, because they can be completed in one clock cycle. In general, a computer instruction can be expressed as a sequence of micro-operations.

The rate at which the CPU completes operations such as $R \leftarrow R + M$ is determined by its bus structure inside the CPU. The cost of the CPU increases with the complexity of the bus structure. Three types of bus structures are typically used: single-bus, two-bus, and three-bus architectures. The simplest of all bus structures is the single-bus organization shown in Figure 6.22. At any time, data may be transferred between any two registers or between a register and the ALU. If the ALU requires two operands such as in response to an ADD instruction, the data can only be transferred one at a time. In single-bus architecture, the bus must be multiplexed among various data. Also, the ALU must have buffer registers to hold the transferred data.

In Figure 6.22, an add operation such as $R_0 \leftarrow R_1 + R_2$ is completed in three clock cycles as follows:

First clock cycle:	The contents of R_1 are moved to buffer register B_1 of the ALU.
Second clock cycle:	The contents of R_2 are moved to buffer register B_2 of the ALU.
Third clock cycle:	The sum generated by the ALU is loaded into R_0.

A single-bus structure slows down the speed of instruction execution even though data may already be in the CPU registers. The instruction's execution time is longer if the operands are in memory; two clock cycles may be required to retrieve the operands into the CPU registers from external memory.

To execute an instruction such as ADD between two operands already in register, the control logic in a single-bus structure must follow a three-step sequence. Each step represents a control state. Therefore, a single-bus architecture requires a large number of states in the control logic; hence, more hardware may be needed to design the control unit. Because all data transfers take place through the same bus one at a time, the design effort to build the control logic is greatly reduced.

Next, consider a two-bus architecture, shown in Figure 6.23. All general-purpose registers are connected to both buses (bus A and bus B) to form a two-bus architecture. The two operands required by the ALU are, therefore, routed in one clock cycle. Instruction execution is faster because the ALU does not have to wait for the second operand, unlike the single-bus architecture. The information on a bus may be from a general-purpose register or a special-purpose register. The special-purpose registers are often divided into two groups. Each group is connected to one of the buses. Data from two special-purpose registers of the same group cannot be transferred to the ALU at the same time.

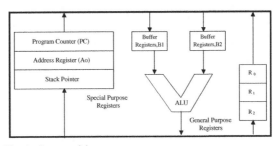

FIGURE 6.22   Single-bus architecture.

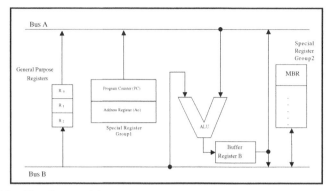

FIGURE 6.23   Two-bus architecture.

In the two-bus architecture, the contents of the program counter are always transferred to the right input of the ALU because it is connected to bus A. Similarly, the contents of the special register MBR (memory buffer register, to hold up data retrieved from external memory) are always transferred to the left input of the ALU because it is connected to bus B.

In Figure 6.23, an add operation such as $R_0 \leftarrow R_1 + R_2$ is completed in two clock cycles as follows:

First clock cycle: The contents of R_1 and R_2 are moved to the inputs of ALU. The ALU then generates the sum in the output register.

Second clock cycle: The sum from the output register is routed to R_0.

The performance of a two-bus architecture can be improved by adding a third bus (bus *C*), at the output of the ALU. Figure 6.24 depicts a typical three-bus architecture. The three-bus architecture perform the addition operation $R_0 \leftarrow R_1 + R_2$ in one cycle as follows:

First cycle: The contents of R_1 and R_2 are moved to the inputs of the ALU via bus A and bus B respectively. The sum generated by the ALU is then transferred to R_0 via bus C.

The addition of the third bus will increase the system cost and also the complexity of the control unit design.

Another important concept required in the design of a control unit is the generation of timing signals. One of the main tasks of a control unit is to properly sequence a set of operations such as a sequence of *n* consecutive clock pulses. To carry out an operation, timing signals are generated from a master clock.

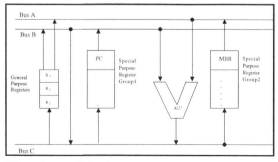

FIGURE 6.24   Three-bus architecture.

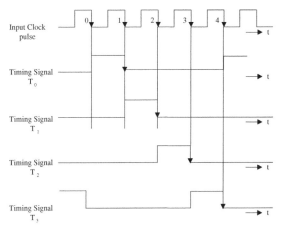

FIGURE 6.25   Timing signals.

Figure 6.25 shows the input clock pulse and the four timing signals T_0, T_1, T_2, and T_3. A ring counter (described in Chapter 5) can be used to generate these timing signals. To carry out an operation P_i at the ith clock pulse, a control unit must count the clock pulses and produce a timing signal T_i.

Hardwired Control Design   The steps involved in hardwired control design are summarized as follows:

1. Derive a flowchart from the problem definition and validate the algorithm by using trial data.
2. Obtain a register transfer description of the algorithm from the flowchart.
3. Specify a processing hardware along with various components.
4. Complete the design of the processing section by establishing the necessary control inputs.
5. Determine a block diagram of the controller.
6. Obtain the state diagram of the controller.
7. Specify the characteristic of the hardware for generating the required timing signals used in the controller.
8. Draw the logic circuit of the controller.

The following example is provided to illustrate the concepts associated with implementation of a typical instruction in a control unit using hardwired control. The unsigned multiplication by repeated addition discussed earlier is used for this purpose. A 4-bit by 4-bit unsigned multiplication will be considered. Assume the result of multiplication is 4 bits.

Step 1:   Derive a flowchart from the problem definition and then validate the algorithm using trial data.

Figure 6.26 shows the flowchart. In the figure, M and Q are two 4-bit registers containing the unsigned multiplicand and unsigned multiplier respectively. Assume that the result of multiplication is 4-bit wide. The 4-bit result of the multiplication called the *product* will be stored in the 4-bit register, R. The contents of R are then output to the outbus. The flowchart in Figure 6.26 is similar to an ASM chart and provides a hardware description of the algorithm. The sequence of events and their

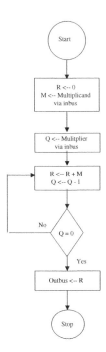

FIGURE 6.26   Flowchart for 4-bit × 4-bit multiplication.

timing relationships are described in the flowchart. For example, the operations, $R \leftarrow 0$ and $M \leftarrow$ multiplicand shown in the same block are executed simultaneously. Note that $M \leftarrow$ multiplicand via inbus and $Q \leftarrow$ multiplier via inbus must be performed serially because both operations use a single input bus for loading data. These operations are, therefore, shown in different blocks. Because $R \leftarrow 0$ does not use the inbus, this operation is overlapped with initialization of M via the inbus. This simultaneous operation is indicated by placing them in the same block.

The algorithm will now be verified by means of a numerical example as shown in Figure 6.27. Suppose $M = 0100_2 = 4_{10}$ and $Q = 0011_2 = 3_{10}$; then $R = $ product $= 1100_2 = 12_{10}$

Step 2: Obtain a register transfer description of the algorithm from the flowchart. Figure 6.28 shows the description of the algorithm.

Step 3: Specify a processing hardware along with various components.

The processing section contains three main components:

• General-purpose registers
• 4-bit adder
• Tristate buffer

Figure 6.29 shows these components. The general-purpose register is a trailing edge-triggered device.

Components of the processing section of 4-bit by 4-bit unsigned multiplication.

Three operations (clear, parallel load, and decrement) can be performed by applying the appropriate inputs at C, L, and D. All these operations are synchronized at the trailing (high to low) edge of the clock pulse.

	R	M	Q
Initialization	0 0 0 0	0 1 0 0	0 0 1 1
Iteration 1 R <-- R + M Q <-- Q - 1	0 1 0 0	0 1 0 0	0 0 1 0
Iteration 2 R <-- R + M Q <-- Q - 1	1 0 0 0	0 1 0 0	0 0 0 1
Iteration 3 R <-- R + M Q <-- Q - 1	1 1 0 0	0 1 0 0	0 0 0 0

Product $= 12_{10}$

FIGURE 6.27   Verification of the unsigned multiplication algorithm.

The 4-bit adder can be implemented using 4-bit adder circuits. The tristate buffer is used to control data transfer to the outbus.

Step 4: Complete the design of the processing section by establishing the necessary control inputs.

Figure 6.30 shows the detailed logic diagram of the processing section, along with the control inputs.

Step 5: Determine a block diagram of the controller. Figure 6.31 shows the block diagram.

The controller has three inputs and seven outputs. The Reset input is an asynchronous input used to reset the controller so that a new computation can begin. The Clock input is used to synchronize the controller's action. All activities are assumed to be synchronized with the trailing edge of the clock pulse.

Step 6: Obtain the state diagram of the controller.

The controller must initiate a set of operations in a specified sequence. Therefore, it is modeled as a sequential circuit. The state diagram of the unsigned multiplier controller is shown in Figure 6.32.

Initially, the controller is in state T_0. At this point, the control signals C_0 and C_1 are HIGH. Operations $R \leftarrow 0$ and $M \leftarrow$ inbus are carried out with the trailing edge

```
Start:  R ← 0, M ← inbus;          Clear Register to 0 and move multiplicand
        Q ← inbus;                 Transfer Multiplier
 Loop:  R ← R + M, Q ← Q -1;       Perform addition, decrement counter
        If Q < > 0 then goto Loop; Repeat if Q ≠ 0
        outbus ← R;
 Halt:  Go to Halt;
```

FIGURE 6.28   Register transfer description 4-bit × 4-bit unsigned multiplication.

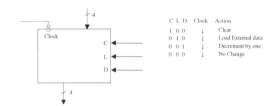

(a)   General Purpose Register

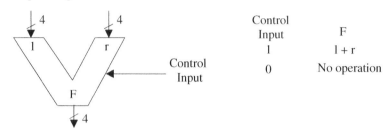

(b)   4-bit Adder

(c)   Tristate Buffer

FIGURE 6.29.

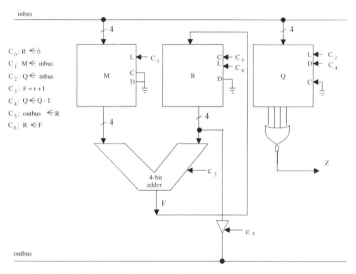

FIGURE 6.30   Detailed logic diagram of the processing section.

of the next clock pulse. The controller moves to state T_1 with this clock pulse. When the controller is in T_2, $R \leftarrow R + M$ and $Q \leftarrow Q - 1$ are performed.

All these operations take place at the trailing edge of the next clock pulse. The controller moves to state T_5 only when the unsigned multiplication is completed.

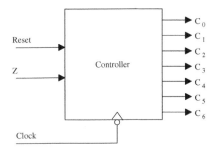

FIGURE 6.31   Block diagram of the unsigned multiplier controller

Control State	Operations Performed	Control Signals to be activated
T_0	$R \leftarrow 0, M \leftarrow$ inbus	C_0, C_1
T_1	$Q \leftarrow$ inbus	C_2
T_2	$R \leftarrow R + M,$ $Q \leftarrow Q - 1$	C_3, C_4, C_6
T_3	None	None
T_4	outbus $\leftarrow$ R	C_5
T_5	None	None

(a) State Diagram   (b) Controller action

FIGURE 6.32   Controller description.

The controller then stays in this state forever. A hardware reset input causes the controller to move to state T_0, and a new computation will start.

In this state diagram, states are selected according to the following guidelines:

- If the operations are independent of each other and can be completed within one clock cycle, they are grouped within one control state. For example, in Figure 6.32, operations $R \leftarrow 0$ and $M \leftarrow$ inbus are independent of each other. With this hardware, they can be executed in one clock cycle. That is, they are microoperations. However, if they cannot be completed within T_0 clock cycle, either clock duration must be increased or the operations should be divided into a sequence of microoperations.

- Conditional testing normally implies introduction of new states. For example, in the figure, conditional testing of Z introduces the new state T_3.

- One should not attempt to minimize the number of states. When in doubt, new states must be introduced. The correctness of the control logic is more important than the cost of the circuit.

Step 7: Specify the characteristics of the hardware for generating the required timing signals.

There are six states in the controller state diagram. Six nonoverlapping timing signals (T_0 through T_5) must be generated so that only one will be high for a clock pulse. For example, Figure 6.33 shows the four timing signals T_0, T_1, T_2, and T_3. A mod-8 counter and a 3-to-8 decoder can be used to accomplish this task. Figure 6.34 shows the mod-8 counter.

Step 8: Draw the logic circuit of the controller.

Figure 6.35 shows the logic diagram of the controller. Figure 6.36 shows the sequence controller (SC) hardware along with its truth table, which sequences the controller according to the state diagram of Figure 6.32.

Consider the logic involved in deriving the entries of the SC truth table. The mod-8 counter is loaded (or initialized) with the specified external data if the counter control inputs C and L are 0 and 1 respectively from Figure 6.34. In this counter, the counter load control input L overrides the counter enable control input E.

From the controller's state diagram of Figure 6.32, the controller counts up automatically in response to the next clock pulse when the counter load control input $L = 0$ because the enable input E is tied to HIGH. Such normal sequencing activity is desirable for the following situations:

- Present control state is T_0, T_1, T_2, T_4.

- Present control state is T_3 and $Z = 1$; the next state is T_4.

The SC must load the counter with appropriate count when the counter is required to load the count out of its normal sequence. For example, from the

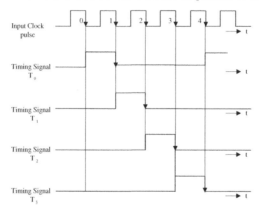

FIGURE 6.33   Timing signals generated by the controller.

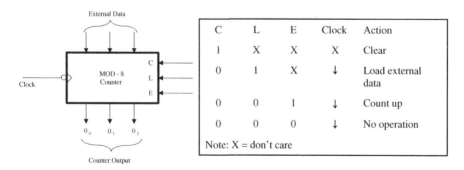

C	L	E	Clock	Action
1	X	X	X	Clear
0	1	X	↓	Load external data
0	0	1	↓	Count up
0	0	0	↓	No operation
Note: X = don't care				

(a) Block Diagram   (b) Function Table

FIGURE 6.34   Characteristics of the counter used in the controller design.

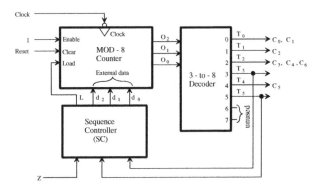

FIGURE 6.35   Logic diagram of the unsigned multiplier controller.

	Inputs			Outputs		
Z	T_3	T_5	L	d_2	d_1	d_0
0	1	x	1	0	1	0
x	x	1	1	1	0	1
Note: x = don't care						

(a)   Truth Table

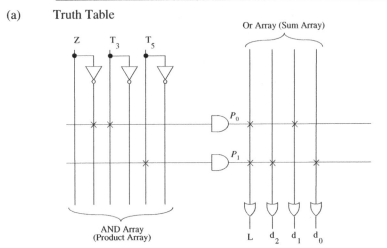

(b)   PLA Implementation

FIGURE 6.36   Sequence controller design.

controller's state diagram of Figure 6.32, if the present control state is T_3 (counter output $O_2O_1O_0 = 011$) and if $Z = 0$, the next state is T_2. When these input conditions occur, the counter must be loaded with external value 010 at the trailing edge of the next clock pulse ($T_2 = 1$ only when $O_2O_1O_0 = 010$. Therefore, the SC generates $L = 1$ and $d_2d_1d_0 = 010$.

Similarly, from the controller's state diagram of Figure 6.32, if the present state is T_5, the next control state is also T_5. The SC must generate the outputs $L = 1$ and $d_2d_1d_0 = 101$. The SC truth table of Figure 6.36(a) shows these out-of-sequence counts.

For each row of the SC truth table of Figure 6.36(a), a product term is generated in the PLA:

$P_0 = \overline{Z}T_3$ and $P_1 = T_5$.

The PLA (Figure 6.36 (b)) generates four outputs: L, d_2, d_1, and d_0. Each output is directly generated by the SC truth table and the product terms. The PLA outputs are as follows:

$$L = P_0 + P_1$$
$$d_2 = P_1$$
$$d_1 = P_0$$
$$d_0 = P_1$$

The controller design is completed by relating the control states (T_0 through T_5) to the control signals (C_0 though C_6) as follows:

$$C_0 = C_1 = T_0$$
$$C_2 = T_1$$
$$C_3 = C_4 = C_6 = T_2$$
$$C_5 = T_4$$

From these equations, when the control is in state T_0 or T_2, multiple micro-operations are performed. Otherwise, when the control is in state T_1 or T_4, a single micro-operation is performed.

The unsigned multiplication algorithm just implemented using hardwired control can be considered as an unsigned multiplication instruction with a typical microcontroller. To execute this instruction, the microcomputer will read (fetch) this multiplication instruction from external memory into the instruction register located inside the CPU. The contents of this instruction register will be input to the control unit for execution. The control unit will generate the control signals C_0 through C_6 as shown in Figure 6.35. These control signals will then be applied to the appropriate components of the processing section in Figure 6.30 at the proper instants of time shown in Figure 6.32. Note that the control signals are physically connected to the hardware elements of Figure 6.30. Thus, the execution of the unsigned multiplication instruction will be completed by a typical microcontroller.

Microprogrammed Control Unit Design   As mentioned earlier, a microprogrammed control unit contains programs written using microinstructions. These programs are stored in a control memory normally in a ROM inside the CPU. To execute instructions, the CPU reads (fetches) each instruction into the instruction register from external memory. The control unit translates the instruction for the CPU. Each control word contains signals to activate one or more microoperations. A program consisting of a set of microinstructions is executed in a sequence of micro-operations to complete the instruction execution. Generally, all microinstructions have two important fields:

- Control word
- Next address

The control field indicates which control lines are to be activated. The next address field specifies the address of the next microinstruction to be executed. The concept of microprogramming was first proposed by W. V. Wilkes in 1951 utilizing a decoder and an 8×8 ROM with a diode matrix. This concept is extended further to include a control memory inside the CPU. The cost of designing a CPU primarily depends on the size of the control memory. The length of a microinstruction, on the other hand, affects the size of the control memory. Therefore, a major design effort is to minimize cost of implementing a microprogrammed CPU by reducing the length of the microinstruction.

The length of a microinstruction is directly related to the following factors:

- The number of micro-operations that can be activated simultaneously. This is called the *degree of parallelism*.

- The method by which the address of the next microinstruction is determined.

All microinstructions executed in parallel can be included in a single microinstruction with a common op-code. The result is a short microprogram. However, the length of the microinstruction increases as parallelism grows.

The control bits in a microinstruction can be organized in several ways. One obvious way is to assign a single bit for each control line. This will provide full parallelism. No decoding of the control field is necessary. For example, consider Figure 6.37 with two registers, X and Y with one outbus.

In figure 6.37, the contents of each register are transferred to the outbus when the appropriate control line is activated:

$$C_0: \text{outbus} \leftarrow X$$
$$C_1: \text{outbus} \leftarrow Y$$

Here, each operation can be performed one at a time because there is only one outbus. A single bit can be assigned to perform each transfer as follows:

Control Bits		Operation Performed
C_0	C_1	
1	0	Outbus $\leftarrow$ X
0	1	Outbus $\leftarrow$ Y
0	0	No operation

This method is called *unencoded format*.

The three operations can be implemented using two bits and a 2-to-4 decoder as shown in Figure 6.38. This is called *encoded format*. The relationship between the encoded and actual control information is as follows:

Encoded Bits		Operation Performed
d_1	d_0	
0	0	No operation
0	1	Outbus $\leftarrow x$
1	0	Outbus $\leftarrow y$

Note that a 5-bit control field is required for five operations. However, three encoded bits are required for five operations using a 3 to 8 decoder. Hence, the encoded format typically provides a short control field and thus results in short microinstructions. However, the need for a decoder will increase the cost. Therefore, there is a trade-off between the degree of parallelism and the cost. Microinstructions can be classified

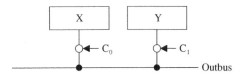

FIGURE 6.37   An example of a register transfer.

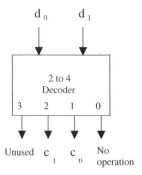

FIGURE 6.38   Encoded format.

into two groups: horizontal and vertical. The horizontal microinstruction mechanism provides long microinstructions, a high degree of parallelism, and little or no encoding. The vertical instruction method, on the other hand, offers short microinstructions, limited parallelism, and considerable decoding.

   Microprogramming is the technique of writing microprograms in a microprogrammed control unit. Writing microprograms is similar to writing assembly language programs. Microprograms are basically written in a symbolic language called *microassembly language*. These programs are translated by a microassembler to generate microcodes, which are then stored in the control memory.

   In the early days, the control memory was implemented using ROMs. However, these days control memories are realized in writeable memories. This provides the flexibility of interpreting different instruction set by rewriting the original microprogram, which allows implementation of different control units with the same hardware. Using this approach, one CPU can interpret the instruction set of another CPU. The design of a microprogrammed control unit is considered next. The 4-bit × 4-bit unsigned multiplication using hardwired control (presented earlier) is implemented by microprogramming. The register transfer description shown in Figure 6.28 is rewritten in symbolic microprogram language as shown in Figure 6.39. Note that the unsigned 4-bit × 4-bit multiplication uses repeated addition. The result (product) is assumed to be 4 bits wide.

Control Memory Address		Operations Performed
0	START	R ← 0, M ← inbus;
1		Q ← inbus;
2	LOOP	R ← R + M, Q ← Q - 1;
3		If Z = 0 then goto Loop;
4		outbus ← R;
5	HALT	Go to HALT

FIGURE 6.39   Symbolic microprogram for 4-bit × 4-bit unsigned multiplication
              using repeated addition.

To implement the microprogram, the hardware organization of the control unit shown in Figure 6.40 can be used. The various components of the hardware of Figure 6.39 are described in the following:

1. **Microprogram Counter (MPC).** The MPC holds the address of the next microinstruction to be executed. It is initially loaded from an external source to point to the starting address of the microprogram. The MPC is similar to the program counter (PC). The MPC is incremented after each microinstruction fetch. If a branch instruction is encountered, the MPC is loaded with the contents of the branch address field of the microinstruction.

2. **Control Word Register (CWR).** Each control word in the control memory in this example is assumed to contain three fields: condition select, branch address, and control function. Each microinstruction fetched from the Control Memory is loaded into the CWR. The organization of the CWR is same for each control word and contains the three fields just mentioned. In the case of a conditional branch microinstruction, if the condition specified by the condition select field is true, the MPC is loaded with the branch address field of the CWR; otherwise, the MPC is incremented to point to the next microinstruction. The control function field contains the control signals.

3. **MUX (Multiplexer).** The MUX is a condition select multiplexer. It selects one of the external conditions based on the contents of the condition select field of the microinstruction fetched into the CWR.

In Figure 6.40, a two-bit condition select field is required as follows:

Condition Select Field		Interpretation
0	0	No branching (no condition)
0	1	Branch if $Z = 0$
1	0	Unconditional branching

From Figure 6.39, six control memory address (addresses 0 through 5) are required for the control memory to store the microprogram. Therefore, a three-bit address

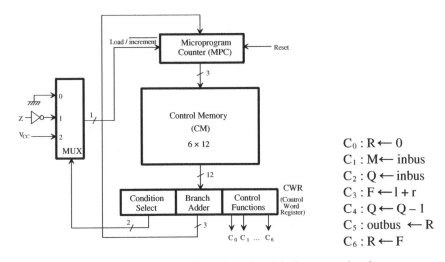

$C_0 : R \leftarrow 0$
$C_1 : M \leftarrow \text{inbus}$
$C_2 : Q \leftarrow \text{inbus}$
$C_3 : F \leftarrow 1 + r$
$C_4 : Q \leftarrow Q - 1$
$C_5 : \text{outbus} \leftarrow R$
$C_6 : R \leftarrow F$

FIGURE 6.40 Microprogrammed unsigned multiplier control unit.

is necessary for each microinstruction. Hence, three bits for the branch address field are required. From Figure 6.40, seven control signals (C_0 through C_6) are required. Therefore, the size of the control function field is 7 bits wide. Thus, the size of each control word can be determined as follows:

$$\text{size of } a \text{ control word} = \text{size of the condition select field} + \text{size of the branch address field} + \text{number of control signals}$$

$$= 2 + 3 + 7$$

$$= 12 \text{ bits}$$

Therefore, the size of the control memory is 6 bits × 12 bits because the microprogram requires six addresses (0 through 5) and each control word is 12 bits wide. The size of the CWR is 12 bits. The complete binary listing of the microprogram is shown in Figure 6.41.

Let us now explain the binary program. Consider the first line of the program. The instruction contains no branching. Therefore, the condition select field is 00. The contents of the branch in this case filled with 000. In the control function field, two micro-operations, C_0 and C_1, are activated. Therefore, both C_0 and C_1 are set to 1; C_2 through C_6 are set to 0.

This results in the following binary microinstruction shown in the first line (address 0) of Figure 6.41:

Condition Select	Branch Address	Control Function
00	000	1100000

Next, consider the conditional branch instruction of Figure 6.40. This microinstruction implements the conditional instruction "If $Z = 0$ then go to address 2." In this case, the microinstruction does not have to activate any control signal of the control function field. Therefore, C_0 through C_6 are zero. The condition select field is 01 because the condition is based on $Z = 0$. Also, if the condition is true ($Z = 0$), the program branches to address 2. Therefore, the branch address field contains 010_2. Thus, the following binary microinstruction is obtained:

Condition Select	Branch Address	Control Function
01	010	000000

ROM Address			Control Word											Comments
In decimal	In binary			Condition Select		Branch Address		Control Function							
								C_0	C_1	C_2	C_3	C_4	C_5	C_6	
0	0	0	0	0	0	0	0	0	1	1	0	0	0	0	0	R←0, M← inbus
1	0	0	1	0	0	0	0	0	0	0	1	0	0	0	0	Q← inbus
2	0	1	0	0	0	0	0	0	0	0	0	1	1	0	1	R ←R + M, Q←Q –1, R←F
3	0	1	1	0	1	0	1	0	0	0	0	0	0	0	0	If Z = 0 then go to address 2 (loop)
4	1	0	0	0	0	0	0	0	0	0	0	0	0	1	0	outbus ←R
5	1	0	1	1	0	1	0	1	0	0	0	0	0	0	0	Go to address 5 (HALT)

FIGURE 6.41   Binary listing of the microprogram for 4-bit × 4-bit unsigned multiplication.

The other lines in the binary representation of the microprogram can be explained similarly. To execute an unsigned multiplication instruction implemented using the repeated addition just described, a microprogrammed microprocessor will fetch the instruction from external memory into the instruction register. To execute this instruction, the microcontroller uses the control unit of Figure 6.40 to generate the control word based on the microprogram of Figure 6.41 stored in the control memory. The control signals C_0 through C_6 of the control function field of the CWR will be connected to appropriate components of Figure 6.30 The instruction will thus be executed by the microcontroller.

By examining the microprogram in Figure 6.41, it is obvious that the control function field contains all zeros in case of branch instructions. In a typical microprogram, there may be several conditional and unconditional branch instructions. Therefore, a lot of valuable memory space inside the control unit will be wasted if the control field is filled with zeros. In practice, the format of the control word is organized in a different manner to minimize its size. This reduces the implementation cost of the control unit. Whenever there are several branch instructions, the microinstructions can be formatted by using a method called *multiple microinstruction format*. In this approach, the microinstructions are divided into two groups: operate and branch instructions.

An operate instruction initiates one or more microoperations. For example, after the execution of an operate instruction, the MPC will be incremented by 1. In the case of a branch instruction, no microoperation will usually be initiated, and the MPC may be loaded with a new value.

This means that the branch address field can be removed from the microinstruction format. Therefore, the control function field is used to specify the branch address itself. Typically, each microinstruction will have two fields, as shown next:

CONDITION-SELECT FIELD		CONTROL FUNCTION FIELD						
S_1	S_0	C_6	C_5	C_4	C_3	C_2	C_1	C_0

If $S_1 S_0 = 00$, the microinstruction is considered as an operate instruction, and the contents of the control function field are treated as the control signals. Assume the Condition Select Field is encoded as follows:

S_1	S_0	
0	0	No branch
0	1	Branch if cond-1 = 1
1	1	Branch if cond-2 = 1
1	0	Unconditional branch

If $S_1 S_0 = 01$, the instruction is regarded as a branch instruction, and the contents of the control field are assumed to be a 7-bit branch address. In this example, it is assumed that when $S_1 S_0 = 01$, the MPC will be loaded with the appropriate address specified by $C_6 C_5 C_4 C_3 C_2 C_1 C_0$ if the condition $Z = 0$ is satisfied; on the other hand, if $S_1 S_0 = 10$, an unconditional branch to the address specified by the control function / branch address field occurs.

In order to illustrate this concept, the microprogram for 4-bit by 4-bit unsigned multiplication of Figure 6.41 is rewritten using the multiple instruction format as shown in Figure 6.42.

ROM Address	Control Word									Comments	
	Condition select field		Control Function / Branch Address Field							Instruction Type	Operation Performed
	S_1	S_0	C_6	C_5	C_4	C_3	C_2 / br_2	C_1 / br_1	C_0 / br_0		
0 0 0	0	0	0	0	0	0	0	1	1	Operate	$R \leftarrow 0$ $M \leftarrow$ Inbus
0 0 1	0	0	0	0	0	0	1	0	0	Operate	$Q \leftarrow$ Inbus
0 1 0	0	0	1	0	1	1	0	0	0	Operate	$R \leftarrow R+M$ $Q \leftarrow Q-1$ $R \leftarrow F$
0 1 1	0	1	0	0	0	0	0	1	0	Branch	If Z=0 Then go to address 2 (loop)
1 0 0	0	0	0	1	0	0	0	0	0	Operate	Outbus $\leftarrow R$
1 0 1	1	0	0	0	0	0	1	0	1	Branch	Go to Address 5(halt)

FIGURE 6.42   Reduction of the length of microinstruction of Figure 6.41.

It can be seen from the figure 6.42 that the total size of the control store is 54 bits ($6 \times 9 = 54$). In contrast, the control store of figure 6.41 contains 72 bits. For large microprograms with many branch instructions, tremendous memory savings can be accomplished using the multiple microinstructon format. Addresses 0, 1, 2, and 4 contain microinstructions with the contents of the conditional select field as 00, and are considered as operate instructions. In this case, the contents of the control function field are directed to the processing hardware.

Address 3 contains a conditional branch instruction since the contents of the condition select field are 01; while address 5 contains an unconditional branch instruction (halt instruction; that is, jump to the same address) since the condition select field is 10. Hence, the 7-bit control function field directly specifies the desired branch addresses 2 and 5, respectively. Figure 6.43 shows the hardware schematic.

Design of a Microprogrammed CPU   Next, the design of a microprogrammed processor is illustrated. The programming model of this processor is shown in Figure 6.44.
The CPU contains two registers:
1. An 8-bit register A   2. A two-bit flag register F

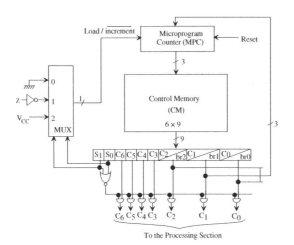

FIGURE 6.43   Microprogrammed Controller for the Microprogram of Figure 6.42.

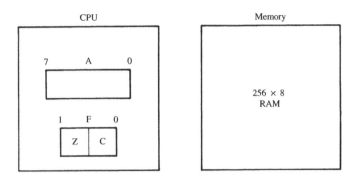

FIGURE 6.44   Programming Model of a Simple Processor.

The flag register holds only zero (Z) and carry (C) flags. All programs and data are stored in the 256×8 RAM. The detailed hardware schematic of the data-flow part of this processor is shown in Figure 6.45.

From the figure, it can be seen that the hardware organization includes four more 8-bit registers, PC, IR, MAR, and BUFFER. These registers are transparent to a programmer. The 8-bit register BUFFER is used to hold the data that is retrieved from memory. In this system, only a restricted number of data paths are available. These paths are controlled by the control inputs C_0 through C_9, as defined in Table 6.4.

The eight ALU operations performed by the CPU are defined by $C_{10}C_{11}C_{12}$ as follows:

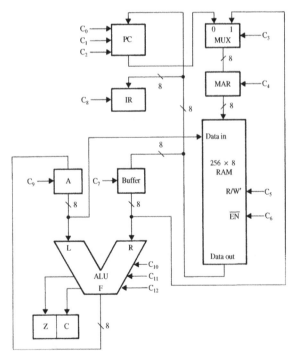

FIGURE 6.45   Hardware Schematic of the Simple Processor (Note: 8-bit PC is connected to eight 2 to 1 MUXs–Not shown above).

TABLE 6.4   Definitions of the Control Inputs C_0-C_9

MICROOPERATION	COMMENT
C_0: PC ← 0	Clear PC to zero.
C_1: PC ← PC + 1	Advance the PC.
$C_2 C_5 \overline{C_6}$: PC ← M ((MAR))	Read the data from the memory and save it in the PC.
$\overline{C_3} C_4$: MAR ← PC	Transfer the contents of the PC into MAR.
$C_5 \overline{C_6} C_7$: BUFFER ← M ((MAR))	Read the data from the memory and save the result in BUFFER.
$C_3 C_4$: MAR ← BUFFER	Transfer the content of the BUFFER into MAR.
$C_5 \overline{C_6} C_8$: IR ← M ((MAR))	Read the data from memory and save the result into IR.
C_9: A ← F	Transfer the ALU output into the A register.
$\overline{C_5} \ \overline{C_6}$: M ((MAR)) ← A	Save contents of register A into memory.

C_{10}	C_{11}	C_{12}	F
0	0	0	0
0	0	1	R
0	1	0	L+R
0	1	1	L-R
1	0	0	L+1
1	0	1	L-1
1	1	0	L AND R
1	1	1	NOT L

Figure 6.46 shows that the proposed instruction set contains 11 instructions. The first 7 instructions are classified as memory reference instructions, since they all require a memory address (which is an 8-bit number in this case). The last 4 instructions do not require any memory address; they are called *nonmemory reference instructions*. Each memory reference instruction is assumed to occupy 2 consecutive bytes in the RAM. The first byte is reserved for the op-code, and the second byte indicates the 8-bit memory address. In contrast, a nonmemory reference instruction takes only one byte of storage. This instruction set supports only two addressing modes: implicit and direct. Both branch instructions are assumed to be absolute mode branch instructions. The op-code encoding for this instruction set is carried out in a logical manner, as explained in Figure 6.47.

The bit I3 of Figure 6.47 decides the instruction type. If I3 = 1, it is a memory reference instruction (MRI), otherwise it is a nonmemory reference instruction (NMRI).

Within the memory reference category, instructions are classified into four groups, as follows:

GROUP NO.	INSTRUCTIONS
0	Load and store
1	Add and subtract
2	Jumps
3	Logical

There are two instructions in the first three groups. Bit I0 is used to determine the desired instruction of a particular group. If I0 of group 0 equals zero, it is the load (LDA) instruction; otherwise it is the store (STA) instruction. Nevertheless, no such classification is required for group 3 and the nonmemory reference instructions.

General Format	Instruction Length in Bytes	Object Code — In binary	Object Code — In hex	Instruction Type	Operation	Comment
LDA ‹addr›	2	0000 1000	08	MRI	A ← M (‹addr›)	Load register A direct
		‹addr8›	‹addrH›			
STA ‹addr›	2	0000 1001	09	MRI	M (‹addr›) ← A	Store register A direct
		‹addr8›	‹addrH›			
ADD ‹addr›	2	0000 1010	0A	MRI	A ← A + M (‹addr›)	Add register A direct
		‹addr8›	‹addrH›			
SUB ‹addr›	2	0000 1011	0B	MRI	A ← A – M (‹addr›)	Subtract register A direct
		‹addr8›	‹addrH›			
JZ ‹addr›	2	0000 1100	0C	MRI	If Z=1 then PC ←‹addr› else PC ← PC + 1	Jump on zero flag set
		‹addr8›	‹addrH›			
JC ‹addr›	2	0000 1101	0D	MRI	If C=1 then PC ←‹addr› else PC ← PC + 1	Jump on carry flag set
		‹addr8›	‹addrH›			
AND ‹addr›	2	0000 1110	0E	MRI	A ← A ^ M (‹addr›)	And register A direct
		‹addr8›	‹addrH›			
CMA	1	0000 0000	00	NMRI	A ← $\overline{A}$	Complement register A
INCA	1	0000 0010	02	NMRI	A ← A + 1	Increment register A
DCRA	1	0000 0100	04	NMRI	A ← A – 1	Decrement register A
HLT	1	0000 0110	06	NMRI	Halt	Halt CPU.

‹addr8›: 8-bit memory address in binary   MRI: memory reference instruction
‹addrH›: 8-bit memory address in hex   NMRI: nonmemory reference instruction.

FIGURE 6.46   Instruction Set to be Implemented.

As mentioned before, the instruction execution involves the following steps:

Step 1: Fetch the instruction.
Step 2: Decode the instruction to find out the required operation.
Step 3: If the required operation is a halt operation, then go to Step 6; otherwise continue.
Step 4: Retrieve the operands and perform the desired operation.
Step 5: Go to Step 1.
Step 6: Execute an infinite LOOP.

The first step is known as the *fetch cycle*, and the rest are collectively known as the *execution cycle*. To decode the instructions, the hardware shown in Figure 6.48 is used.

With this hardware and the status flags (Z and C), a microprogram to implement the instruction set can be written. The symbolic version of this microprogram is shown in Figure 6.49.

Mnemonic	Op-code Bit and Their Interpretations							
					TC	GN		SC
	I7	I6	I5	I4	I3	I2	I1	I0
LDA	0	0	0	0	1	0	0	0
STA	0	0	0	0	1	0	0	1
ADD	0	0	0	0	1	0	1	0
SUB	0	0	0	0	1	0	1	1
JZ	0	0	0	0	1	1	0	0
JC	0	0	0	0	1	1	0	1
AND	0	0	0	0	1	1	1	0
CMA	0	0	0	0	0	0	0	0
INCA	0	0	0	0	0	0	1	0
DCRA	0	0	0	0	0	1	0	0
HLT	0	0	0	0	0	1	1	0

Note:

TC: Type classifier (if I3 = 1, then it is a MRI; otherwise it is a NMRI)

GN: Group number within a type

(I2   I1   Group no.

0   0   0

0   1   1

1   0   2

1   1   3)

SC: Subcategory within a group

FIGURE 6.47   Op-code Encoding Logic.

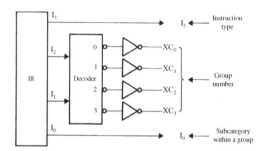

FIGURE 6.48   Instruction-decoding Hardware.

No attempt has been made toward arriving at a minimal microprogram. Rather, the concept was presented. Figure 6.50 shows the hardware for the microprogrammed control unit.

6.2   Memory Organization

The memory system can be divided into three groups:

- Processor memory
- Primary or main memory
- Secondary memory

Symbolic Microprogram:

ROM Address

0		PC ←0;	These operations constitute the fetch cycle.
1	FETCH	MAR PC;	
2		IR ←M ((MAR)), PC ←PC + 1;	
3		IF I$_3$ = 1 then go to MEMREF;	
4		IF XC$_0$ = 1 then go to CMA;	Here we decode the instructions.
5		IF XC$_1$ = 1 then go to INCA;	
6		IF XC$_2$ = 1 then go to DCRA;	
7		Go to HALT;	
8	CMA	A←$\overline{A}$;	Execute CMA instructions.
9		Go to FETCH;	
10	INCA	A ←A + 1;	Execute INCA instruction.
11		Go to FETCH;	
12	DCRA	A ←A - 1;	Execute DCRA instruction.
13		Go to FETCH;	
14	MEMREF	IF XC$_0$ = 1 then go to LDSTO;	Here we branch to the various groups of the memory
15		IF XC$_1$ = 1 then go to ADSUB;	reference instruction.
16		IF XC$_2$ = 1 then go to JMPS;	
17	AND	MAR←PC;	
18		BUFFER ←M ((MAR)), PC ←PC + 1;	Execute AND instruction.
19		MAR ←BUFFER;	
20		BUFFER ←M ((MAR));	
21		A ←A Λ BUFFER;	
22		Go to FETCH;	
23	LDSTO	MAR ←PC;	
24		BUFFER ←M ((MAR)), PC ←PC + 1;	
25		MAR ←BUFFER;	
26		IF I$_0$ = 1 then go to STO;	
27	LOAD	BUFFER ←M ((MAR));	
28		A ←BUFFER;	
29		Go to FETCH;	
30	STO	M ((MAR)) ←A;	
31		Go to FETCH;	
32	ADSUB	MAR ←PC;	
33		BUFFER←M ((MAR)), PC←PC + 1;	
34		MAR ←BUFFER ;	
35		BUFFER ←M ((MAR));	
36		IF I$_0$ = 1 then go to SUB;	
37	ADD	A ←A + BUFFER;	Execute ADD instruction
38		Go to FETCH;	
39	SUB	A ←A – BUFFER;	Execute SUB instruction
40		Go to FETCH;	
41	JMPS	MAR ←PC;	
42			
43		IF I$_0$ = 1 then go to JOC;	
44	JOZ	IF Z = 1 then go to LOADPC;	Execute JZ instruction
45		PC←PC + 1;	
46		Go to FETCH;	
47	JOC	IF C = 1 then go to LOADPC;	Execute JC instruction
48		PC←PC + 1;	
49		Go to FETCH;	
50	LOADPC	PC ←M((MAR));	
51		Go to FETCH;	
52	HALT	Go to HALT;	Execute HALT instruction

FIGURE 6.49   Symbolic Microprogram that implements the instruction set of
figure 6.46.

Processor memory refers to a set of CPU registers described before. These registers are used to hold temporary results when a computation is in progress. Also, there is no speed disparity between these registers and the CPU because they are fabricated using the same technology. However, the cost involved in this approach limits a microcontroller architect to include only a few registers in the CPU.

Main memory is the storage area in which all programs are executed. The CPU can directly access only those items that are stored in main memory. Therefore, all programs must be within the main memory prior to execution. CMOS technology

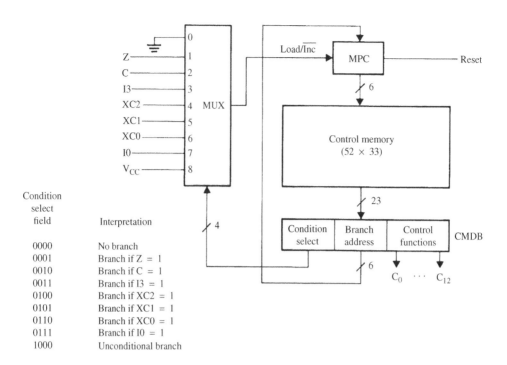

Condition select field	Interpretation
0000	No branch
0001	Branch if $Z = 1$
0010	Branch if $C = 1$
0011	Branch if $I3 = 1$
0100	Branch if $XC2 = 1$
0101	Branch if $XC1 = 1$
0110	Branch if $XC0 = 1$
0111	Branch if $I0 = 1$
1000	Unconditional branch

FIGURE 6.50   Microprogrammed Controller for the CPU.

is normally used these days in main memory design. The size of the main memory is usually much larger than processor memory and its operating speed is slower than the processor registers. Main memory normally includes ROMs and RAMs.

Electromechanical memory devices such as hard disks are extensively used as computer's secondary memory and allow storage of large programs at a low cost. These secondary memory devices access stored data serially. Hence, they are significantly slower than the main memory. Popular secondary memories include hard disk. Programs are stored on the disks in files. Secondary memory stores programs in excess of the main memory. Secondary memory is also referred to as *auxiliary* or *virtual* memory. The CPU cannot directly execute programs stored in the secondary memory, so in order to execute these programs, the CPU must transfer them to its main memory by a program called the *operating system*.

Programs in hard disk memories are stored in tracks. A track is a concentric ring of programs stored on the surface of a disk. Each track is further subdivided into several sectors. Each sector typically stores 512 or 1024 bytes of information. All secondary memories use magnetic media except the optical memory, which stores programs on a plastic disk. CD-ROM is an example of a popular optical memory used with microcomputer systems. The CD-ROM is used to store large programs such as a C compiler. Other optical memories include CD-RAM, DVD-ROM and DVD-RAM.

The main memory (also called *primary memory*) is used to store instructions and data. The main memory is an integral part of a microcontroller, and is contained in the same chip as the CPU.

Primary or main memory is the storage area in which all programs are executed. The microcontroller can directly access only those items that are stored in main memory. Therefore, all programs must be in the main memory prior to execution. CMOS technology is normally used in main memory design. Typically microcontrollers such as the PIC18F contain main memory consisting of Flash memory (program memory) and SRAM (data memory).

For 8-bit microcontrollers, the memory is divided into a number of 8-bit units called *memory words*. An 8-bit unit of data is termed a *byte*. Therefore, for an 8-bit microcontroller, *memory word* and *memory byte* mean the same thing. For 16-bit microcontrollers, a word contains two bytes (16 bits). A memory word is identified in the memory by an address. For example, the PIC18F 4321 is an 8-bit microcontroller, and can directly address a maximum of two Megabytes (2^{21}) of program memory space. The data memory address, on the other hand, is 12-bit wide. Hence, the PIC18F family members can directly address data memory of up to 4 Kbytes (2^{12}). This provides a maximum of 2^{12} = 4096 bytes of data memory addresses, ranging from 000 to FFF in hexadecimal.

An important characteristic of a memory is whether it is volatile or nonvolatile. The contents of a volatile memory are lost if the power is turned off. On the other hand, a nonvolatile memory retains its contents after power is switched off. ROM is a typical example of nonvolatile memory. RAM is a volatile memory unless backed up by batteries.

Large areas of data memory require an efficient addressing scheme to make rapid access to any address possible. Ideally, this means that an entire address does not need to be provided for each read or write operation.

For PIC18F, this is accomplished with a RAM banking scheme. This divides the data memory into 16 contiguous banks (bank 0 through 15) of 256 bytes. Depending on the instruction, each location can be addressed directly by its full 12-bit address, or an 8-bit low-order address and a 4-bit Bank Pointer.

Figure 6.51 shows a simplified data memory layout of the PIC18F. In the figure, the high 4 bits of an address specify the bank number. As an example, consider address 0x105 of segment 1. The high 4 bits, 0001, of this address define the location as in bank 1, and the low 8 bits, 0x05, specify the particular address in bank 1.

6.2.1 Types of Main memory

The main memory can be categorized into two types: read-only memory (ROM) and random-access memory (RAM). As shown in Figure 6.52, ROMs and RAMs are then divided into a number of subcategories, which are discussed next.

FIGURE 6.51   PIC18F data memory.

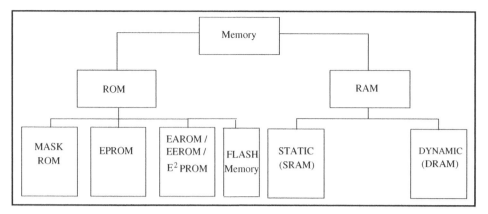

FIGURE 6.52   Summary of available semiconductor memories for typical
microcontrollers.

Read-Only Memory   ROMs (Read-Only memories) can only be read, so is nonvolatile memory. CMOS technology is used to fabricate ROMs. ROMs are divided into two common types: mask ROM and Erasable Programmable ROM (EPROM) such as the 2732, and EAROM (Electrically Alterable ROM) [also called *EEPROM* or *E²PROM* (electrically erasable PROM)] such as the 2864.

Mask ROMs are programmed by a masking operation performed on a chip during the manufacturing process. The contents of mask ROMs are permanent and cannot be changed by the user. EPROMs can be programmed, and their contents can also be altered by using special equipment, called an *EPROM programmer*. When designing a microcontroller for a particular application, permanent programs are stored in ROMs.

EPROMs can be reprogrammed and erased. The EPROM chip must be removed from the system before programming. This memory is erased by exposing the chip to ultraviolet light via a lid or window on the chip. Typical erase times vary between 10 and 20 minutes. The EPROM can be programmed by inserting the EPROM chip into a socket of the EPROM programmer and providing proper addresses and voltage pulses at the appropriate pins of the chip.

EEPROMs can be programmed without removing the memory from the ROM's sockets. These memories are also called *read-mostly memories* (RMMs), because they have much slower write times than read times. Therefore, these memories are usually suited for operations when mostly reading rather that writing will be performed.

Flash Memory   Another type of memory, called *Flash memory* (nonvolatile), invented in the mid-1980s by Toshiba, is designed using a combination of EPROM and E²PROM technologies. Flash memory can be reprogrammed electrically while embedded on the board. One can change multiple bytes at a time. An example of flash memory is the Intel 28F020 (256K × 8-bit). Flash memory is typically used in cellular phones and digital cameras. Note that the PIC18F uses flash memory as its program memory.

Random-Access Memory   There are two types of RAM: static RAM (SRAM), and dynamic RAM (DRAM). *Static RAM* stores data in flip-flops. Therefore, this memory does not need to be refreshed. RAMs are volatile unless backed up by battery. The PIC18F uses SRAM for its data memory.

Dynamic RAM stores data in capacitors. That is, it can hold data for a few milliseconds. Hence, dynamic RAMs are refreshed typically by using external refresh circuitry. Dynamic RAMs (DRAMs) are used in applications requiring large memory. DRAMs have higher densities than static RAMs (SRAMs). Typical examples of DRAMs are the 4464 (64K × 4-bit), 44256 (256K × 4-bit), and 41000 (1M × 1-bit). DRAMs are inexpensive, occupy less space, and dissipate less power than SRAMs. Two enhanced versions of DRAM are EDO DRAM (extended data output DRAM) and SDRAM (synchronous DRAM).

The EDO DRAM provides fast access by allowing the DRAM controller to output the next address at the same time the current data is being read. An SDRAM contains multiple DRAMs (typically, four) internally. SDRAMs utilize the multiplexed addressing of conventional DRAMs. That is, like DRAMs, SDRAMs provide row and column addresses in two steps. However, the control signals and address inputs are sampled by the SDRAM at the leading edge of a common clock signal (133 MHz maximum). SDRAMs provide higher densities than conventional DRAMs by further reducing the need for support circuitry and faster speeds. The SDRAM has been used in PCs (personal computers).

6.2.2   READ and WRITE Timing Diagrams

To execute an instruction, the CPU of the microcontroller reads or fetches the op-code via the data bus from a memory location in the ROM/RAM external to the CPU. It then places the op-code (instruction) in the instruction register. Finally, the CPU executes the instruction. Therefore, the execution of an instruction consists of two portions, instruction fetch and instruction execution. We consider the instruction fetch, memory READ, and memory WRITE timing diagrams in the following using a single clock signal. Figure 6.53 shows a typical instruction fetch timing diagram.

In Figure 6.53, to fetch an instruction, when the clock signal goes to HIGH, the CPU places the contents of the program counter on the address bus via address pins A_0–A_{15} on the chip. Note that since each of lines A_0–A_{15} can be either HIGH or LOW, both transitions are shown for the address in Figure 6.53.

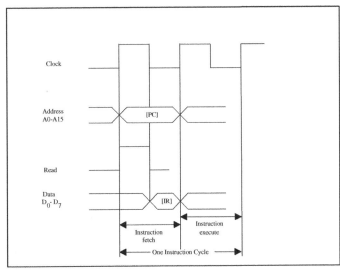

FIGURE 6.53   Typical instruction fetch timing diagram for an 8-bit microcontroller.

The instruction fetch is basically a memory READ operation. Therefore, the CPU raises the signal on the READ pin to HIGH. As soon as the clock goes to LOW, the logic external to the CPU gets the contents of the memory location addressed by A_0–A_{15} and places them on the data bus D_0–D_7. The CPU then takes the data and stores it in the instruction register so that it gets interpreted as an instruction. This is called *instruction fetch*. The CPU performs this sequence of operations for every instruction.

We now describe the READ and WRITE timing diagrams. A typical READ timing diagram is shown in Figure 6.54. Memory READ is basically loading the contents of a memory location of the main ROM/RAM into an internal register of the CPU. The address of the location is provided by the contents of the memory address register (MAR). Let us now explain the READ timing diagram of Figure 6.54.

1. The CPU performs the instruction fetch cycle as before to READ the op-code.
2. The CPU interprets the op-code as a memory READ operation.
3. When the clock pin signal goes HIGH, the CPU places the contents of the memory address register on the address pins A_0–A_{15} of the chip.
4. At the same time, the CPU raises the READ pin signal to HIGH.
5. The logic external to the CPU gets the contents of the location in the main ROM/RAM addressed by the memory address register and places it on the data bus.
6. Finally, the CPU gets this data from the data bus via pins D_0 – D_7 and stores it in an internal register.

Memory WRITE is basically storing the contents of an internal register of the CPU into a memory location of the main RAM. The contents of the memory address register provide the address of the location where data is to be stored. Figure 6.55 shows a typical WRITE timing diagram.

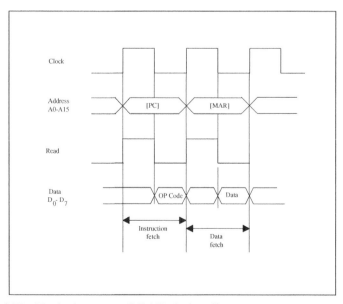

FIGURE 6.54   Typical memory READ timing diagram.

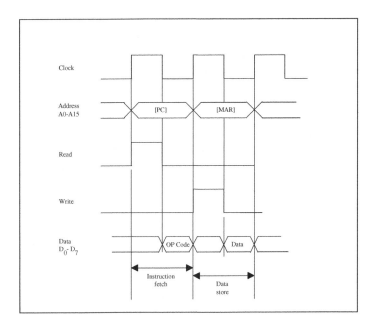

FIGURE 6.55   Typical memory WRITE timing diagram.

1.   The CPU fetches the instruction code as before.
2.   The CPU interprets the instruction code as a memory WRITE instruction and then proceeds to perform the DATA STORE cycle.
3.   When the clock pin signal goes HIGH, the CPU places the contents of the memory address register on the address pins A_0–A_{15} of the chip.
4.   At the same time, the CPU raises the WRITE pin signal to HIGH.
5.   The CPU places data to be stored from the contents of an internal register onto data pins D_0–D_7.
6.   The logic external to the CPU stores the data from the register into a RAM location addressed by the memory address register.

6.2.3   Main Memory Organization

Typical microcontroller on-chip main memory also called *memory module* may include ROM/EPROM/EPROM, Flash, and SRAM. As mentioned earlier, the PIC18F main memory consists of Flash memory (program memory) and SRAMs (data memory). A microcontroller system designer is normally interested in how the microcontroller memory is organized or, in other words, how to connect the memory units to the CPU, and then determine the memory map of the microcontroller. That is, the PIC18F designer would be interested in finding out what memory locations are assigned to the Flash memory and SRAMs.

Main Memory Array Design   In a typical microcontroller, the designer has to implement the required capacity by interconnecting several memory circuits to the CPU. This concept is known as *memory array design*. We address this topic in this section and show how to interface a data memory system (SRAM) with a typical CPU. In the following, we will use common signals associated with the CPU and memory units internal to typical microcontrollers.

Now let us discuss how to design SRAM arrays. In particular, our discussion is focused on the design of memory arrays for a hypothetical CPU. The pertinent signals of a typical CPU necessary for main memory interfacing are shown in Figure 6.56. There are 16 address lines, A_{15}-A_0, with A_0 being the least significant bit. This means that this CPU can address directly a maximum of $2^{16} = 65,536$ or 64K bytes of memory locations.

The control line $M/\overline{IO}$ goes LOW if the CPU executes an I/O instruction; it is held HIGH if the CPU executes a memory instruction. Similarly, the CPU drives control line $R/\overline{W}$ HIGH for READ operation; it is held LOW for WRITE operation. Note that all 16 address lines and the two control lines ($M/\overline{IO}$, $R/\overline{W}$) described so far are unidirectional in nature; that is, information can always travel on these lines from the processor to external units. Eight bidirectional data lines, D_7-D_0 (with D_0 being the least significant bit) are also shown in Figure 6.56. These lines are used to allow data transfer from the CPU to memory module, and vice versa.

The block diagram of a typical 1K × 8 RAM SRAM is shown in Figure 6.57. In this circuit, there are 10 address lines, A_9-A_0, so one can read or write 1024 ($2^{10} = 1024$) different memory words. Also, in this chip there are eight bidirectional data lines, D_7-D_0 so that information can travel back and forth between the CPU and the memory module. The three control lines $\overline{CS1}$, CS2, and $R/\overline{W}$ are used to control the SRAM unit according to the truth table shown in Table 6.5 from which it can be concluded that the RAM chip is enabled only when $\overline{CS1} = 0$ and CS2 = 1. Under this condition, $R/\overline{W} = 0$ and $R/\overline{W} = 1$ imply write and read operations, respectively.

To connect a CPU to the memory module, two address decoding techniques are commonly used for each memory type: linear decoding and full decoding. Let us

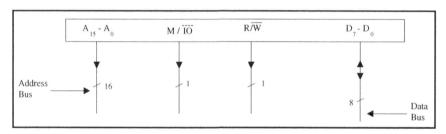

FIGURE 6.56   Pertinent signals of a typical CPU required for main memory interfacing.

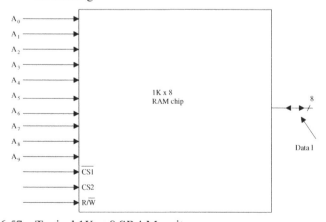

FIGURE 6.57   Typical 1K × 8 SRAM unit.

discuss first how to interconnect a CPU with a 4K SRAM array comprised of the four 1K SRAMs of Figure 6.57 using the linear decoding technique. Figure 6.58 uses linear decoding to accomplish this. In this approach, address lines A_9-A_0 of the CPU are connected to all SRAM units. Similarly, the control lines $M/\overline{IO}$ and $R/\overline{W}$ of the CPU are connected to control lines CS2 and $R/\overline{W}$ respectively to each of the SRAM unit. The high-order address bits A_{10}-A_{13} act directly for selecting memory units. In particular, address lines A_{10} and A_{11} select SRAM units I and II, respectively. Similarly, the address lines A_{12} and A_{13} select the SRAM units III and IV, respectively. A_{15} and A_{14} are don't cares and are assumed to be zero.

TABLE 6.5   Truth Table for the SRAM

$\overline{CS1}$	CS2	$R/\overline{W}$	Function
0	1	0	Write Operation
0	1	1	Read Operation
1	X	X	The chip is not selected
X	0	X	The chip is not selectd

X means Don't Care

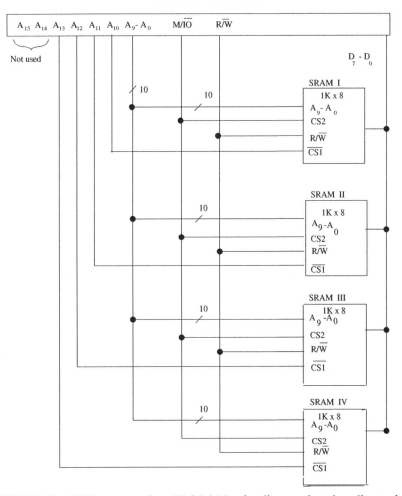

FIGURE 6.58   CPU connected to 4K SRAM using linear select decoding technique.

Table 6.6 shows the memory (address) map for the four 1K SRAMs. The primary advantage of this method, known as *linear select decoding*, is that it does not require decoding hardware. However, if two or more of lines A_{10}-A_{13} are low at the same time, more than one SRAM units are selected, and this causes a bus conflict.

Because of this potential problem, the software must be written such that it never reads into or writes from any address in which more than one of bits A_{13}-A_{10} are low. Another disadvantage of this method is that it wastes a large amount of address space. For example, whenever the address value is B800 or 3800, SRAM chip I is selected. In other words, address 3800 is the mirror reflection of address B800 (this situation is also called *memory foldback*). This technique is therefore limited to a small system. The system of Figure 6.58 can be expanded up to a total capacity of 6K using A_{14} and A_{15} to select two more 1K SRAM units.

To resolve problems with linear decoding, the full decoded memory addressing is used. In this technique, a decoder is used. The 4K memory designed using this technique is shown in Figure 6.59. In Figure 6.59, the decoder output selects one of the four 1K SRAMs, depending on the values of A_{12}, A_{11}, and A_{10} (Table 6.7).

Note that the decoder output will be enabled only when $\overline{E3} = \overline{E2} = 0$ and $E1 = 1$. Therefore, in the organization of Figure 6.59, when any one of the high-order bits A_{15}, A_{14}, or A_{13} is 1, the decoder will be disabled, and thus none of the SRAM's will be selected. In this arrangement, the memory addresses are assigned as shown in Table 6.8.

This approach does not waste any address space since the unused decoder outputs (don't cares) can be used for memory expansion. For example, the 3-to-8 decoder of Figure 6.59 can select eight 1K SRAM's. Also, this method does not generate any bus conflict. This is because the decoder output selected ensures enabling of one memory unit at a time.

6.3 Input/Output (I/O)

The technique of data transfer between a microcontroller and an external device is called *input/ output* (I/O). One communicates with a microcontroller via the I/O devices interfaced to it. The user can enter programs and data using the keyboard on a terminal and execute the programs to obtain results. Therefore, the I/O devices connected to a microcontroller provide an efficient means of communication between the microcontroller and the outside world. These I/O devices, commonly called *peripherals* and include keyboards, seven-segment displays, and LCD's (Liquid Crystal Displays).

There are two ways of transferring data between a microcontroller and I/O devices. These are programmed I/O, and interrupt I/O. Using *programmed I/O*, the CPU executes a program to perform all data transfers between the CPU and the external

TABLE 6.6   Memory (Address Map) of the Memory Organization of Figure 6.58

Address Range (Hex)	SRAM Number
3800-3BFF	I
3400-37FF	II
2C00-2FFF	III
1C00-1FFF	IV

device. The main characteristic of this type of I/O technique is that the external device carries out the functions dictated by the program contained in the microcontroller memory. In other words, the CPU controls all transfers completely.

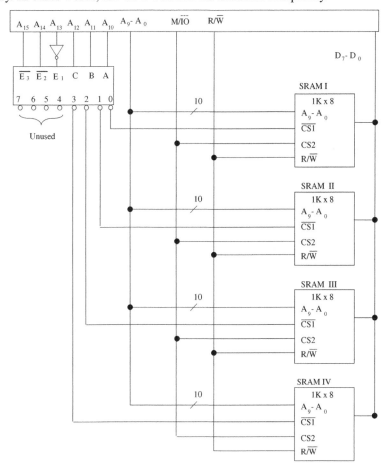

FIGURE 6.59   Interconnecting a CPU with four 1K × 8 RAMs (4K × 8) using full decoded memory addressing.

TABLE 6.7   Decoding Guide

A_{12}	A_{11}	A_{10}	SRAM Number
0	0	0	I
0	0	1	II
0	1	0	III
0	1	1	IV

TABLE 6.8   Address Map of the Memory Organization of Figure 6.59

Address Range (Hex)	SRAM Number
0000-03FF	I
0400-07FF	II
0800-0BFF	III
0C00-0FFF	IV

In *interrupt I/O*, an external device can force the CPU to stop executing the current program temporarily so that it can execute another program known as an *interrupt service routine*. This routine satisfies the needs of the external device. After completing this program, a return from interrupt instruction can be executed at the end of the service routine to return control at the right place in the main program.

The interrupt is initiated externally via hardware or internally via occurrence of events such as completion of ADC (Analog to Digital Conversion). Once the interrupt is recognized, the microcontroller internally saves the return address and the contents of some other registers , and automatically branches to an address predefined by the manufacturer. The user writes a program called *interrupt service routine* at this address. This program is similar to functions in C.

6.3.1   Simple I/O Devices

A simple input device such as a DIP switch can be connected to a microcontroller's I/O port as shown in Figure 6.60. The figure shows a switch circuit that can be used as a single bit input into an I/O port. When the DIP switch is open, V_{IN} is HIGH. When the switch is closed, V_{IN} is LOW. V_{IN} can be used as an input bit for performing laboratory experiments. Note that unlike TTL, a 1K ohm resistor is connected between the switch and the input of the MOS gate. This provides protection against static discharge.

For performing simple I/O experiments using programmed I/O, light-emitting diodes (LEDs) and seven-segment displays can be used as output devices. An LED is typically driven by low voltage and low current, which makes it a very attractive device for use with microcontrollers.

Basically, an LED will be ON, generating light, when its cathode is sufficiently negative with respect to its anode. A microcontroller can therefore light an LED either by grounding the cathode (if the anode is tied to +5 V) or by applying +5 V to the anode (if the cathode is grounded) through an appropriate resistor value. A typical hardware interface between a microcontroller and an LED is depicted in Figure 6.61. In Figure 6.61 (a), a '1' from the microcontroller will turn the LED ON while a '0' will turn it

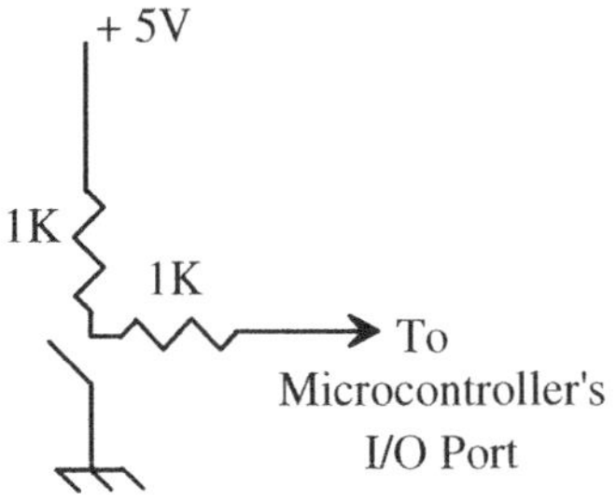

FIGURE 6.60   Typical switch for a microcontroller's input.

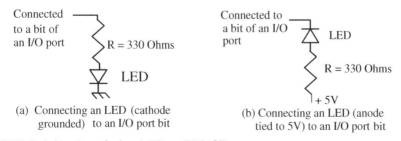

(a) Connecting an LED (cathode grounded) to an I/O port bit

(b) Connecting an LED (anode tied to 5V) to an I/O port bit

FIGURE 6.61   Interfacing LED to PIC18F.

OFF. In Figure 6.61 (b), on the other hand, a '0' from the microcontroller will turn the LED ON while a '1' will turn it OFF. This interface hardware is included in a typical microcontroller's I/O port, such as the PIC18F. Hence, the PIC18F microcontroller outputs adequate current to directly turn an LED ON or OFF; the hardware interface does not need to be connected to the port.

From Chapter 1, a red LED requires 10 mA current at 1.7V. In Figure 6.61 (a), a HIGH at the microcontroller output will turn the LED ON. This will allow a path of current to flow from the +5 V source through R and the LED to the ground. In Figure 6.61(b), a LOW at the microcontroller output will turn the LED ON. This will allow a path of current to flow from the +5V source through R and the LED to the ground (microcontroller I/O port). The appropriate value of R needs to be calculated to satisfy the voltage and current requirements of the LED. The value of R can be calculated as follows:

$$R = \frac{5-1.7}{10 \text{ mA}} = \frac{5-1.7}{10 \text{ mA}} = 330 \text{ } \Omega$$

Therefore, the interface design is complete, and a value of R = 330 Ω is required. A seven-segment display can be used with programmed I/O to display, for example, decimal numbers from 0 to 9. The name *seven segment* is based on the fact that there are seven LEDs, one in each segment of the display. Figure 6.62 shows a typical seven-segment display. In the figure, each segment contains an LED. All decimal numbers from 0 through 9 can be displayed by turning the appropriate segment ON or OFF.

For example, a '0' can be displayed by turning the LED in segment *g* OFF and turning the other six LEDs in segments *a* through *f* ON. There are two types of seven-segment displays: send a HIGH to light a segment and a LOW to turn it off. In a common-anode configuration, on the other hand, the microcontroller sends a LOW to light a segment and a HIGH to turn it off.

Seven-segment displays can be interfaced to typical microcontrollers using programmed I/O. BCD to seven-segment code converter chips such as 7447 or 7448 can be replaced by a lookup table. This table can be stored in a microcontroller's memory. An assembly language program can be written to read the appropriate code for a BCD digit stored in this table. This data can be output to display the BCD digit on a seven-segment display connected to an I/O port of the microcontroller. Programs to accomplish this are written in C language as shown in Chapter 8.

6.3.2 Programmed I/O

A microcontroller communicates with an external device via one or more registers called *I/O ports* using programmed I/O. Each bit in the port can be configured individually as either input or output. Each port can be configured as an input or

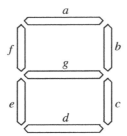

FIGURE 6.62 A seven-segment display.

output port by another register usually called the *Data Direction Register* (DDR). The port contains the actual input or output data. The data direction register is an output register and can be used to configure the bits in the port as inputs or outputs.

Each bit in the port can be set up as an input or output, normally by writing a 0 or a 1 using programming languages such as C in the corresponding bit of the DDR. The PIC18F microcontroller makes an I/O port bit an input by writing a '1' in the corresponding bit in DDR. On the other hand, writing a '0' in a particular bit in DDR will configure the corresponding bit in the port as an output.

For example, if an 8-bit DDR in the PIC18F contains 0xCB (CB Hex), the corresponding port is defined as shown in Figure 6.63. In this example, because 0xCB (1100 1011) is stored in the data-direction register, bits 0, 1, 3, 6, and 7 of the port are set up as inputs, and bits 2, 4, and 5 of the port are defined as outputs. The microcntroller can then send output to external devices, such as LEDs, connected at bits 2, 4, and 5 through a proper interface.

Similarly, the microcontroller can input the status of external devices, such as switches, through bits 0, 1, 3, 6, and 7. To input data from the input switches, the microcontroller inputs the complete byte, including the bits to which LEDs are connected. While receiving input data from an I/O port, however, the microcontroller places a value, probably 0, at the bits configured as outputs and the program must interpret them as "don't cares." At the same time, the microcontroller's outputs to bits configured as inputs are disabled.

I/O ports are addressed using either standard I/O or memory-mapped I/O techniques. *Using Standard I/O* or sometimes called *port I/O* (also called *isolated I/O* by Intel), the CPU outputs an internal signal such as the M/$\overline{\text{IO}}$ for memory and I/O units on the microcontroller chip. The CPU outputs a HIGH on M/$\overline{\text{IO}}$ to indicate to memory and the I/O that a memory operation is taking place. A LOW output from the CPU to M/$\overline{\text{IO}}$ indicates an I/O operation. Execution of an IN or OUT instruction makes the M/$\overline{\text{IO}}$ LOW, whereas memory-oriented instructions, such as MOVE, drive the M/$\overline{\text{IO}}$ to HIGH.

In standard I/O, the CPU uses the M/$\overline{\text{IO}}$ output signal to distinguish between I/O and memory. Intel microcontrollers such as the Intel 8051 uses standard I/O.

In *memory-mapped I/O*, the CPU does not use the M/$\overline{\text{IO}}$ control signal. Instead, the CPU uses an unused address pin to distinguish between memory and I/O. The PIC18F uses memory-mapped I/O.

Unconditional and Conditional Programmed I/O   There are typically two ways in which programmed I/O can be utilized: unconditional I/O and conditional I/O. The microcontroller can send data to an external device at any time using *unconditional Programmed I/O*. The external device must always be ready for data transfer. A typical example is that of a microcontroller outputting a 7-bit code through an I/O port to drive a seven-segment display connected to this port.

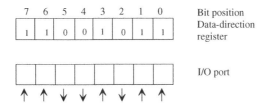

FIGURE 6.63   I/O port with the corresponding data-direction register.

In *conditional Programmed I/O*, the microcontroller waits for a particular condition (such as conversion complete signal for Analog-to-Digital converter) to happen. Upon occurrence of the condition, the microcontroller can be programmed to input the Analog-to-Digital converter's converted data. Conditional Programmed I/O is sometimes called *Polled I/O*.

6.3.3 Interrupt I/O

A disadvantage of conditional programmed I/O is that the CPU needs to check the status bit (Conversion complete output for the Analog-to-Digital Converter) by waiting in a loop. This type of I/O transfer is dependent on the occurrence of the external condition. This waiting may slow down the CPU's ability to process other data. The interrupt I/O technique is efficient in this type of situation.

Interrupt I/O is a device-initiated I/O transfer. The external device is connected to a pin called the *interrupt* (INT) pin on the microcontroller chip. When the device needs an I/O transfer with the microcontroller, it activates its interrupt pin. The microcontroller usually completes the current instruction and saves the return address and contents of certain registers internally.

The microcontroller then automatically branches to a program called the *interrupt service routine*. This program is written by the user. The external device wants the microcontroller to execute this program to transfer data.

The basic characteristics of interrupt I/O have been discussed so far. The main features of interrupt I/O provided with a typical microcontroller are discussed next.

Interrupt Types There are typically two types of interrupts: external interrupts and internal interrupts. *External interrupts* are initiated through a microcontroller's interrupt pins by external devices. External interrupts can be divided further into two types: maskable and nonmaskable. The nonmaskable interrupt cannot be enabled or disabled by instructions, whereas a microcontroller's instruction set typically contains instructions to enable or disable maskable interrupt. A nonmaskable interrupt has a higher priority than a maskable interrupt. If maskable and nonmaskable interrupts are activated at the same time, the processor will service the nonmaskable interrupt first.

A nonmaskable interrupt is typically used as a power failure interrupt. Microcontrollers normally use +5 V dc, which is transformed from 110 V ac. If the power falls below 90 V AC, the DC voltage of +5 V cannot be maintained. However, it will take a few milliseconds before the AC power drops below 90 V AC. In these few milliseconds, the power-failure-sensing circuitry can interrupt the processor. The interrupt service routine can be written to store critical data in nonvolatile memory such as battery-backed CMOS RAM, and the interrupted program can continue without any loss of data when the power returns.

Internal interrupts are usually nonmaskable, and cannot be disabled by instructions. They are activated internally by conditions such as completion of A/D (Analog to Digital) conversion, Timer interrupt or interrupt due to serial I/O. Interrupts are handled in the same way as external interrupts. The user writes a service routine to take appropriate action to handle the interrupt. Some microcontrollers include software interrupt instructions. When one of these instructions is executed, the microcontroller is interrupted and serviced similarly to external or internal interrupts.

Some microcontrollers such as the Motorola/Freescale HC11/HC16 provide both external (maskable and nonmaskable) and internal (exceptional conditions and software instructions). The PIC18F provides external maskable interrupts only. The PIC18F does not have any external nonmaskable interrupts. However, the PIC18F provides internal interrupts. The internal interrupts are activated internally by conditions such as timer interrupts, completion of analog to digital conversion, and serial I/O.

Interrupt Address Vector   The technique used to find the starting address of the service routine (commonly known as the *interrupt address vector*) varies from one processor to another. The microcontroller manufacturers typically define the fixed starting address for each interrupt.

6.4   CPU design using Verilog

Memory can be modeled in Verilog as an array of registers. The following are some of the typical examples of specifying memory in Verilog:
```
reg addr [0:2047]; // Memory with 2K 1-bit words (Addresses addr[0]
                 // through addr[2047]).
reg [15:0] addr [0:4095]; // Memory with 4K 16-bit words (Addresses
                      // addr[0] through addr[4095]).
reg [22:0] mem [52:0]; // Memory of size 53X23 bits (Addresses mem[0]
                    // through mem[52]).
data = mem[loc]   //Memory read operation. Read the contents of a memory
         //location addressed by loc into a register called data.
mem[loc] = data //Memory write operation. Write the contents of a register
          //called data into a memory location addressed by loc.
```

Example 6.1

Write a Verilog description for the ALU of Table 6.9 (re-drawn from Table 6.3 for convenience).

Solution

```
//The verilog code for 4-bit ripple carry adder is:


module Add4(c_out, Sum, A, B, c_in);
//Add 2 4-bit numbers A & B with carry in
//output Sum and c_out
```

TABLE 6.9   Truth table selecting the operations of the ALU of Example 6.1

Select Lines		Output Z	Comment
s_1	s_0		
0	0	X plus Y	Addition
0	1	X plus $\overline{Y}$ Plus 1	2's Complement subtraction
1	0	X $\wedge$ Y	Boolean AND
1	1	X $\oplus$ Y	Exclusive-OR

```
 output c_out;
 output [3:0] Sum;
 input [3:0] A, B;
 input c_in;
 wire [2:0] carry;
//need 4 full adders

 FA fa0(carry[0], Sum[0], A[0], B[0], c_in);
 FA fa1(carry[1], Sum[1], A[1], B[1], carry[0]);
 FA fa2(carry[2], Sum[2], A[2], B[2], carry[1]);
 FA fa3(c_out, Sum[3], A[3], B[3], carry[2]);
endmodule

//The included code for full adder is:

module FA(c_out, sum, a, b, c_in);
//Full Adder
 input       a, b, c_in;
 output      sum, c_out;
 assign{c_out, sum} = a + b + c_in;
endmodule

//The code for a 2-to-1 multiplexer is:

module mux2to1(x, select, A0, A1);
 output x;
 input select, A0, A1;
 assign x = (select)? A1: A0;
endmodule

//description: 4-bit ALU

module ALU(F, C_out, X, Y, fCode);
 output [3:0] F;
 output C_out;
 input [3:0] X, Y;
 input [1:0] fCode;
 wire [3:0] B, Y_not, AU, LU, LU_0, LU_1;
 wire carry;

//Structure of Arithmetic unit
//Prep inverted Y
 not(Y_not[0], Y[0]);
 not(Y_not[1], Y[1]);
 not(Y_not[2], Y[2]);
 not(Y_not[3], Y[3]);

//Prep input B to adder
 mux2to1 B0( B[0], fCode[0], Y[0], Y_not[0]);
 mux2to1 B1( B[1], fCode[0], Y[1], Y_not[1]);
```

```
 mux2to1 B2( B[2], fCode[0], Y[2], Y_not[2]);
 mux2to1 B3( B[3], fCode[0], Y[3], Y_not[3]);

//Feed signal to adder
 Add4 Adder(carry, AU, X, B, fCode[0]);
//Only when S1 = 0, we need carry
//otherwise carry should be 0
 and(C_out, carry, ~fCode[1]);

//Structure of logic unit;
//Input when S0 == 0
 and(LU_0[0], X[0], Y[0]);
 and(LU_0[1], X[1], Y[1]);
 and(LU_0[2], X[2], Y[2]);
 and(LU_0[3], X[3], Y[3]);
//Input when S0 == 1
 xor(LU_1[0], X[0], Y[0]);
 xor(LU_1[1], X[1], Y[1]);
 xor(LU_1[2], X[2], Y[2]);
 xor(LU_1[3], X[3], Y[3]);

//calc output of logic unit
 mux2to1 G0(LU[0], fCode[0], LU_0[0], LU_1[0]);
 mux2to1 G1(LU[1], fCode[0], LU_0[1], LU_1[1]);
 mux2to1 G2(LU[2], fCode[0], LU_0[2], LU_1[2]);
 mux2to1 G3(LU[3], fCode[0], LU_0[3], LU_1[3]);
//Connect arithmethic and logic unit together
 mux2to1 F0(F[0], fCode[1], AU[0], LU[0]);
 mux2to1 F1(F[1], fCode[1], AU[1], LU[1]);
 mux2to1 F2(F[2], fCode[1], AU[2], LU[2]);
 mux2to1 F3(F[3], fCode[1], AU[3], LU[3]);
endmodule
```

Waveform:

In the above, when opcode is 2, L is 15, R is 13. A Boolean AND operation between L and R is performed, and the answer is 13 (D) as expected. For opcode 0, operation L plus R is performed generating an answer of 12 with 1 carry out as expected.

Example 6.2

Write a Verilog description for the microprogrammed CPU of Section 6.1.4 (Figures 6.44 through 6.50).

Solution

Xlinix ModelSim simulator can be used to simulate the CPU. A test bench can then be written to instantiate and test the simulation for correctness. Seven modules are created in the Verilog program to implement the microprogrammed CPU. The modules are **memcntrl, reg_8bit, alu_8bit, mux_8bit, ram, processor and cpu**. The design is created using hierarchical method. The **cpu** module is at the top of the hierarchy, **processor** and **memcntrl** are under **cpu** module, and finally the rest of the modules are under the **processor**.

The **memcntrol** contains the ROM, filled with a 23-bit value, which contains a 4-bit condition select, a 6-bit branch address, and 13-bit control input (C12 - C0) for the registers, ALU, and RAM. It also has the conditional statement that will make the Microprogram Counter (MPC) to count up by one if the load/!increment is LOW, or will load the branch address passed by the control memory buffer if load/!increment is HIGH. The **processor** module connects mux, ALU, registers (regA, regIR, regMAR, regPC, regBUFF), and the RAM. It also includes the instruction decoder and performs the following (Figure 6.50) : If condition select field = 0, load/!increment = 0, no branch. If condition select = 1 and Z = 1, branch. If condition select = 2 and C =1, branch. If condition select = 3 and I3 = 1, branch. If condition select = 4 and XC2 = 1, branch. If condition select = 5 and XC1 = 1, branch. If condition select = 6 and XC0 = 1, branch. If condition select = 7 and I0 = 1, branch.

The 256 × 8 RAM holds program instructions and data. The program is stored beginning at RAM address 0. This program tests two instructions, LOAD and ADD of the CPU.

The program will first load a value into register A from RAM address 100, add it to itself and store the result in register A.
The CPU module has only two inputs. These are reset and clock. It connects the **processor** module with the **memory control** module to complete the hierarchy of the microporgrammed CPU design.

Verilog code for the microprogrammed CPU is provided in the following:

```
// Microprogrammed Controller Module for the CPU
// Port declarations
module memcntrl (C_fn, Z, C, I3, XC2, XC1, XC0,I0, reset,
clk);
input Z, C, I3, XC2, XC1, XC0, I0, reset, clk;
output [12:0] C_fn;
reg [22:0] mem [52:0];
reg [12:0] C_fn;
reg [22:0] regCMDB;
reg [5:0] regMPC;
reg ld_inc;
// Binary microprogram
```

```
// The size of the control memory is 53 x 23 bits. The
// 23-bit control word consists of 13-bit control function
// containing C0 through C12 with C0 as bit 12 and C12 as
// bit 0. The condition select field is 4-bit wide (bits 19-22.
// For example, consider the code for line 0 with the
// operation PC <- 0 in the following. Since there is no
// condition in this operation, condition select field ( CS )
// bits are 0's. The branch address field ( Brn )bits are
// assumed as don't cares arbitrarily. To clear PC to 0,
// C0 = 1 (bit 12). To disable RAM, C6 = 1. C1, C2, C4,
// C7, C8 and C9 are initialized to 0's. Other bits are
// arbitrarily initialized as don't cares.
initial
begin

// 23-bit value contains a 4-bit condition select, a 6-bit
// branch address,and 13-bit control. input ( C12 - C0 ) for
// the registers, ALU, and RAM.
//      22 19    12        0
//       cs  Brn  Cntrl Func
mem[0]  = 23'b0000xxxxxx100x0x1000xxx;
mem[1]  = 23'b0000xxxxxx00001x1000xxx;
mem[2]  = 23'b0000xxxxxx010x010010xxx;
mem[3]  = 23'b0011001110000x0x1000xxx;
mem[4]  = 23'b0110001000000x0x1000xxx;
mem[5]  = 23'b0101001010000x0x1000xxx;
mem[6]  = 23'b0100001100000x0x1000xxx;
mem[7]  = 23'b1000110100000x0x1000xxx;
mem[8]  = 23'b0000xxxxxx000x0x1001111;
mem[9]  = 23'b1000000001000x0x1000xxx;
mem[10] = 23'b0000xxxxxx000x0x1001100;
mem[11] = 23'b1000000001000x0x1000xxx;
mem[12] = 23'b0000xxxxxx000x0x1001101;
mem[13] = 23'b1000000001000x0x1000xxx;
mem[14] = 23'b0110010111000x0x1000xxx;
mem[15] = 23'b0101100000000x0x1000xxx;
mem[16] = 23'b0100101001000x0x1000xxx;
mem[17] = 23'b0000xxxxxx00001x1000xxx;
mem[18] = 23'b0000xxxxxx010x010100xxx;
mem[19] = 23'b0000xxxxxx00011x1000xxx;
mem[20] = 23'b0000xxxxxx000x010100xxx;
mem[21] = 23'b0000xxxxxx000x0x1001110;
mem[22] = 23'b1000000001000x0x1000xxx;
mem[23] = 23'b0000xxxxxx00001x1000xxx;
mem[24] = 23'b0000xxxxxx010x010100xxx;
mem[25] = 23'b0000xxxxxx00011x1000xxx;
mem[26] = 23'b0111011110000x0x1000xxx;
mem[27] = 23'b0000xxxxxx000x010100xxx;
```

```
mem[28]  = 23'b0000xxxxxx000x0x1001001;
mem[29]  = 23'b1000000001000x0x1000xxx;
mem[30]  = 23'b0000xxxxxx000x000000xxx;
mem[31]  = 23'b1000000001000x0x1000xxx;
mem[32]  = 23'b0000xxxxxx00001x1000xxx;
mem[33]  = 23'b0000xxxxxx010x010100xxx;
mem[34]  = 23'b0000xxxxxx00011x1000xxx;
mem[35]  = 23'b0000xxxxxx000x010100xxx;

mem[36]  = 23'b0111100111000x0x1000xxx;
mem[37]  = 23'b0000xxxxxx000x0x1001010;
mem[38]  = 23'b1000000001000x0x1000xxx;
mem[39]  = 23'b0000xxxxxx000x0x1001011;
mem[40]  = 23'b1000000001000x0x1000xxx;
mem[41]  = 23'b0000xxxxxx00001x1000xxx;
mem[42]  = 23'b0000xxxxxx000x0x1000xxx;
mem[43]  = 23'b0111101111000x110000xxx;
mem[44]  = 23'b0001110010000x0x1000xxx;
mem[45]  = 23'b0000xxxxxx010x0x1000xxx;
mem[46]  = 23'b1000000001000x0x1000xxx;
mem[47]  = 23'b0010110010000x0x1000xxx;
mem[48]  = 23'b1000000001000x0x1000xxx;
mem[49]  = 23'b0000xxxxxx010x0x1000xxx;
mem[50]  = 23'b0000xxxxxx001x010000xxx;
mem[51]  = 23'b1000000001000x0x1000xxx;
mem[52]  = 23'b1000110100000x0x1000xxx;
end
always @( reset )
     if ( reset )
          begin                     // when reset is active and
                                    // reset is high
               regMPC = 6'b000000;// initialize MPC to zero
          end

// conditional statement that will make the Microprogram
// Counter (MPC) count up by one if the load !increment
// is low, or will load the branch address passed by the
// control memory buffer.

always @ ( posedge clk )   // when clock is at positive edge
     begin
            regCMDB = mem[regMPC];
// register regCMDB contains 23-bit contents of memory
addressed by regMPC
            C_fn = regCMDB [12:0];
     // control function equals to first 13 bits of register CMDB
     // if condition select field = 0, load /!increment= 0, no branch.
     // if condition select = 1 and Z = 1, branch
```

```
        // if condition select = 2 and C =1, branch
        // if condition select = 3 and I3 = 1, branch
        // if condition select = 4 and XC2 = 1, branch
        // if condition select = 5 and XC1 = 1, branch
        // if condition select = 6 and XC0 = 1, branch
        // if condition select = 7 and I0 = 1, branch
        // if condition select = 8 and load /!increment= 1, branch
              assign ld_inc =
              ( regCMDB [22:19] == 0 )?1'b0: // if cmdb= 0 ld_inc = 0
              ( regCMDB [22:19] == 1 )?Z: // if cmdb= 1 ld_inc = Z

( regCMDB [22:19] == 2 )?C:    // if cmdb= 2 ld_inc = C
   ( regCMDB [22:19] == 3 )?I3:   // if cmdb= 3 ld_inc = I3
   ( regCMDB [22:19] == 4 )?XC2:  // if cmdb= 4 ld_inc = XC2
   ( regCMDB [22:19] == 5 )?XC1:// if cmdb= 5 ld_inc =XC1
   ( regCMDB [22:19] == 6 )?XC0: // if cmdb= 6 ld_inc = XC0
   ( regCMDB [22:19] == 7 )?I0: // if cmdb= 7 ld_inc = I0
   ( regCMDB [22:19] == 8 )?1'b1: // if cmdb= 8 ld_inc = 1
      1'bx;                         // else ld_inc = x
      if (ld_inc)
              regMPC = regCMDB [18:13]; // load branch address
      else
              regMPC = regMPC + 1;   // increment MPC by 1
   end
endmodule
```

//Register 8 bit module

```
// General Purpose Register (GPR)
module reg_8bit (b, a, sel, clk);
input [7:0] a;
input [2:0] sel;
input clk;
output [7:0] b;
reg [7:0] b;
      always @ (sel)
            begin
                  b <=  (sel==0)?b:     // b = b if sel = 0
                        (sel==1)?0 :    // b= 0 if sel = 1
                        (sel==2)?b+1 :  // b= b+1 if sel = 2
                        (sel==4)?a:     // b= a if sel = 4
                        8'bx;           // else b=xxxxxxxx
            end
endmodule
```
//ALU module

```
// ALU with zero and carry flags
module alu_8bit ( f, z_flag, c_flag, a, b, sel);
input [2:0] sel;
input [7:0] a, b;
output [7:0] f;
output z_flag, c_flag;
reg z_flag, c_flag;
     initial
       begin
             z_flag = 1'b0;   // initialize zero and carry flag to
zero
             c_flag = 1'b0;   //
       end
       assign f =(sel==0)?0 :     // f=0 if sel=0
                 (sel==1)?b:          // f=b if sel=1

                 (sel==2)?a+b:  // f=a+b if sel=2
                 (sel==3)?a-b:  // f=a-b if sel=3
                 (sel==4)?a+1 : // f=a+1 if sel=4
                 (sel==5)?a-1 ://f=a-1 if sel=5
                 (sel==6)?a&b://f=a&b if sel=6
                 (sel==7)?~a://f=~a if sel=7
                 8'bx;      // else f=xxxxxxxx

//Carry and Zero Flag registers
       always @ ( f )
             begin
                   if (f==0)    // if alu output = 0, zero flag = 1
                      assign z_flag =1;
           else if ( f != 0 & ( sel != 3'bxxx )) // if f not zero
                                 // and
                                          // sel not xxx
                       assign z_flag = 0; // zero flag = 0

             end

       always@ ( f )

       begin
             if(sel==4 | sel==2)
              carry = (a[7]+b[7])*f[7]+a[7]*b[7];
             if ( carry )        // if alu outputs carry, carry
flag = 1
               assign c_flag = 1;
             else if ( !carry & ( sel != 3'bxxx )) // if not carry and
              assign c_flag = 0; // sel not xxx, carry = 0
```

```
        end
endmodule
```
//Processor module (Figures 6.45 and 6.48)
```
// Processor

module processor (I3, XC0, XC1, XC2, XC3, I0, z_flag, c_
flag, clock, c0, c1, c2, c3, c4, c5, c6, c7, c8, c9, c10, c11,
c12);
 input clock;
 input c0, c1, c2, c3, c4, c5, c6, c7, c8, c9, c10, c11,
c12;
 output I3, XC0, XC1, XC2, XC3, I0, z_flag, c_flag;
 wire [7:0] IR_out;
 wire [7:0] F_out, BUFF_out, RAM_dataout, RAM_addr, MAR_in, PC_out;
 reg [7:0] regA_out;
 reg I0, I3, XC0, XC1, XC2, XC3;

//module mux_8bit(z, sel, mux_in0, mux_in1);
mux_8bit Mux1(MAR_in, c3, PC_out, BUFF_out);
//module alu_8bit(f, z_flag, c_flag, a, b, sel);
alu_8bit ALU1(F_out, z_flag, c_flag, regA_out, BUFF_out, {c10, c11,
c12});

//module reg_8bit(b, a, sel, clk);
//reg_8bit regA(regA_in, F_out, {c9, 1'b0, 1'b0}, clock);
 reg_8bit regIR(IR_out, RAM_dataout, {c8, 1'b0, 1'b0}, clock);
 reg_8bit regMAR(RAM_addr, MAR_in, {c4, 1'b0, 1'b0}, clock);
 reg_8bit regPC(PC_out, RAM_dataout, {c2, c1, c0}, clock);
 reg_8bit regBUFF(BUFF_out, RAM_dataout, {c7, 1'b0, 1'b0}, clock);

//module ram(dataout, memeaddr, datain, rw, en);
ram RAM1(RAM_dataout, RAM_addr, regA_out, c5, c6);
 initial
 begin
      XC0 <= 0;   //initialize control signals to zero
      XC1 <= 0;
      XC2 <= 0;
      XC3 <= 0;
      I0 <= 0;
      I3 <= 0;
 end

 always@(clock)
 begin

      I3 <= IR_out[3];  // instruction decoder
      I0 <= IR_out[0];  // I3= irout[3] , I0 = irout[0]
```

```
   case ( {IR_out[2], IR_out[1]} )

 2'd0:begin XC0 =1;XC1 =0; XC2 = 0; end//if irout[2:1]=0,XC0 =1,others zero
 2'd1:begin XC1 =1;XC0 =0; XC2 = 0; end// if irout[2:1]=1,XC1=1,others zero
 2'd2:begin XC2 =1;XC0 =0; XC1 =0; end // if irout[2:1]=2,XC2=1,others zero
 2'd3:begin XC3 =1;XC0 =0; XC1=0; XC2= 0;end//ifirout[2:1]=3, XC3=1,other
0
 default:
 begin
      XC0 =1'bx; XC1 = 1'bx; XC2 = 1'bx; XC3 =1'bx; end//
else everything x
  endcase
 end
      always @ (posedge clock)
        begin

              regA_out <=(c9==0)?regA_out: // if c9=0, regA_out=regA_out
                    (c9==1)?F_out: // if c9 =1, regA_out = F_out
                    8'bx;        // else regA_out= xxxxxxxx
        end
endmodule
```

//**Mux 8 bit module**
```
module mux_8bit (z, sel, mux_in0, mux_in1);

 input sel;

 input [7:0] mux_in0, mux_in1;
 output[7:0]     z;

//    The output is defined as register
 reg[7:0]   z;


// The output changes whenever any of the inputs changes
      always @(sel or mux_in0 or mux_in1)
          // Check the control signal
        case (sel)
            1'b0:
                 z = mux_in0;  // if sel= 0 , z = in0
            1'b1:
                 z = mux_in1; // if sel=1, z = in 1
        endcase
endmodule
```

//**256 x 8 Ram**
```
module ram ( dataout, memaddr, datain, rw, en );
//--------------Input Ports----------------------
 input [7:0] memaddr;
```

```
 input [7:0] datain;
 input rw, en;
 output [7:0] dataout;
//--------------Internal variables----------------
 reg [7:0] dataout ;
 reg [7:0] mem [0:255];
//--------------Code Starts Here-----------------
 initial
 mem[0] = 8'b00001000;  // LDA mem <addr>
 mem[1] = 100;          // <addr> = 100, A<-5
 mem[2] = 8'b00001010;  // ADD A <- A + MEM<addr>
 mem[3] = 100;          // <addr> = 100, A<-10
 mem[100] = 8'b00000101;       // init data = 5
      always @ (memaddr or datain or rw)
  begin : MEM_WRITE
      if ( !en && !rw )
    mem[memaddr] = datain;
  end
      always @ (memaddr or rw or en)
  begin : MEM_READ
       if (!en && rw )
    dataout = mem[memaddr];
  end
endmodule
```

//CPU module has only two inputs (system clock and system reset)
```
module cpu ( clock, reset );
 input clock, reset;
 wire xc2, xc1, xc0, i3, i0, z, c;
 wire [12:0] cfn;
 processor p1(.clock(clock), .XC2(xc2), .XC1(xc1),
      .XC0(xc0), .I3(i3), .I0(i0), .z_flag(z), .c_flag(c),
      .c0(cfn[12]), .c1(cfn[11]), .c2(cfn[10]),.c3(cfn[9]),
      .c4(cfn[8]), .c5(cfn[7]), .c6(cfn[6]),.c7(cfn[5]),.
      c8(cfn[4]), .c9(cfn[3]), .c10(cfn[2]), .c11(cfn[1]),
      .c12(cfn[0]) );
 memcntrl memc(.clk(clock), .reset(reset), .XC2(xc2),
      .XC1(xc1), .XC0(xc0), .I3(i3), .I0(i0), .Z(z), .C(c),
      .C_fn(cfn));
endmodule
```

//Test Bench for CPU module
```
module test_cpu;
 reg clock, rst;
 cpu dut (clock, rst);
 initial // Clock generator
 begin          // generating clock with period of 2ns
  clock = 0;
  #1001 forever
```

```
  #1000 clock = !clock;
 end
 initial    // Test stimulus
 begin
  rst = 1;   // reset goes high for 3.5 ns then goes low
  #3500 rst = 0;
 end
endmodule
```

Timing Diagram

All eleven instructions are tested successfully by simulating a sample program. Timing diagrams are generated accordingly. The following simple program inside the 256×8 RAM is simulated for testing the proper operation of two (LDA,ADD) of the eleven instructions. The timing diagram of Figure 6.64 is generated. Note that PC is the program counter for the sample program in the RAM, and MPC is the microprogram counter for the symbolic program in the ROM (Figure 6.49) inside the memory control module.
Program for testing LDA and ADD:

```
mem[0]   = LDA            // A<- MEM <addr>
mem[1]   = 100;           // <addr> = 100, A<-5
mem[2]   = ADD            // A <- A + MEM<addr>
mem[3]   = 100;           // <addr> = 100,A<-10
mem[100] = 8'b00000101;   // init data = 5
```

LDA (PC=0) instruction with reference address 100, goes through the subroutines in the symbolic program (Figure 6.49): FETCH (MPC=1 at t=2ns), branching to MEMREF(MPC=14 at t=8ns), then to LDSTO(MPC=23 at t=10ns), all the way through

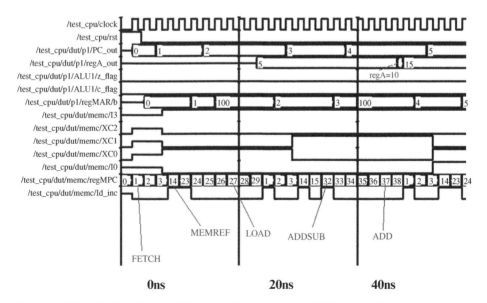

Figure 6.64   Verilog Timing Diagram (Top diagram-CPU clock, Next-Reset,
              Next-PC, Next-reg A, Next-Zflag, Next-Cflag, Next-regMAR, Next-I3,
              Next-XC2, Next-XC1, Next-XC0, Next-I0, Next-mpc, Next-ld_ !inc)

LOAD (MPC = 27 at t=18ns), and back to FETCH. At t=23ns, register A holds 05H, showing that it has loaded the contents of RAM memory address 100 (See Figure 6.64). Next, ADD (PC=2) operation is performed using reference address 100. At this point, ADD goes through the following subroutines in the symbolic program: FETCH (MPC=1 at t=24ns), branching to MEMREF(MPC=14 at t=30ns), then to ADDSUB(MPC=32 at t=34ns), all the way through ADD(MPC=37 at t=44ns), then back to FETCH (See Figure 6.64). At t=46ns, register A and BUFFER hold the contents of memory address 100. They are now the inputs to the ALU. The ALU will add these two values and its output will then go to register A, as commanded by the ADD<addr> instruction. At t=47ns, one can see that the contents of register A have changed to 0AH (10) (Figure 6.64).

Questions and Problems

6.1 Design a combinational logic shifter with 4-bit input and 4-bit output as follows:

$\overline{OE}$	Shift Count		4 - bit output
	S_1	S_0	
1	X	X	High Impedance output lines
0	0	0	No Shift
0	0	1	Right Shift once
0	1	0	Right Shift twice
0	1	1	Right Shift three times

where X means don't care. Using multiplexers and tristate buffers, draw a logic diagram.

6.2 Draw a logic diagram for a 4 × 4 barrel shifter.

6.3 Using a 4-bit CLA as the building block, design an 8-bit adder.

6.4 Design:
(a) a 16-bit adder whose worst-case add-time is 10Δ using a 4-bit CLA as a building block.
(b) the fastest 64-bit adder using a 4-bit CLA as the building block. Estimate the worst-case add-time of your design.
(c) a combinational circuit to compute the function where $(3/8)x$ is a 4-bit 2's complement number.

6.5 Design an arithmetic logic unit to perform the following functions:

S_1	S_0	F
0	0	*A* plus *B*
0	1	*A* minus *B*
1	0	*A* AND *B*
1	1	*A* OR *B*

Use multiplexers, binary adders, and gates as needed. Assume that A and B are 4-bit numbers. Draw a logic circuit.

6.6 Design a 4-bit ALU to perform the following operations:

S	F
0	Logical Left Shift *A* once
1	0

Assume that A is a 4-bit number. Draw a logic diagram using a binary adder, multiplexers, and inverters as necessary.

6.7 Design a 4-bit arithmetic unit as follows:

S	F
0	*A* plus *B*
1	*A* plus *1*

Assume that A and B are 4-bit numbers

6.8   Design an ALU to perform the following operations:

S_1	S_0	F
0	0	x plus y
0	1	x
1	0	B
1	1	x $\oplus$ y

Assume that x and y are 4-bit numbers, and $B = \bar{y}_3\,\bar{y}_2\,\bar{y}_1\,\bar{y}_0$. Draw a logic diagram.

6.9   What is the purpose of the control unit in the CPU?

6.10   Draw a logic diagram to implement the following register transfers:
         (a)   If the content of the 8-bit register R is odd, then
                     $x \leftarrow x \oplus y$, else $x \leftarrow x$ AND y
               Assume x and y are 4 bits wide.

         (b)   If the number in the 8-bit register R is negative, then
               $x \leftarrow x - 1$ else $x \leftarrow x + 1$. Assume x and y are 4 bits wide.

6.11   Discuss briefly the advantages and disadvantages of single-bus, two-bus, and three-bus architectures inside a control unit.

6.12   What is the basic difference between hardwired control, and microprogramming? Name the technique used for designing the control unit of the PIC18F and the HC11 microcontrollers.

6.13   Using the following components: 4-bit general-purpose register, 4-bit adder/subtractor, and tristate buffer, and assuming the inbus and outbus are 4 bits wide, design a control unit using hardwired control to perform the operations shown in

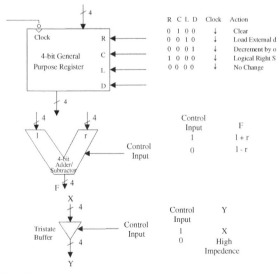

FIGURE P 6.13.

Figure P6.13. You may use counters, decoders, and PLAs as required.

(a) Outbus ← 4 × A. Assume A is a 4-bit unsigned number and the result is 4 bits wide.

(b) If the 4-bit number in register B is odd, outbus ← 0; otherwise outbus ← A + (B/2). Assume A and B are unsigned 4 bit numbers. Also, assume data is already loaded into B.

(c) If the content of a 4-bit register Q = 0, perform R ← M and then transfer the 4-bit result to outbus. On the other hand, if the content of the 4-bit register Q = 0, perform R ← 0 and then transfer the 4-bit result to the outbus. Assume M and R are 4 bits wide.

6.14   Repeat Problem 6.13 using microprogramming.

6.15   Discuss the basic differences between CISC and RISC.

6.16   Design and implement a combinational circuit that will work as follows:

S1	S0	F
0	0	A plus B
0	1	Shift left (A)
1	0	A plus B plus 1
1	1	Shift left (A) + 1

Note that A and B are 4-bit operands.

6.17 i) Design a combinational circuit that will work as follows:

S1	S0	Y_i
0	0	0
0	1	X_i
1	0	$\overline{X_i}$
1	1	1

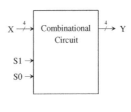

ii) Using the results of part i), design a 4-bit, 8-function arithmetic unit that will function as follows:

S2	S1	S0	F
0	0	0	A
0	0	1	A plus B
0	1	0	A plus $\overline{B}$
0	1	1	A minus 1
1	0	0	A plus 1
1	0	1	A plus B plus 1
1	1	0	A plus $\overline{B}$ plus 1
1	1	1	A

6.18   Design a 4-bit, 8-function arithmetic unit that will meet the following specifications:

S2	S1	S0	F
0	0	0	2A
0	0	1	A plus $\overline{B}$
0	1	0	A plus B
0	1	1	A minus 1
1	0	0	2A plus 1
1	0	1	A plus $\overline{B}$ plus 1
1	1	0	A plus B plus 1
1	1	1	A

6.19   (a)   Using a 4-bit binary adder with inputs (A, B, and C_{in}), outputs (F and C_{out}), and one selection bit (S0), design an arithmetic circuit as follows:

S0	FUNCTION TO BE PERFORMED
0	A plus B
1	B plus 1

        (b)   Using another selection bit S1, modify the circuit of i) to include the arithmetic and logic functions as follows:

S1	S0	FUNCTION TO BE PERFORMED
0	0	F = A plus B
0	1	F = B
1	0	F = shift left (logical) A
1	1	F = $\overline{A}$

        (c)   Design a 4-bit logic unit that will function as follows:

S1	S0	F
0	0	A + B
0	1	A • B
1	0	$\overline{A}$
1	1	A ⊕ B

6.20   Design and implement a 6 × 6 array multiplier.

6.21   Design an unsigned 8 × 4 non-additive multiplier using additive-multiplier-module whose block diagram representation is as follows:

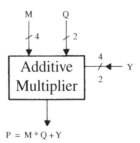

$$P = M * Q + Y$$

Assume that M, Q, and Y are unsigned integers.

6.22   Consider the registers and ALU shown in Figure P6.22:

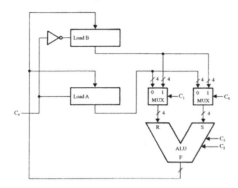

The interpretation of various control points are summarized as follows:

C_3	C_2	F
0	0	R plus S
0	1	R minus S
1	0	R and S
1	1	R EX-OR S

C_1	C_0	R-INPUT	S-INPUT
0	0	A	A
0	1	A	B
1	0	B	A
1	1	B	B

C_4	ACTION
0	B ← F
1	A ← F

FIGURE P 6.22.

Answer the following questions by writing suitable control word(s). Each control word must be specified according to the following format: C_4 C_3 C_2 C_1 C_0 For example:

$$C_4 \quad C_3 \quad C_2 \quad C_1 \quad C_0$$
$$1 \quad \ 0 \quad \ 0 \quad \ 0 \quad \ 1 \ ; A \leftarrow A \text{ plus } B$$

(a)   How will the A register be cleared? (Suggest at least two possible ways.) DIRECT CLEAR input is not available.
(b)   Suggest a sequence of control words that exchanges the contents of A and B registers (exchange means A ← B and B ← A).

6.23   Consider the following algorithm:

Declare registers A [8], B [8], C [8];
START: A ← 0; B ← 00001010;
LOOP: A ← A + B; B ← B – 1;
            If B < > 0 then go to LOOP
            C ← A;
HALT: Go to HALT

Design a hardwired controller that will implement this algorithm.

6.24   What is the basic difference between main memory and secondary memory?

6.25   A microcontroller has 24 address pins. What is the maximum size of the main memory?

6.26   What is the basic difference between: (a) EPROM and EEPROM? (b) SRAM and DRAM?

6.27   Given a memory with a 14-bit address and an 8-bit word size.

    (a)   How many bytes can be stored in this memory?

    (b)   If this memory were constructed from 1K × 1 RAMs, how many memory chips would be required?

    (c)   How many bits would be used for chip select?

6.28   Draw a block diagram showing the address and data lines for the 2732, and 2764 EPROM chips.

6.29   (a)   How many address and data lines are required for a 1M × 16 memory chip?

    (b)   What is the size of a decoder with one chip enable ($\overline{\text{CE}}$) to obtain a 64K × 32 memory from 4K × 8 chips? Where are the inputs and outputs of the decoder connected?

6.30   A microcontroller with 24 address pins and eight data pins is connected to a 1K × 8 memory with one enable. How many unused address bits of the microcontroller are available for interfacing other 1K × 8 memory units? What is the maximum directly addressable memory available with this microcontroller?

6.31   Name the methods used in main memory array design. What are the advantages and disadvantages of each?

6.32   The block diagram of a 512 × 8 RAM is shown in Figure P6.32. In this arrangement the memory unit is enabled only when $\overline{\text{CS1}}$ = L and CS2 = H. Design a 1K × 8 RAM system using the 512 × 8 RAM as the building block. Draw a neat logic diagram of your implementation. Assume that the CPU can directly address 64K with a $\text{R}/\overline{\text{W}}$ and eight data pins. Using linear decoding and don't-care conditions as 1's, determine the memory map in hexadecimal.

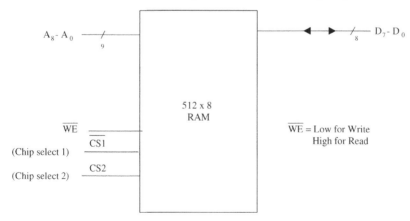

FIGURE P6.32.

6.33   Consider the hardware schematic shown in Figure P6.33.

    (a)   Determine the address map of this system. *Note:* $\overline{\text{MEMR}}$ = 0 for read, $\overline{\text{MEMR}}$ = 1 for write, M/$\overline{\text{IO}}$ = 0 for I/O and M/$\overline{\text{IO}}$ = 1 for memory.

(b) Is there any possibility of bus conflict in this organization? Clearly justify your answer.

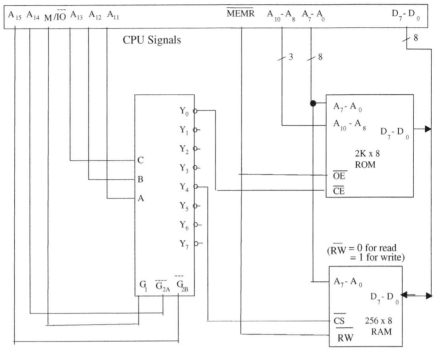

FIGURE P6.33.

6.34 Interface a CPU with 16-bit address pins and 8-bit data pins and a pin to a $1K \times 8$ EPROM and two $1K \times 8$ RAM's such that the memory map shown in Table P6.34 is obtained: Assume that both EPROM and RAM contain two enable pins: $\overline{CE}$ and $\overline{OE}$ for the EPROM, $\overline{CE}$ and $\overline{WE}$ for each RAM. Note that $\overline{WE} = 1$ and $\overline{WE} = 0$ indicate read and write operations for the RAM chip. Use a decoder block identical to the 74138.

TABLE P6.34

Device	Size	Address Assignment (Hex)
EPROM	$1K \times 8$	8000–83FF
RAM chip 0	$1K \times 8$	9000–93FF
RAM chip 1	$1K \times 8$	C000–C3FF

6.35 Repeat Problem 6.34 to obtain the memory map shown in Table P6.35 using a decoder block identical to the 74138.

TABLE P6.35

Device	Size	Address Assignment in hex
EPROM	$1K \times 8$	7000–73FF
RAM 0	$1K \times 8$	D000–D3FF
RAM 1	$1K \times 8$	F000–F3FF

6.36   What is meant by *foldback* in linear decoding?

6.37   Define the three types of I/O. Identify each as either CPU-initiated or device-initiated.

6.38   What is the basic difference between standard I/O and memory-mapped I/O? Identify the programmed I/O technique used by the PIC18F.

6.39   What is the difference between memory mapping in a microcontroller and memory-mapped I/O?

6.40   Discuss the basic difference between polled I/O and interrupt I/O.

6.41   What is the difference between subroutine and interrupt I/O?

6.42   What is an interrupt address vector?

6.43   Summarize the basic difference between maskable and nonmaskable interrupts. Describe how power failure interrupt is normally handled.

6.44   Discuss the basic difference between internal and external interrupts.

7

MICROCONTROLLER BASICS

In this chapter we describe the fundamental material needed to understand the basic characteristics of microcontrollers. It includes topics such as typical microcontroller architectures, timing signals. An overview of pipelining and RISC vs. CISC, PIC18F functional block diagram is included. Finally, an introduction to programming languages is also covered.

7.1 Basic Blocks of a Microcontroller

In order to understand the functions performed by typical modules contained in a microcontroller, it is necessary to cover the basic blocks of a microcomputer.

A microcomputer has three basic blocks: a microprocessor (CPU on a chip), a memory unit, and an input/output (I/O) unit. Figure 7.1 shows the basic blocks of a microcomputer. A system bus (comprised of several wires) connects these blocks. The *CPU* executes all the instructions and performs arithmetic and logic operations on data. The CPU of the microcomputer contains all the registers and the control unit, as well as arithmetic-logic circuits of the microcomputer.

A *memory unit* stores both data and instructions. The memory section typically contains ROM and RAM chips. The ROM can only be read and is nonvolatile; that is, it retains its contents when the power is turned off. A ROM is typically used to store instructions and data that do not change. For example, it might store a table of seven-segment codes for outputting data to a display external to the microcomputer for turning on a digit from 0 through 9.

One can read from and write into a RAM. The RAM is volatile; that is, it does not retain its contents when the power is turned off. A RAM is used to store instructions and data that are temporary and might change during the course of executing a program. An *I/O unit* transfers data between the microcomputer and the external devices via I/O ports (registers).

In a single-chip microcomputer, these three elements are on one chip, whereas in a single-chip microprocessor, separate chips are required for memory and I/O. Microcontrollers, which evolved from single-chip microcomputers, are typically used for dedicated applications such as automotive systems, home appliances, and home entertainment systems. Microcontrollers include a CPU, memory, and IOP (I/O and Peripherals) on a single chip. Note that a typical IOP contains I/O unit of a microcomputer, timers, A/D (analog-to-digital) converter, analog comparators, serial I/O, and other peripheral functions (to be discussed later). Two typical microcontrollers are Microchip Technology's PIC18F (8-bit) and Texas Instruments MSP 430 (16-bit).

Since the microcomputer is an integral part of a microcontroller, it is necessary to investigate a typical microcomputer in detail. Once such a clear understanding

Fundamentals of Digital Logic and Microcontrollers, Sixth Edition. M. Rafiquzzaman.
© 2014 John Wiley & Sons, Inc. Published 2014 by John Wiley & Sons, Inc.

is obtained, it will be easier to work with any specific microcontroller. Figure 7.2 illustrates a very simplified version of a typical microcomputer and shows the basic blocks of a microcomputer system. The various buses that connect these blocks are also shown. Although this figure looks very simple, it includes all of the main elements of a typical microcomputer system.

7.1.1 System Bus

The microcomputer's system bus (internal to the microcontroller) contains three buses, which carry all of the address, data, and control information involved in program execution. These buses connect the CPU to each of the ROM, RAM, and I/O chips so that information transfer between the CPU and any of the other elements can take place. In a microcomputer, typical information transfers are carried out with respect to the memory or I/O. When a memory or an I/O unit receives data from the CPU, it is called a *WRITE operation*, and data are written into a selected memory location or an I/O port (register). When a memory or an I/O unit sends data to the CPU, it is called a *READ operation*, and data are read into the CPU from a selected memory location or an I/O port.

In the *address bus*, information transfer takes place in only one direction, from the CPU to the memory or I/O elements.

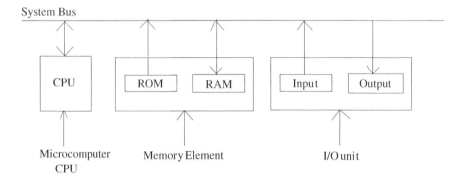

FIGURE 7.1   Basic blocks of a microcomputer.

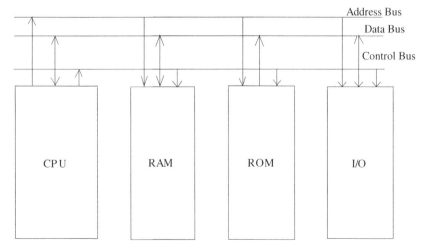

FIGURE 7.2   Simplified version of a typical microcomputer.

This is therefore called a *unidirectional bus*. The size of the address bus determines the total number of memory addresses available in which programs can be executed by the microprocessor. The address bus is specified by the total number of address bits required by the CPU. This also determines the direct addressing capability or the size of the main memory of the microcontroller. The microcontroller's CPU can execute programs located only in the main memory. For example, a CPU with 16 address bits can generate $2^{16} = 65,536$ bytes [64 kilobytes (kB)] of different possible addresses (combinations of 1's and 0's) on the address bus. The CPU includes addresses from 0 to $65,535_{10}$ (0000_{16} through $FFFF_{16}$). A memory location can be represented by each of these addresses. For example, an 8-bit data item $2B_{16}$ can be stored at 16-bit address 0200_{16}.

When a CPU with a 16-bit address bus wants to transfer information between itself and a certain memory location, it generates the 16-bit address from an internal register on its 16 address pins, $A_0–A_{15}$, which then appears on the address bus. These 16 address bits are decoded to determine the desired memory location. The decoding process normally requires hardware (decoders) not shown in Figure 7.2. In the *data bus*, data can flow in both directions, that is, to or from the CPU. This is therefore a bidirectional bus.

The *control bus* consists of a number of signals that are used to synchronize operation of the individual microcomputer elements. The CPU sends some of these control signals to the other elements to indicate the type of operation being performed. Each microcontroller has a unique set of control signals. However, some control signals such as Read and Write are common to all microcontrollers.

7.1.2   Clock Signals

The system clock signals are contained in the control bus. These signals generate the appropriate clock periods during which instruction executions are carried out by the CPU. Typical microcontrollers have an internal clock generator circuit to generate a clock signal. Figure 7.3 shows a typical clock signal.

The number of cycles per second (Hertz, abbreviated as Hz) is referred to as the *clock frequency*. The CPU clock frequencies of typical microcontrollers vary from 1MHz (1×10^6Hz) to 40MHz (40×10^6Hz). The clock defines the speed of the microcontroller. Note that one clock cycle = $1/f$, where f is the clock frequency. The execution times of microcontroller instructions are provided in terms of the number of clock cycles.

For example, suppose that execution time for the addition instruction by a microcontroller is one cycle. This means that a microcontroller with a 40MHz clock will execute the ADD instruction in 25 nanoseconds [clock cycle = $1/(40 \times 10^6)$ = 25 nanoseconds]. On the other hand, for a 4MHz microcontroller, the addition instruction will be executed in 250 nanoseconds [clock cycle = $1/(4 \times 10^6)$ = 250 nanoseconds]. This implies that the higher the clock frequency, the faster the microcontroller can execute the instructions.

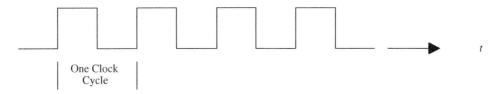

FIGURE 7.3   Typical clock signal.

7.2   Microcontroller Architectures

The microcontroller requires memory to store programs and data. The various microcontrollers available today are basically the same in principle. The main variations are in the number of memory units, and the address and data buses they use. As mentioned in Chapter 1, two types of CPU architectures are used for designing microcontrollers. They are von Neumann (Princeton) and Harvard.

In von Neumann architecture, a single memory system with the same address and data buses is used for accessing both programs and data. This means that programs and data cannot be accessed simultaneously. This may slow down the overall speed. Texas Instrument's MSP 430 uses von Neumann architecture. Figure 7.4 shows a block diagram of the von Neumann architecture.

Harvard architecture is a type of microcontroller architecture that uses separate program and data memory units along with separate buses for instructions and data. This means that these microcontrollers can execute instructions and access data simultaneously. Microcontrollers designed with this architecture require four buses for program memory and data memory. These are: one data bus for instructions, one address bus for addresses of instructions, one data bus for data, and one address bus for addresses of data. The sizes of the address and data buses for instructions may be different from those of the address and data buses for data. Several microcontrollers including the PIC18F are designed using the Harvard architecture. Figure 7.5 shows a block diagram of the Harvard architecture.

Most microcontrollers use the Harvard architecture. This is because it is inexpensive to implement these buses inside the chip since both program and data memories are internal to the chip.

Although processors designed using the von Neumann architecture are slower compared to the Harvard architecture since instructions and data cannot be accessed simultaneously because of the single bus, typical microprocessors such as the Pentium use this architecture. This is because memory units such as ROMs and RAMs are external to the microprocessor. This will require almost half the number of wires on the mother board since address and data pins for only two buses rather than four buses (Harvard architecture) are required. This is the reason Harvard architecture would be very expensive if utilized in designing microprocessors. Note that microcontrollers using Harvard architecture internally will have to use von Neumann architecture externally.

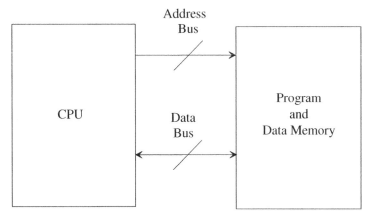

FIGURE 7.4   von Neumann architecture.

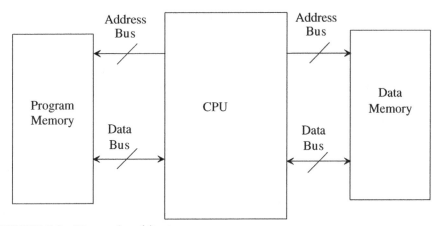

FIGURE 7.5   Harvard architecture.

Finally, depending on the register section, the CPU can be classified either as an accumulator-based or general-purpose register-based CPU. In an accumulator-based CPU, data are assumed to be held in a register called the accumulator. All arithmetic and logic operations are performed using the accumulator as one of the data sources. Microchip Technology's PIC18F is accumulator-based. The general-purpose register-based CPU, on the other hand, contains several general purpose registers. Any general-purpose register can be used as an accumulator. Texas Instruments MSP430 is general-purpose register-based.

7.3   Basic Concept of Pipelining

Since typical microcontrollers such as the PIC18F implement pipelining, basic concepts associated with pipelining will be covered in this section. As mentioned in Chapter 6, in order to execute a program, a conventional CPU repeats the following three steps for completing each instruction:

1. *Fetch.* The CPU fetches (instruction read) the instruction from the program memory (external to the CPU) into the instruction register.
2. *Decode.* The CPU decodes or translates the instruction using the control unit. The control unit inputs the contents of the instruction register, and then decodes (translates) the instruction to determine the instruction type.
3. *Execute.* The CPU executes the instruction using the control unit. To accomplish the task, the control unit generates a number of enable signals required by the instruction.

For example, suppose that it is desired to add the contents of two registers, X and Y, and store the result in register Z. To accomplish this, the conventional CPU performs the following steps:

1. The CPU fetches the instruction into the instruction register.
2. The control unit (CU) decodes the contents of the instruction register.
3. The CU executes the instruction by generating enable signals for the register and ALU sections to perform the following:
   a. The CU transfers the contents of registers X and Y from the Register section into the ALU.
   b. The CU commands the ALU to ADD.

    c. The CU transfers the result from the ALU into register Z of the register
       section.

Hence, the conventional CPU executes a program by completing one instruction
at a time and then proceeds to the next. This means that the control unit would have
to wait until the instruction is fetched from memory. Also, the ALU would have to wait
until the required data are obtained. Since the speeds of microcontrollers are increasing
at a more rapid rate than memory speeds, the control unit and ALU will be idle while
the conventional CPU fetches each instruction and obtains the required data. Typical
microcontrollers such as the PIC18F utilize the control unit and ALU efficiently by
prefetching the next instruction(s) and the required data before the control unit and ALU
require them. As mentioned earlier, conventional CPUs execute programs in sequence;
typical microcontrollers such as the PIC18F, on the other hand, implement the feature
called *pipelining* to prefetch the next instruction while the control unit is busy executing the
current instruction. Hence, PIC18F implements pipelining to increase system throughput.

The basic concepts associated with pipelining will be considered next. Assume
that a task T is carried out by performing four activities: Al, A2, A3, and A4, in that
order. Hardware Hi is designed to perform activity Ai. Hi is referred to as a *segment*,
and it essentially contains combinational circuit elements. Consider the arrangement
shown in Figure 7.6.

In this configuration, a latch is placed between two segments so the result
computed by one segment can serve as input to the following segment during the next
clock period. The execution of four tasks Tl, T2, T3, and T4 using the hardware of
Figure 7.6 is described using the space-time chart shown in Figure 7.7.

Initially, task Tl is handled by segment 1. After the first clock, segment 2 is busy
with Tl while segment 1 is busy with T2. Continuing in this manner, task Tl is completed at
the end of the fourth clock. However, following this point, one task is shipped out per clock.
This is the essence of the pipelining concept. A pipeline gains efficiency for the same reason
as an assembly line does. Several activities are performed but not on the same material.

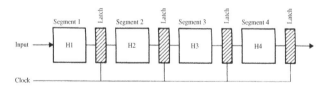

FIGURE 7.6   Four-segment pipeline.

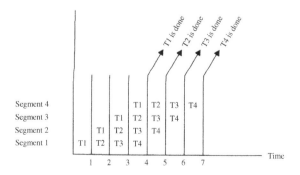

FIGURE 7.7   Overlapped execution of four tasks using a pipeline.

The PIC18F implements a two-stage pipeline. As mentioned earlier, the execution of an instruction by a typical CPU is completed in two stages. During the first stage, the instruction is fetched from program memory. During the second stage, the PIC18F CPU while executing the instruction, fetches the next instruction. This is called *two-stage instruction pipelining*, and is used by the PIC18F to increase the speed of instruction execution. It should be mentioned that when the PIC18F fetches a branch instruction, it clears or flushes the pipeline and executes a new sequence of instructions starting at the new branch address.

7.4 RISC vs. CISC

There are two types of CPU architectures: RISC (reduced instruction set computer) and CISC (complex instruction set computer). A RISC microcontroller such as the PIC18F emphasizes simplicity and efficiency. RISC designs start with a necessary and sufficient instruction set. The purpose of using RISC architecture is to maximize speed by reducing clock cycles per instruction. Almost all computations can be obtained from a few simple operations. The goal of RISC architecture is to maximize the effective speed of a design by performing infrequent operations in software and frequent functions in hardware, thus obtaining a net performance gain. The following list summarizes the typical features of a RISC CPU:

1. The RISC CPU is designed using hardwired control with little or no microcode. Note that variable-length instruction formats generally require microcode design. All RISC instructions have fixed formats, so microcode design is not necessary.
2. A RISC CPU executes most instructions in a single cycle.
3. The instruction set of a RISC CPU typically includes only register, load, and store instructions. All instructions involving arithmetic operations use registers, and load and store operations are utilized to access memory.
4. The instructions have a simple fixed format with few addressing modes.
5. A RISC CPU processes several instructions simultaneously and thus includes pipelining.
6. Software can take advantage of more concurrency. For example, jumps occur after execution of the instruction that follows. This allows fetching of the next instruction during execution of the current instruction.

RISC CPUs are suitable for embedded applications. *Embedded controllers* are embedded in the host system. This means that the presence and operation of these controllers are basically hidden from the host system. Typical embedded control applications include office automation systems such as printers.

RISC CPUs are well suited for applications such as image processing, robotics, and instrumentation. The key features of the RISC CPUs that make them ideal for these applications are their relatively low level of integration in the chip and instruction pipeline architecture. These characteristics result in low power consumption, fast instruction execution, and fast recognition of interrupts.

CISC CPUs such as the Motorola /Freescale HC16 CPU contain a large number of instructions and many addressing modes. In contrast, RISC CPUs such as the PIC18F include a simple instruction set with a few addressing modes. Almost all computations can be obtained from a few simple operations. RISC basically supports a small set of commonly used instructions that are executed at a fast clock rate compared to CISC, which contains a large instruction set (some of which are rarely used) executed at a slower clock rate. To implement the fetch/execute cycle for supporting a large instruction set for CISC, the clock is typically slower.

In CISC, most instructions can access memory whereas RISC contains mostly load/store instructions. The complex instruction set of CISC requires a complex control unit, thus requiring microprogrammed implementation. RISC utilizes hardwired control which is faster. CISC is more difficult to pipeline; RISC provides more efficient pipelining. An advantage of CISC over RISC is that complex programs require fewer instructions in CISC with fewer fetch cycles, while RISC requires a large number of instructions to accomplish the same task with several fetch cycles. However, RISC can significantly improve its performance with a faster clock, more efficient pipelining, and compiler optimization.

7.5   Functional Representation of a Typical RISC Microcontroller— The PIC18F4321

Figure 7.8 depicts the functional block diagram of the PIC18F4321 microcontroller. The block diagram can be divided into three sections, namely, CPU, memory, and I/O (input/output). A brief description of these blocks will be provided in the following.

The PIC18F4321 CPU contains registers, ALU, an instruction decode and control unit, along with the oscillator blocks. Typical CPU registers include IR (Instruction Register), W (data register called *Working register* or *Accumulator*), PC (Program Counter), three memory address registers (FSR0 through FSR2). An on-chip hardware multiplier is also included for performing unsigned multiplication. The on-chip memory contains program memory and data memory. As mentioned before, the PIC18F4321 is designed using the Harvard architecture; a separate 21-bit address bus for program memory and a separate 12-bit address bus for data memory are shown in the figure.

The on-chip I/O block includes five I/O ports (Port A through Port E), four hardware timers, a 10-bit ADC (analog-to-digital Converter), and CCP (capture, compare, PWM) and associated modules. Note that the PIC18F4321 can perform functions such as capture, compare, and pulse width modulation (PWM) using the timers and CCP modules. The PIC18F4321 can compute the period of an incoming signal using the capture module. The PIC18F4321 can produce a periodic waveform or time delays using the compare module. The PIC18F4321's on-chip PWM can be used to obtain pulse waveforms with a particular period and duty cycle, ideal for applications such as motor control.

7.6   Basics of Programming Languages

Microcontrollers are typically programmed using semi-English-language statements (assembly language). In addition to assembly languages, micrococontrollers use a more understandable human-oriented language called *high-level language*. No matter what type of language is used to write programs, microcontrollers understand only binary numbers. Therefore, all programs must eventually be translated into their appropriate binary forms.

Programming languages can typically be divided into three main types: machine language, assembly language, and high-level language. A *machine language program* consists of either binary or hexadecimal op-codes. Programming a microcontroller with either one is relatively difficult, because one must deal only with numbers. The CPU architecture of the microcontroller determines all of its instructions. These

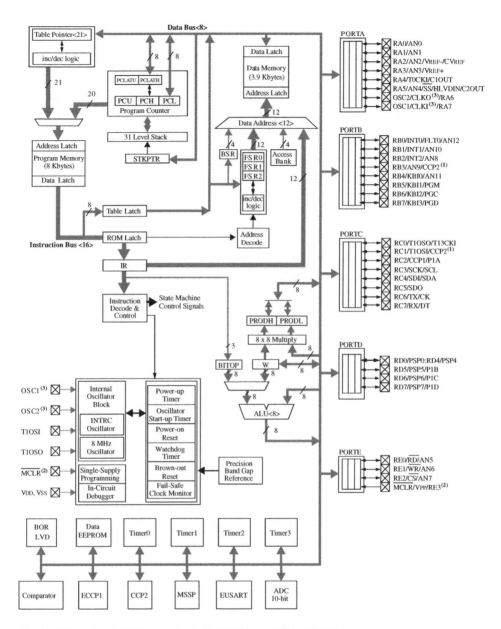

FIGURE 7.8   PIC18F4321 Functional block diagram.

instructions are called the microcontroller's *instruction set*. Programs in *assembly* and *high-level languages* are represented by instructions that use English-language-type statements. The programmer finds it relatively more convenient to write programs in assembly or high-level language than in machine language. However, a translator must be used to convert such programs into binary machine language so that the microcontroller can execute the programs. This is shown in Figure 7.9.

FIGURE 7.9   Translating assembly or high-level language into binary machine language.

An *assembler* translates a program written in assembly language into a machine language program. A *compiler* or *interpreter*, on the other hand, converts a high-level language program such as C into a machine language program. Assembly or high-level language programs are called *source codes*. Machine language programs are known as *object codes*. A *translator* converts source codes to object codes. Next, we discuss the three main types of programming language in more detail.

7.6.1   Machine Language

A microcontroller has a unique set of machine language instructions defined by its manufacturer. No two microcontrollers by different manufacturers have the same machine language instruction set. For example, Microchip Technology's PIC18F microcontroller uses the code $D7FF_{16}$ for the assembly language statement "HERE BRA HERE" (branch always to HERE), whereas the Motorola/Freescale HC11 microcontroller uses the code $20FE_{16}$ for the same statement with its BRA instruction. Therefore, a machine language program for one microcontroller will not run on the microcontroller of a different manufacturer.

At the most elementary level, a microcontroller program can be written using its instruction set in binary machine language. As an example, the following program adds two numbers using the PIC18F machine language:

0000111000000010

0000111100000011

0110111001000000

1110111100000011

Obviously, the program is very difficult to understand unless the programmer remembers all of the PIC18F codes, which is impractical. Because one finds it very inconvenient to work with 1's and 0's, it is almost impossible to write an error-free program on the first try. Also, it is very tiring for a programmer to enter a machine language program written in binary into the microcontroller's RAM. For example, the programmer needs a number of binary switches to enter the binary program. This is definitely subject to error.

To increase the programmer's efficiency in writing a machine language program, hexadecimal numbers rather than binary numbers are used. The following is the same addition program in hexadecimal using the PIC18F instruction set:

0E02

0F03

6E40

EF03

It is easier to detect an error in a hexadecimal program, because each byte contains only two hexadecimal digits. One would enter a hexadecimal program using a hexadecimal keyboard. A keyboard monitor program in ROM provides interfacing of the hexadecimal keyboard with the microcontroller. This program must convert each key actuation into binary machine language in order for the microcontroller to understand the program. However, programming in hexadecimal is not normally used.

7.6.2 Assembly Language

The next programming level uses assembly language. Each line in an assembly language program includes four fields:

- label field
- instruction, mnemonic, or op-code field
- operand field
- comment field

As an example, a typical program for adding two 8-bit numbers written in PIC18F assembly language is as follows:

Label	Mnemonic	Operand	Comment
	MOVLW	1	;Move 1 into accumulator
	ADDLW	2	;Add 2 with 3 , store result in accumulator
	SLEEP		;Halt

Obviously, programming in assembly language is more convenient than programming in machine language, because each mnemonic gives an idea of the type of operation it is supposed to perform. Therefore, with assembly language, the programmer does not have to find the numerical op-codes from a table of the instruction set, and programming efficiency is improved significantly.

An assembly language program is translated into binary via a program called an *assembler*. The assembler program reads each assembly instruction of a program as ASCII characters and translates them into the respective binary op-codes. For example, the PIC18F assembler translates the SLEEP (places the PIC18F in sleep mode; same as HALT instruction in other processors) instruction into its 16-bit binary op-code as 0000 0000 0000 0011 (0003 in hex), as depicted in Table 7.1.

7.6.3 High-Level Language

As mentioned earlier, a programmer's efficiency increases significantly with assembly language compared to machine language. However, the programmer needs to be well acquainted with the CPU architecture and its instruction set. Further, the programmer has to provide an op-code for each operation that the CPU has to carry out in order to execute a program. As an example, for adding two numbers, the programmer would instruct the CPU to load the first number into a register, add the second number to the register, and then store the result in memory. However, the programmer might find it tedious to write all the steps required for a large program. Also, to become a reasonably good assembly language programmer, one needs to have a lot of experience.

TABLE 7.1 Conversion of PIC18F SLEEP instruction into its binary op-code

Assembly code	Binary form of ASCII codes as seen by the assembler		Binary op-code created by the MPLAB PIC18F assembler
S	0101	0011	
L	0100	1000	0000 0000 0000 0011
E	0100	0101	
E	0100	0101	
P	0101	0000	

High-level language programs composed of English-language-type statements rectify all these deficiencies of machine and assembly language programming. The programmer does not need to be familiar with the internal microcontroller structure or its instruction set. Also, each statement in a high-level language corresponds to a number of assembly or machine language instructions. For example, consider the statement "f = a + b;" written in a high-level language called C. This single statement adds the contents of 'a' with 'b' and stores the result in f. This is equivalent to a number of steps in machine or assembly language, as mentioned before. It should be pointed out that the letters a, b, and f do not refer to particular registers within the CPU. Rather, they are memory locations.

C is a very popular language used and is widely used by microcontrollers. A high-level language is a problem-oriented language. The programmer does not have to know the details of the architecture of the microcontroller and its instruction set. Basically, the programmer follows the rules of the particular language being used to solve the at hand. A second advantage is that a program written in a particular high-level language can be executed by two different microcontrollers, provided that they both understand that language. For example, a program written in C for a PIC18F microcontroller will run on a Texas Instrument's MSP 430 microcontroller because both microcontrollers have a compiler to translate the C language into their particular machine language; minor modifications are required for I/O programs.

Typical microcontrollers are also provided with a program called an *interpreter*. This is provided as part of the software development package. The interpreter reads each high-level statement such as F = A + B and directs the microcontroller to perform the operations required to execute the statement. The interpreter converts each statement into machine language codes but does not convert the entire program into machine language codes prior to execution. Hence, it does not generate an object program. Therefore, an interpreter is a program that executes a set of machine language instructions in response to each high-level statement in order to carry out the function. A compiler, however, converts each statement into a set of machine language instructions and also produces an object program that is stored in memory. This object program must then be executed by the CPU to perform the required task in the high-level program.

In summary, an interpreter executes each statement as it proceeds, without generating an object code, whereas a compiler converts a high-level program into an object program that is stored in memory. This program is then executed. Compilers for microprocessors normally provide inefficient machine codes because of the general guidelines that must be followed for designing them. However, modern C compilers for microcontrollers generate very tight and efficent codes. Also, C is a high-level language that includes input/output instructions. Hence, C is a very populer programming language with microcontrollers.

7.7 Choosing a Programming Language

Compilers used to provide inefficient machine codes because of the general guidelines that must be followed for designing them. However, modern C compilers for microcontrollers generate very tight and efficient codes. Hence, C is widely used with microcontrollers these days. Assembly language programming, on the other hand , is important in the understanding the internal architecture of a microcontroller, and may be useful for debugging programs. A brief coverage of C is provided in the next section.

7.8 Introduction to C Language

As mentioned before, a programmer's efficiency increases significantly with assembly language compared to machine language. However, the programmer needs to be well acquainted with the microcontroller's architecture and its instruction set. Furthermore, the programmer has to provide an opcode for each operation that the microcontroller has to carry out in order to execute a program. As an example, for adding two numbers, the programmer would instruct the microcontroller to load the first number into a register, add the second number to the register, and then store the result in memory. However, the programmer might find it tedious to write all of the steps required for a large program. Also, to become a reasonably good assembly language programmer, one needs to have a lot of experience. Also, it takes a long time to debug assembly code.

High-level language programs composed of English-language-type statements rectify all of these deficiencies of machine and assembly language programming. The programmer does not need to be familiar with the internal microcontroller structure or its instruction set. Also, each statement in a high-level language corresponds to a number of assembly or machine language instructions. For example, consider the statement c = a + b; written in a high-level language such as C. This single statement adds the contents of a with b and stores the result in c. This is equivalent to a number of steps in machine or assembly language, as mentioned before. It should be pointed out that the letters a, b, and c do not refer to particular registers within the microcontroller. Rather, they are memory locations.

The C language is widely used at present. Typical microcontrollers such as the PIC18F family can be programmed using this high-level language. A high-level language is a problem-oriented language. The programmer does not have to know the details of the architecture of the microcontroller and its instruction set. Basically, the programmer follows the rules of the particular language being used to solve the problem at hand. A second advantage is that a program written in a particular high-level language can be executed by two different microcontrollers, provided that they both understand that language. For example, a program written in C for a PIC18F microcontroller will run on the HC16 microcontroller because both microcontrollers have a compiler to translate the C language into their particular machine language; minor modifications are required for I/O programs. C is a high-level language that includes I/O instructions.

Compilers for microprocessors normally provide inefficient machine codes because of the general guidelines that must be followed for designing them. However, modern C compilers for microcontrollers generate very tight and efficient codes. C is widely used these days for writing programs for both real-time and non-real-time applications with microcontrollers. *Real time* indicates that the task required by the application must be completed before any other input to the program can occur that would change its operation.

The C Programming language was developed by Dennis Ritchie of Bell Labs in 1972. C has become a very popular language for many engineers and scientists, primarily because it is portable except for I/O and, however, can be used to write programs requiring I/O operations with minor modifications. This means that a program written in C for the PIC18F4321 will run on the Texas Instruments MSP430 with some modifications related to I/O as long as C compilers for both microcontrollers are available.

C is a general-purpose programming language and is found in numerous applications as follows:

- **Systems Programming.** Many operating systems (such as UNIX and its variant LINUX), compilers, and assemblers are written in C. Note that an operating system typically is included with the personal computer when it is purchased. The operating system provides an interface between the user and the hardware by including a set of commands to select and execute the software on the system.

- **Computer-Aided Design (CAD) Applications.** CAD programs are written in C. Typical tasks to be accomplished by a CAD program are logic synthesis and simulation.

- **Numerical Computation.** Software written in C is used to solve mathematical problems such as solving linear system of equations and matrix inversion. Industry standard MATLAB software is written in C.

- **Other Applications.** These include programs for printers and digital control algorithms such as PI (Proportional Integral) and PID (Proportional Integral Derivative) algorithms using microcontrollers.

A C-program may be viewed as a collection of functions. Execution of a C program will always begin by a call to the function called *main*. This means that all C programs should have its main program named as **main**. However, one can give any name to other functions.

A simple C program that prints "I wrote a C-program" is

```
/* First C-program */
# include <stdio.h>
void main ( )
{
printf ("I wrote a C-program") ;
}
```

Here, `main ( )` is a function of no arguments, indicated by `( )`. The parenthesis must be present even if there are no arguments. The braces { } enclose the statements that make up the function. The line `printf("I wrote a C-program");` is a function call that calls a function named `printf`, with the argument `I wrote a C-program`. `printf` is a library function that prints output on the terminal. Note that `/* */` is used to enclose comments. These are not translated by the compiler. C++ compilers are used to compile C programs these days. Hence, // followed by comment can be used instead of /* */. Note that the comments in C++ are written after // and it spans until the end of the line.
A variation of the C program just described is

```
// Another C program
# include <stdio.h>
void main ( )
{
printf ("I wrote");
printf ("a C-");
printf ("program);
printf ("\n");
}
```

Here, #include is a preprocessor directive for the C language compiler. These directives give instructions to the compiler that are performed before the program is compiled. The directive `#include <stdio.h>` inserts additional statements in the program. These statements are contained in the file `stdio.h`. The file `stdio.h` is included with the standard C library. The `stdio.h` file contains information related to the input/output statement.

The `\n` in the last line of the program is C notation for the newline character. Upon printing, the cursor moves forward to the left margin on the next line. `printf` never supplies a newline automatically. Therefore, multiple `printf`'s may be used to output "I wrote a C-program" on a single line in a few steps. The escape sequence `\n` can be used to print three statements on three different lines. An illustration is given in the following:

```
# include <stdio.h>
void main (void)
{
printf ("I wrote a C-program \n");
printf ("This will be printed on a new line \n");
printf ("So also is this line \ n");
}
```

All variables in C must be declared before their use. The compiler provides an error message if one forgets a declaration. A declaration includes a type and a list of variables that have that type. For example, the declaration `int a,b` implies that the variables a and b are integers.

Next, write a program to add two integers a and b where a = 100 and b = 200. The C program is

```
# include <stdio.h>
void main ( )
{int a = 0x64, b = 0xc8; // a and b are specified in hex
printf ("The sum is %x\n", a+b);
}
```

This program shows how to declare two integers and initialize them with hexadecimal numbers. The format specifier %x allows the sum to be printed as a hexadecimal number. This program will print the sum as 0x12C which is 300 in decimal.

The scanf allows the programmer to enter data from the keyboard. A typical expression for `scanf` is `scanf("%d%d", &a, &b);`

This expression indicates that the two values to be entered via the keyboard are in decimal. These two decimal numbers are to be stored in addresses a and b. Note that the symbol `&` is an address operator.

The C program for adding and subtracting two integers a and b using `scanf` is

```
// C program that performs basic I/O
# include <stdio.h>
void main ( )
{
int a, b;
printf ("Input two integers: ");
scanf ("%d%d" , &a, &b) ;
printf ("Their sum is: %d\n", a+b);
printf ("The difference is: %d\n" , a-b);
}
```

In summary, writing a working C program involves four steps as follows:

Step 1: Using a text editor, prepare a file containing the C code. This file is called the *source file*.

Step 2: Preprocess the code. The preprocessor makes the code ready for compiling. The preprocessor looks through the source file for lines that start with a **#**. In the previous programming examples, #include <stdio.h> is a preprocessor directive. This directive copies the contents of the standard header file stdio.h into the source code. This header file stdio.h describes typical input/output functions such as scanf ( ) and printf ( ) functions.

Step 3: The compiler translates the preprocessed code into machine code. The output from the compiler is called *object code*.

Step 4: The linker combines the object file with code from the C libraries. For instance, in the examples shown here, the actual code for the library function printf ( ) is inserted from the standard library to the object code by the linker. The linker generates an executable file. Thus, the linker makes a complete program.

Before writing C programs, the programmer must make sure that the computer runs either the UNIX or MS-DOS operating system. Two essential programming tools are required. These are a text editor and a C compiler. The text editor is a program provided with a computer system to create and modify compiler files. The C compiler is also a program that translates C code into machine code.

In summary, the C language offers the following features:

- provides support to structured programming

- is portable and small size

- includes many operators for *low and high level operations*

- provides data structures such as arrays, strings, structures, and unions

7.8.1   Data Types

The data types in C language include char, int, float, and double. A variable declared as a char (character) usually holds eight bits of data. A variable of int (integer) type, on the other hand, can hold 16 or 32 bits of data. The type float specifies a 32-bit single precision floating point number. The type double can be used to declare a data as a 64-bit double precision floating point number. We will use only char and int data types in this book. Note that float and double data types are not needed in most of the microcontroller-based applications. In addition, floating-point computations are too costly in terms of space and time.

The qualifiers unsigned and signed can be used with char and int data types. The unsigned char is always positive, and covers a range of values from 0 to 255. Typical examples of unsigned char include age and memory address which are always positive.

The signed char covers a range of values from −128 to +127 (0 being positive). The C compilers use signed char as default. Hence, using char instead of signed char will specify the data as a signed character. Typical examples of signed char include voltage and temperature which can be positive and negative.

The unsigned integer covers a range of values from 0 to 65535 while the signed integer covers a range of values from −32,768 to + 32767 (0 being positive). The signed int or simply int (default for C compilers) can be used to specify values from -32,768 to +32,767 (0 being positive).

The lengths of integers can be modified using the qualifiers `shortlong` and `long`. For typical C compilers, shortlong is 24-bit while long is 32-bit.

Examples of declaring `char` and `int` data types are provided below:

```
unsigned char i; /* specifies i as an unsigned 8-bit
number*/
char x; /*declares x as a signed 8-bit number */
unsigned int b; /*declares b as ufnsigned 16-bit integer*/
int a; /* defines variable as 16-bit signed integer*/
```

7.8.2 Bit Manipulation Operators

C provides six bit manipulation operators as shown in Table 7.2. The applications of these bit manipulation operators are discussed in Chapter 4.

Typical examples for each of these operators are provided below:

```
0x24 & 0x0F = 0x04
0x70 | 0x02 = 0x72
0xE1 ^ 0xFF = 0x1E
~ 0x25 = 0xDA
0x27 >>2 = 0x09
0xA1 << 3 = 0x08
```

The bit manipulation operators are very common in I/O operations. Hence, some examples showing their applications in bit manipulation for PIC18F I/O ports are provided in the following.

The AND operator is typically used for clearing one or more bits to 0. For example, the C statement `PORTC = PORTC & 0x7F;` will clear bit 7 of PORTC to 0 without changing the other bits of PORTC. The OR operator is typically used to set one or more bits to 1.

For example, the C statement `PORTD = PORTD | 0x05;` will insert 1's at bits 0 and 2 of PORTD without changing the other bits of PORTD. The XOR is typically used to find one's complement of (toggle) one or more bits. For example, the C statement `PORTC = PORTC ^ 0xFF;` will toggle all bits of PORTC.

The above three statements can also be specified in a compact form as shown below:

$$PORTC \,\&= 0x7F;$$
$$PORTD \,|= 0x05;$$
$$PORTC \,^\wedge= 0xFF;$$

Next, let us discuss some of the applications of the logic operators.

The left shift operation is very useful for multiplying an unsigned number by 2^n by shifting it n times to left provided a '1' is not shifted out of the most significant bit. For example, consider $y = 10*x$. Note that 10_{10} in binary is 1010_2.

Also, $10*x = 8*x + 2*x$. This means that $8*x = x<<3$, and $2*x = x<<1$.

Hence, $y = (x<<3) + (x<<1)$.

The right shift operation, on the other hand, is very convenient for dividing

TABLE 7.2   Bit manipulation operations in C

Logic operators	Operation performed
&	AND
|	OR
^	XOR
! or ~	NOT
>>	Right Shift
<<	Left Shift

an unsigned number by 2^n by shifting it n times to the right, provided a '1' is not shifted out of the least significant bit. As an example, consider y = (a + b)/2.

Note that y = (a + b)/2 = (a+b) >>1

Similarly, y = (a + b +c +d)/4 = (a + b + c + d) >>2

In order illustrate the Exclusive-OR (XOR) operator, consider X = Y. This expression is the same as !(X^Y). When X equals Y then X^Y = 0x00, and !(0x00) evaluates to true. The XOR can also be used to swap two variables without the need for a temporary variable. The following example illustrates this:

SWAP with temporary variable
temp = x;
x = y;
y = temp;
SWAP without temporary variable
y = x ^ y;
x = x ^ y;
y = x ^ y;

Let us verify the above using a numerical example. With x = 1001 and y = 0111, y = x ^ y = (1001)^(0111) = 1110, x = x ^ y = (1001) ^ (1110) = 0111, y = x ^ y = (0111)^(1110) = 1001. Hence, y = x. The above swapping of x with y works because of the following identity,

p /(p /q) = q

Also, the above statements can be represented in compact form as:

y ^ = x;
x ^ = y;
y ^ = x;

Example 7.1

Write a C program to convert a 16-bit number, each byte containing an ASCII digit into a packed BCD byte.

Solution

```
# include <p18f4321.h>
void main ()
{
unsigned char a, b, c;
unsigned char addr1 = 0x32;
unsigned char addr2 = 0x31;
unsigned char addr3;
addr1 = addr1 & 0x0f     // Mask off upper four bits of the low
                         //byte
addr2 = addr2 & 0x0f;    // Mask off upper four bits of the high
                         //byte
addr2 = addr2 << 4;      // Shift high byte 4 times to left
addr3 = addr2 | addr1; // Packed BCD byte in addr3
}
```

7.8.3 Control Structures

Control structures allow programmers to modify control flow which is a sequential flow by default. Structures allow one to make decisions and create loops which make the hardware to replicate execution of statements. Typical structured control structures in C include `if-else`, `switch`, `while`, `for` and, `do-while`.

The `if-else` Construct The syntax for the `if-else` construct is as follows:

```
if (cond)
statement1;
else
statement2;
```

Figure 7.10 shows the flowchart for the `if-else` construct.

This is a one-entry-one-exit structure in that if the condition is true, the statement1 is executed; else (if the condition is false), statement1 is skipped, and the statement2 is executed. An example of the `if-else` structure (flowchart in Figure 7.11) is provided in the following:

```
unsigned char x, y, z;
if (x < y)
z = x + y;
else
z = x -y;
```

In the above, if x is less than y, the unsigned 8-bit numbers x and y are added, and the 8-bit result is stored in z. On the other hand, if x is not less than y, then the statement z = x -y is executed. As another example, consider the following. This code

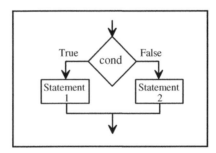

FIGURE 7.10   The `if-else` construct.

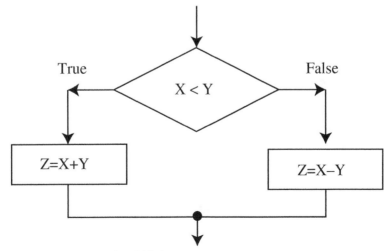

FIGURE 7.11   An example of if-else structure.

finds the larger of the two 8-bit unsigned numbers, a and b, and saves the result in big as follows :

```
unsigned char a, b, big;
if (a>b)
big = a;
else
big = b;
```

The flowchart in Figure 7.12 illustrates the above example.

Braces are required if multiple statements need to be executed if the condition is true or false. Consider x and y as two unsigned numbers. If x>y, then find x+y, and x/y and store them in u and v respectively. Otherwise, find x-y and increment w by 4, and store them in z and w respectively. Finally, compute x*y, and store in t. The following program illustrates this:

```
unsigned char t, u, v, w, x, y, z;
if (x < y) {
u = x + y; // Add x with y, and store in u
v = x / y; // Divide x by y and store in v
}
else {
z = x-y; // Subtract y from x and store result in z
w+ = 4; // Value of w is incremented by 4
}
t = x * y // Multiply x by y and store in t
```

In the above, if x < y, then the statements for u and v in the braces are executed. The statement t = x * y is then executed. If x is not greater than y, then only the statements for z and w+ in the braces are executed. The statement for t is then executed. In either case, t = x * y is executed. Figure 7.13 shows the flowchart for the above example.

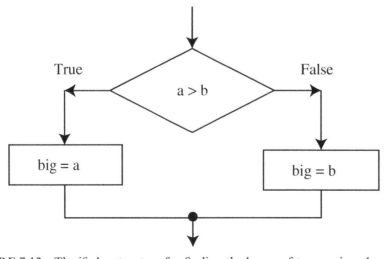

FIGURE 7.12   The if-else structure for finding the larger of two unsigned numbers.

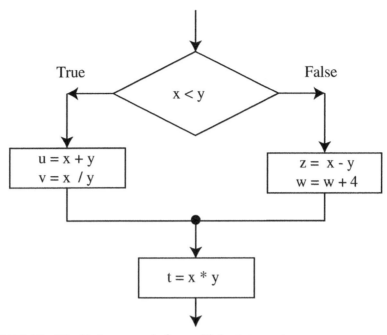

FIGURE 7.13   The if-else example for multiple statements.

Finally, the following example will illustrates the use of the `if-else` construct in an I/O application. Consider ports C and port D of the PIC18F microcontroller. These are 8-bit ports. TRISC and TRISD are 8-bit data direction registers of port C and port D respectively. A '0' written at a particular bit position of the TRISx (x = C or D) will configure the specified bit in the corresponding port (port C or port D) as an output while a '1' will configure it as an input. Hence, port C and port D of the PIC18F4321 can be configured as inputs and outputs, respectively, using the following C language statements:

```
TRISC = 0xFF; // Configure PORTC as an input port
TRISD = 0x00; // Configure PORTD as an output port
```

Next, suppose that there are two switches connected to bits 0 and 1 of PORTC, and an LED connected to bit 0 of port D. Using the else-if construct, a program is written in the following to turn the LED ON if the switches are either both LOW or both HIGH; otherwise, turn the LED OFF:

```
unsigned char X, Y;
TRISC = 0xFF;        //Configure PORTC as an input port
TRISD = 0x00;        //Configure PORTD as an output port
X = PORTC;           //Input switches via PORTC
X& = 0x03;           //Mask all bits except bits 0, 1 and retain
                     //switch values
if (X == 0)          //If both switches are LOW, turn LED ON Y = 1;
else if (X == 1)     //If switch at bit 1 is LOW and bit 0 is
                     //HIGH, turn LED OFF
Y = 0;
```

```
else if (X == 2)    //If switch at bit 1 is HIGH and bit 0 is
                    //LOW, turn LED OFF
Y = 0;
else Y = 1;         //If both switches are HIGH, turn LED ON
PORTD = Y;          //Output Y to PORTD
```

7.8.4 The `switch` Construct

The syntax for the `switch` expression is
```
switch (integer){
case 1:
statements;
break;
case 2:
statements;
break:
---
---
case n:
statement;
}
```

In the above, the integer included with the switch statement is compared with the each of the integers included with the case statements. If the values match, the statements associated with that case statement are executed. For example, consider switch (2). If integer = 2, then the statements with case 2 are executed. The break exits from the switch construct. The if-else statements of the switch/LED example of the last section can be replaced using the switch construct as follows:

```
switch (X) {
Case 0: Y = 1; break;
Case 1: Y = 0; break;
Case 2: Y = 0; break;
Case 3: Y = 1;
}
```

7.8.5 The `while` Construct

The `while` construct allows a programmer to describe loops. The syntax for the `while` construct is provided below:

```
while (condition) {
statements
}
next statement;
```

In the above, if the condition is true, the statements are executed, and then control is returned to the top of the loop. The condition is tested again. If the condition is true again, statements are executed, and control is returned to the top of the loop. The process is repeated as long as the condition is true. However, if the condition is false at the start or during repeating the process when the condition is checked, the statements in the braces are not executed. The next statement following the second brace is executed. Note that braces are not required for a single statement. Figure 7.14 depicts the flowchart for the while construct.

As an example, consider the following with a single statement in the loop:

```
int n = 4;
while (n <= 16)
n + = 4;
z = n;
```

the above, n is 4 before entering the while loop. The condition n <= 16 is true the first time. The expression n = n + 4 continuously changes the value of n in increments of 4 to 8, 12, 16, 20. As soon as n = 16, the condition n <= 16 is false, and the next statement z = n is executed, which assigns z with the value of 20.

Finally, note that infinite loop occurs when the condition in the while construct is always true. Note that

```
while (1)
```

describes an infinite loop in C.

The while loops can also be used to write software delay routines. A simple delay loop using the while construct is provided below:

```
unsigned int k = 1000; // initialize k to 1000
while (k>10)
k --;
```

A nested delay loop using the while construct is provided below:

```
unsigned int i, j, k ;
i = 0; k = 1000;
while (i < k) {
j = 1;
while (j < 100)
j++;
i++;
}
```

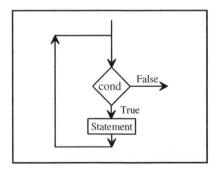

FIGURE 7.14   The while construct.

7.8.6 The `for` Construct

The `for` construct is another loop structure supported by the C language. It is more flexible than the while construct. Hence, the `for` construct is often a preferred choice. The syntax of a `for` construct is as follows:

```
for (e1; e2; e3)
s;
```

where e1, e2, and e3 are C expressions and 's' is a valid C statement.

Figure 7.15 shows the flowchart for the `for` loop. In Figure 7.15, the expression e1 is evaluated first. The expression e2 is then evaluated; if it is false, the loop terminates. Otherwise, the statement 's' is executed. The expressions e3 and e2 are then evaluated, and the process continues. The expression e1 normally contains code to initialize the loop. The expression e2 describes the exit condition. The purpose of the expression e3 is to modify the exit condition so that the loop terminates at some point. The following 'while' loop

```
int n = 4;
while (n <= 16)
n = n + 4;
z = n;
```

described earlier can be replaced with the 'for' loop provided below:

```
int n;
for (n = 4; n<= 16; n+ = 4)
z = n;
```

The single loop and nested loop delay routines can be written using the `for` construct provided in the following:
Simple delay routine using `for` with a single loop:

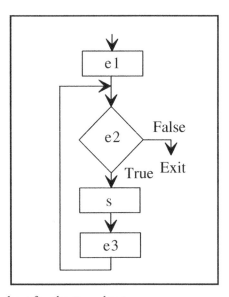

FIGURE 7.15   Flowchart for the `for` loop.

```
unsigned int k;
for (k = 1000 ; k>0 ; k--)
```
Delay routine using 'for' with nested loop:

```
unsigned int i, j, k = 1000;
for (i = 0 ; i < k ; i++)
for (j = 0 ; j<100 ; j++)
;
```

7.8.7 The do-while Construct

The C language provides another loop structure called the do-while. The syntax for the do-while construct is provided below:

```
do {
statements
} while (condition);
```

Figure 7.16 shows the flowchart for the do-while construct. The do-while loop is a post-checked loop in the sense that the exit condition is evaluated after executing all statements. If the condition evaluates to false, the loop terminates. Otherwise, the process continues until the condition becomes false. Whatever is accomplished by a do-while construct can be achieved by a for loop construct. Hence, this loop structure is not as popular as while or for loop structures.

7.8.8 Structures and Unions

In addition to built-in data types such as char and int, C supports user-defined nonhomogeneous collections. A structure permits the programmer to access a group of different data types using a common user-defined name. The structure can be declared using the keyword struct followed by a user-defined name. An example of the structure declaration is provided below:

```
struct struct_name {
int a;
char b ;
} my_struct ;
```

Note that in the above, the name is optional. However, if it is present, it defines the tag or user-defined name of the structure, and the tag can be used later. The variable name my_struct is of type struct_name. Also, Size of the structure = Sum of the sizes of its componenets. Hence, Size of the above structure = Size(int) + Size (char) = 2 bytes + 1 byte = 3 bytes.

Union is a space-saving structure. The memory referenced by the keyword union can store different types of data with the restriction that at any one time, the memory holds a single type of data. Note that different data types of the union share the same memory space. The data in a union must be referenced by a member of the proper data type. As an example, a union can be declared as follows:

```
union num {
char x ;
int a ;
} ;
```

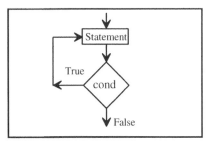

FIGURE 7.16   Figure for the `do-while` construct.

Note that Size of a union = Size of the largest data type held by the union. In the above, Size of the union "num" = Max {size of (int), size of (char)} = Max {2, 1} = 2 bytes.

In the above code, any variable of type `union` num can hold a char or an int. The names 'x' and 'a' specify which type of data is referenced. To specify a variable y of type `union` num, the following statement can be used:

        `union num y;`

Now, a character B can be assigned to x using the following:

`y . x = 'B' ;`

Next, an integer 2756 can be assigned to y using the following:

`y . a = 2756 ;`

Since at any one time y holds data of a single type, the assignment of 2756 to y cancels the assignment of 'B' to y.

As mentioned before, Union offers a space saving structure. For union variable, memory space is allocated to hold the largest number. This memory can then be used to hold the smaller numbers. As an example. consider the following:

```
union {
int a ;
char c[2] ;
} myu ;
```

In the above union, 16 bits of memory space are allocated to integer 'a'; the same 16- bit storage space is shared by 8-bit characters c[0] and c[1], as shown in Figure 7.17. This union allows a prgrammer to perform byte swap operation very efficiently.

7.8.9   Functions in C

A function in C allows a programmer to encapsulate a task. Hence, a function is a task-specialized module. A C-program is often comprised of a collection of functions. Functions are written and tested separately before they are added to the library. The end user can use the functions in a library as many times as needed. Since a function is tested thoroughly prior to placing in the library, its use not only reduces program development time, but also increases the software reliability.

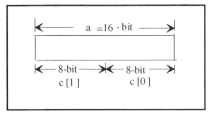

FIGURE 7.17   Example of space saving with union.

As an example, consider the add function shown below:

```
int add (int p , int q) {
int t;
t = p + q;
return t;
}
```

This function inputs two integers (p and q), and returns the sum of p and q via the local variable "t" of the function. Note that the function add (10, 12) will return 22 as the answer.

As another example, consider the following function which returns the number of ones in a given byte x:

```
unsigned char count (unsigned char x) {
unsigned char c;
for (c=0 ; x! = 0 ; x>>1)
if (x & 0x01)
c++;
return c;
}
```

In many situations, a function does not have to return a value. A typical example is the software delay loop provided below:

```
void delay (unsigned int p) {
unsigned int i ;
for (i = 0; i<p; i++);
}
```

The reserve word void indicates that this function does not return any result to the caller.

7.8.10 Macros

Macros can make programming in C easier by reducing the amount of code the programmer has to actually write. This is accomplished by letting the compiler produce those redundant pieces of program that are used routinely in various places in the program. Basically a macro is a set of codes used repeatedly throughout the program. When the macro is written, it is assigned a name. Rather than writing the same sequence of codes each time they are needed, the name of the macro is inserted in their place. During compiling, whenever the name of the macro is encountered, the compiler will insert the sequence of codes that this name represents.

The macro is in some way similar to a function, in that the usage is repeated. As an example, consider the following macro:

#define MPY10 (X) (X << 3) + (X<< 1)

This macro multiplies X by 10 using two left shift operations discussed earlier. A typical macro call is Z = MPY10 (Y);

Note that there is no transfer of control in macro. In contrast, a function call is performed via transfer of control in the program. The C compiler expands the macro in the program. For example,

Z = MPY10 (A) + MPY10 (B) is expanded as
Z = (A<<3) + (A<< 1) + (B<<3) + (B<<1);

Macros are faster than functions or subroutines since there is no overhead in macro associated with transfer of control.

Questions and Problems

7.1   What is the basic difference between a microprocessor, a single-chip microcomputer and a microcontroller?

7.2   What is meant by an 8-bit microcontroller? Name one commercially available 8-bit microcontroller.

7.3   What is the difference between
        (a) a program counter and a memory address register?
        (b) General-purpose register and instruction register?

7.4   How many bits are needed to access a 4 MB main memory? What is the hexadecimal value of the last address in this memory?

7.5   If the last address of an on-chip memory of a microcontroller is 0x7FF, determine its size.

7.6   What is the difference between von Neumann and Harvard CPU architectures? Provide an example of a commercially available microcontroller using each type of CPU.

7.7   What is the basic difference between program execution by a conventional CPU and the PIC18F CPU?

7.8   Discuss briefly the purpose of the functional units (CCP, ADC, serial communication) implemented in the PIC18F?

7.9   What is meant by pipelining?

7.10   Summarize the basic features of PIC18F pipelining.

7.11   What is the basic difference between assembly and high-level languages? Why would you choose one over the other?

7.12   Assume that two microcontrollers, the PIC18F and the HC16, have C compilers. Will a program written in C language run on both microcontrollers?

7.13   Will a program written in Microchip's PIC18F assembly language run on microcontrollers from other manufacturers?

7.14   What are the basic differences between assembly language and C language? Which programming language is popular with microcontrollers? Why?

7.15   Find the results after each of the following operations:

            (a) 0x24 & 0x81
            (b) 0x60 | 0x90
            (c) 0x81 ^ 0xA2
            (d) ! 0x25
            (e) 0x1F <<4

7.16   What is the advantage of using union?

7.17   Why would you use Macros?

8

PIC18F HARDWARE AND INTERFACING USING C: PART 1

In this chapter, we describe the first part of hardware aspects of the PIC18F4321. Topics include PIC18F4321 pins and signals, clock and reset circuits, programmed and interrupt I/O.

8.1 PIC18F Pins and Signals

The PIC18F4321 is contained in three types of packaging as follows:
- 40-pin plastic dual in-line package (PDIP)
- 44-pin quad flat no-lead plastic package (QFN)
- 44-pin thin plastic quad flat pack package (TQFP)

Figure 8.1 shows the PIC18F4321 pin diagram for a PDIP. A brief description of all pins and signals for the PIC18F4321 contained in the 40-pin PDIP is provided in Table 8.1. There are two VDD (Vcc) pins and two VSS (ground) pins which are not shared (multiplexed) with other pins. The range of voltages for the VDD pins are from + 4.2 V to +5.5 V. However, the VDD pins are normally connected to +5 V. The VSS pins are connected to ground. The maximum power dissipation for the PIC18F4321 is one watt. Note that multiple pins for power and ground are used in order to distribute the power and reduce noise problems at high frequencies.

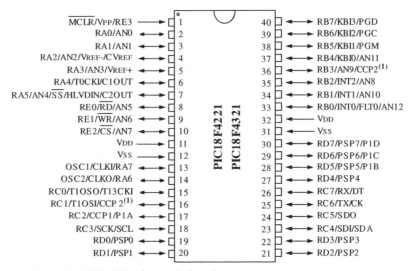

FIGURE 8.1   PIC18F4321 pins and signals.

All other 36 pins are multiplexed (shared) with other signals. There are 36 pins assigned to five I/O ports, namely, Port A (8-bit, RA0-RA7), Port B (8-bit, RB0-RB7), Port C (8-bit, RC0-RC7), Port D (8-bit, RD0-RD7), and Port E (4-bit, RE0-RE3). These pins are multiplexed with other signals such as the clock/oscillator, reset, external interrupt, analog inputs, and CCP (Capture/Compare/Pulse Width Modulation).

The PIC18F pins associated with clock, reset, and I/O will be discussed next.

8.1.1 Clock

Upon hardware reset, the PIC18F4321 operates from an internal clock frequency of one MHz (default). However, after hardware reset, the internal clock (crystal) frequency can be changed from 31 KHz to 8 MHz by writing appropriate data into the OSCCON register (page 31 of Microchip's PIC18F4321 data sheet) as shown in Figure 8.2. Note that internal clock means that the PIC18F4321 operates from an on-chip clock provided by Microchip, and no external clock circuit is required.

TABLE 8.1   PIC18F4321 pinout description

Pin number	Pin name	Pin type	Description
1	$\overline{\text{MCLR}}$	Input	Master clear reset ; active low input
	Vpp	Input	Programming voltage input for the flash memory
	RE3	Input	Digital input; Port E bit 3
2	RA0	Input/Output	Digital I/O; Port A bit 0
	AN0	Input	Analog input 0
3	RA1	Input/Output	Digital I/O; Port A bit 1
	AN1	Input	Analog input 1
4	RA2	Input/Output	Digital I/O; Port A bit 2
	AN2	Input	Analog input 2
	VREF-	Input	A/D reference voltage (low) input
	CVREF	Output	Comparator reference voltage output
5	RA3	Input/Output	Digital I/O; Port A bit 3
	AN3	Input	Analog input 3
	VREF+	Input	A/D reference voltage (high) input
6	RA4	Input/Output	Digital I/O; Port A bit 4
	T0CKI	Input	Timer0 external clock input
	C1OUT	Output	Comparator 1 output
7	RA5	Input/Output	Digital I/O; Port A bit 5
	AN4	Input	Analog input 4
	SS	Input	SPI slave select input
	HLVDIN	Input	High/low-voltage detect input
	C2OUT	Output	Comparator 2 output
8	RE0	Input/Output	Digital I/O; Port E bit 0
	$\overline{\text{RD}}$	Input	Read control for parallel slave port (see also $\overline{\text{WR}}$ and $\overline{\text{CS}}$ pins).
	AN5	Input	Analog input 5
9	RE1	Input/Output	Digital I/O; Port E bit 1
	$\overline{\text{WR}}$	Input	Write control for parallel slave port (see $\overline{\text{CS}}$ and $\overline{\text{RD}}$ pins).
	AN6	Input	Analog input 6

TABLE 8.1   PIC18F4321 pinout description (continued)

Pin number	Pin name	Pin type	Description
10	RE2	Input/Output	Digital I/O; Port E bit 2
	$\overline{\text{CS}}$	Input	Chip Select control for parallel slave port (see related $\overline{\text{RD}}$ and $\overline{\text{WR}}$).
	AN7	Input	Analog input 7
11	VDD	Power	Positive supply for logic and I/O pins.
12	VSS	Ground	Ground reference for logic and I/O pins.
13	OSC1	Input	Oscillator crystal input or external clock source input.
	CLKI	Input	External clock source input. Always associated with pin function OSC1. (See related OSC1/CLKI, OSC2/CLKO pins.)
	RA7	Input/Output	Digital I/O; Port A bit 7
14	OSC2	Output	Oscillator crystal output. Connects to crystal or resonator in crystal oscillator mode. In RC, EC, and INTIO modes, OSC2 pin outputs.
	CLKO	Output	CLKO which has one-fourth the frequency of OSC1 and denotes the instruction cycle rate.
	RA6	Input/Output	Digital I/O; Port A bit 6
15	RC0	Input/Output	Digital I/O; Port C bit 0
	T1OSO	Output	Timer1 oscillator analog output
	T13CKI	Input	Timer1/Timer3 external clock input
16	RC1	Input/Output	Digital I/O; Port C bit 1
	T1OSI	Input	Timer1 oscillator analog input
	CCP2	Input/Output	Capture 2 input/Compare 2 output/ PWM 2 output; default assignment for CCP2 when Configuration bit, CCP2MX, is set.
17	RC2	Input/Output	Digital I/O; Port C bit 2
	CCP1	Input/Output	Capture 1 input/Compare 1 output/ PWM 1 output
	P1A	Output	Enhanced CCP1 output
18	RC3	Input/Output	Digital I/O; Port C bit 3
	SCK	Input/Output	Synchronous serial clock input/output for SPI mode.
	SCL	Input/Output	Synchronous serial clock input/output for I²C™ mode.
19	RD0	Input/Output	Digital I/O; Port D bit 0
	PSP0	Input/Output	Parallel Slave Port data
20	RD1	Input/Output	Digital I/O; Port D bit 1
	PSP1	Input/Output	Parallel slave port data
21	RD2	Input/Output	Digital I/O; Port D bit2
	PSP2	Input/Output	Parallel slave port data
22	RD3	Input/Output	Digital I/O; Port D bit 3
	PSP3	Input/Output	Parallel slave port data

TABLE 8.1   PIC18F4321 pinout description (continued)

Pin number	Pin name	Pin type	Description
23	RC4	Input/Output	Digital I/O
	SDI	Input	SPI data in
	SDA	Input/Output	I²C data I/O
24	RC5	Input/Output	Digital I/O; Port C bit 5
	SDO	Output	SPI data out
25	RC6	Input/Output	Digital I/O; Port C bit 6
	TX	Output	EUSART asynchronous transmit
	CK	Input/Output	EUSART synchronous clock (see related RX/DT).
26	RC7	Input/Output	Digital I/O; Port C bit 7
	RX	Input	EUSART asynchronous receive
	DT	Input/Output	EUSART synchronous data (see related TX/CK)
27	RD4	Input/Output	Digital I/O; Port D bit 4
	PSP4	Input/Output	Parallel slave port data
28	RD5	Input/Output	Digital I/O; port D bit 5
	PSP5	Input/Output	Parallel slave port data
	P1B	Output	Enhanced CCP1 output
29	RD6	Input/Output	Digital I/O; port D bit 6
	PSP6	Input/Output	Parallel slave port data
	P1C	Output	Enhanced CCP1 output
30	RD7	Input/Output	Digital I/O; Port D bit 7
	PSP7	Input/Output	Parallel slave port data
	P1D	Output	Enhanced CCP1 output
31	VSS	Ground	Ground reference for logic and I/O pins
32	VDD	Power	Positive supply for logic and I/O pins
33	RB0	Input/Output	Digital I/O; Port B bit 0
	INT0	Input	External interrupt 0
	FLT0	Input	PWM fault input for enhanced CCP1
	AN12	Input	Analog input 12
34	RB1	Input/Output	Digital I/O; Port B bit 1
	INT1	Input	External interrupt 1
	AN10	Input	Analog input 10
35	RB2	Input/Output	Digital I/O; Port B bit 2
	INT2	Input	External interrupt 2
	AN8	Input	Analog input 8
36	RB3	Input/Output	Digital I/O; Port B bit 3
	AN9	Input	Analog input 9
	CCP2	Input/Output	Capture 2 input/compare 2 output/ PWM 2 output; alternate assignment for CCP2 when configuration bit CCP2MX, is cleared.
37	RB4	Input/Output	Digital I/O; Port B bit 7
	KBI0	Input	Interrupt-on-change pin
	AN11	Input	Analog input 11
38	RB5	Input/Output	Digital I/O; Port B bit 5
	KBI1	Input	Interrupt-on-change pin

TABLE 8.1   PIC18F4321 pinout description (continued)

Pin number	Pin name	Pin type	Description
	PGM	Input/Output	Low-voltage programming enable pin
39	RB6	Input/Output	Digital I/O; Port B bit 6
	KBI2	Input	Interrupt-on-change pin
	PGC	Input/Output	In-circuit debugger and programming clock pin
40	RB7	Input/Output	Digital I/O; Port B bit 7
	KBI3	Input	Interrupt-on-change pin
	PGD	Input/Output	In-circuit debugger and ICSP programming data pin.

From Figure 8.2, the following C code will set the internal clock frequency to 4 MHz:

```
OSCCON = 0xEC // Set the internal crystal to 4 MHz
```

The PIC18F4321 can also be operated in ten different oscillator modes. The user can program the Configuration bits, FOSC3:FOSC0, in Configuration Register (page 257 of the PIC18F4321 data sheet) to select one of these ten modes.

7	6	5	4	3	2	1	0	
IDLEN	IRCF2	IRCF1	IRCF0	OSTS	IOFS	SCS1	SCS0	OSCCON

bit 7 bit 0

bit 7 **IDLEN:** Idle Enable bit
1 = Device enters an Idle mode when a SLEEP instruction is executed
0 = Device enters Sleep mode when a SLEEP instruction is executed
bit 6-4 **IRCF2:IRCF0:** Internal Oscillator Frequency Select bits(5)
111 = 8 MHz (INTOS C drives clock directly)
110 = 4 MHz
101 = 2 MHz
100 = 1 MHz(3)
011 = 500 kHz
010 = 250 kHz
001 = 125 kHz
000 = 31 kHz (from either INTOSC/256 or INTRC directly) (2)
bit 3 **OSTS:** Oscillator Start-up Time-out Status bit(1)
1 = Oscillator Start-up Timer (OST) time-out has expired; primary oscillator is running
0 = Oscillator Start-up Timer (OST) time-out is running; primary oscillator is not ready
bit 2 **IOFS:** INTOSC Frequency Stable bit
1 = INTOSC frequency is stable
0 = INTOSC frequency is not stable
bit 1-0 **SCS1:SCS0:** System Clock Select bits(4)
1x = Internal oscillator block
01 = Secondary (Timer1) oscillator
00 = Primary oscillator
Note 1: Reset state depends on state of the IESO Configuration bit.
2: Source selected by the INTSRC bit (OSCTUNE<7>), see text.
3: Default output frequency of INTOSC on Reset.
4: Modifying the SCSI:SCSO bits will cause an immediate clock source switch.
5: Modifying the IRCF3:IRCF0 bits will cause an immediate clock frequency switch if the internal oscillator is providing the device clocks.
Legend:
R = Readable bit W = Writable bit U = Unimplemente d bit, read as '0'
-n = Value at POR '1' = Bit is set '0' = Bit is cleared x = Bit is unknown

FIGURE 8.2   OSCCON (Oscillator Control Register).

8.1.2   PIC18F Reset

Upon reset, the PIC18F loads '0' into program counter. Thus the PIC18F reads the first instruction from the contents of address 0 in the program memory. Most registers are unaffected by a Reset. All I/O ports are configured as inputs. The PIC18F4321 can be reset in several different ways. For simplicity, the two most commonly used RESET techniques are power-on and manual resets. These two resets will be discussed in the following.

Power-On Reset (POR)   A power-on reset pulse is generated on-chip upon power-up whenever VDD rises above a certain threshold. This allows the device to start in the initialized state when VDD is adequate for operation. The reset circuit in Figure 8.3 provides a simple power-on reset circuit with a pushbutton (manual) switch. When the power is turned ON, the resistors in Figure 8.3 with the switch open will provide power-on reset. When the PIC18F exits the reset condition, and starts normal operation, a program can be executed by pressing the pushbutton, and program execution can be restarted upon activation of the pushbutton.

Manual Reset   Figure 8.3 shows a typical circuit for manual reset. The $\overline{\text{MCLR}}$/Vpp pin is normally HIGH. Upon activating the push button, the $\overline{\text{MCLR}}$/Vpp pin is driven from HIGH to LOW. The internal on-chip circuitry connected to the $\overline{\text{MCLR}}$/Vpp pin ensures that the pin must be LOW for at least 2 μsec (minimum requirement for reset). Since it can be shown that the circuit of Figure 8.3 satisfies the minimum timing requirement for reset, this circuit can be used to reset the PIC18F manually.

8.1.3   A Simplified Setup for the PIC18F4321

Figure 8.4 shows a simplified setup for the PIC18F4321 microcontroller using the default clock of one MHz. Appendix D shows the hardware and software aspects of how to interface the PIC18F4321 to a personal computer or a laptop using PicKit3. This setup can be used for performing inexpensive meaningful experiments in laboratories using a breadboard.

As mentioned before, upon activating the reset, the PIC18F loads '0' into program counter. Thus, the PIC18F reads the first instruction from the contents of address 0 in the program memory. Most registers are unaffected by a Reset. The $\overline{\text{MCLR}}$/Vpp pin is normally HIGH. Upon activating the push button, the $\overline{\text{MCLR}}$/Vpp pin is driven from HIGH to LOW. The internal on-chip circuitry connected to the $\overline{\text{MCLR}}$/Vpp pin ensures that the pin must be LOW for at least 2 μsec (minimum requirement for reset).

In Figure 8.4, there are two pairs of pins on the PIC18F4321 that must be connected to power and ground; pins 11 (VDD) and 32 (VDD) should be connected directly to +5V and pins 12 (VSS) and 31 (VSS) must be connected directly to ground.

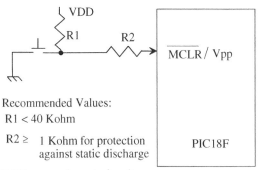

FIGURE 8.3   PIC18F manual reset circuit.

Note that the operating voltage for VDD is between 4.2V and 5.5V. The manual reset circuit is connected to pin 1 of the PIC18F4321 chip. When the push button is activated, the PIC18F4321 is reset. This also allows for an easy way to restart a program in the PIC18F4321. The manual reset circuit is connected to pin 1 of the PIC18F4321 chip.

8.2 PIC18F4321 programmed I/O using C

In this section, basics of PIC18F programmed I/O will be described. In addition, interfacing the PIC18F4321 to simple I/O devices such as LEDs, switches, and seven-segment displays will also be covered. Note that the PIC18F ports are buffered. This means that the PIC18F ports provide adequate currents to drive devices such as LEDs and seven-segment displays.

8.2.1 PIC 18F4321 I/O ports

The PIC18F4321 contains five ports namely Port A (8-bit), Port B (8-bit), Port C (8-bit), Port D (8-bit), and Port E (4-bit). Note that all bits of Ports A through D are available for general I/O operation. On the other hand, only three bits of Port E (bits 0, 1, 2) are available for general I/O. The fourth pin (bit 3) of PORT E (MCLR/VPP/RE3) is an input only pin. Its operation is controlled by the MCLRE Configuration bit in a special register called *CONFIG3H* (Configuration Register 3 High, Page 257 of Microchip's PIC18F4321 Data Sheet). For simplicity, Port A through Port D will be considered in this discussion.

It is relatively easier to configure Port C and Port D as input or output port. Each port is associated with a corresponding data direction register called *TRISx* in the PIC18F4321. Table 8.2 lists Ports A through E along with their corresponding TRISx registers. A bit in a TRISx register is used to configure the corresponding bit in

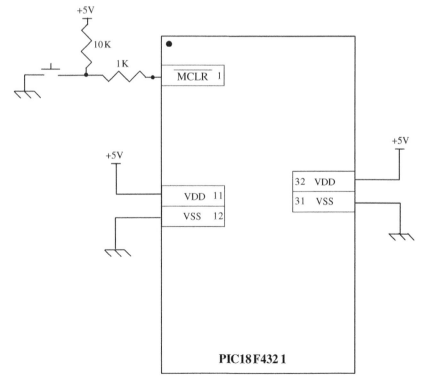

FIGURE 8.4   Simplified PIC18F4321 setup.

the associated port as an input or output. For example, writing a '1' at bit 0 of TRISC will make bit 0 of Port C as an input bit. On the other hand, writing a '0' at bit 1 of TRISC will make bit 1 of Port C as an output bit. Note that upon hardware reset, all ports are automatically configured as inputs. The following C language statements will make all bits of Port C and Port D as inputs and outputs respectively:

```
TRISC = 0xFF; // Configure PORT C as an input port
TRISD = 0; // Configure PORT D as an output port
```

As another example, consider the following C-language statement:

```
TRISD = 0x34; // Configure PORT D
```

In the above instruction sequence, data 34 (hex) is moved into TRISD (8-bit data direction register for PORTD), PORTD is then defined as shown in Figure 8.5. In this example, because 34H (0011 0100) is written into TRISD, bits 0, 1, 3, 6, and 7 of the port are set up as outputs, and bits 2, 4, and 5 of the port are defined as inputs. The microcontroller can then send output to external devices, such as LEDs, connected to bits 0, 1, 3, 6, and 7 through a proper interface. Similarly, the PIC18F4321 can input the status of external devices, such as switches, through bits 2, 4, and 5. To input data from the input switches, the PIC18F4321 inputs the complete byte, including the bits to which output devices such as LEDs are connected. While receiving input data from an I/O port, however, the PIC18F4321 places a value, probably 0, at the bits configured as outputs and the program must interpret them as "don't cares." At the same time, the PIC18F4321's outputs to bits configured as inputs are disabled.

The MPLAB C18 compiler provides built-in unions for configuring a port bit. This allows the programmer to address a single bit in a port without changing the other bits in the port. For example, bit 2 of PORT C can be configured as an output by writing a '0' at bit 2 of TRISC as follows:

```
# define portbit PORTCbits.RC2 // Declare a bit (bit 2) of PORT C
TRISCbits.TRISC2 = 0 ; // Configure bit 2 of PORT C as an output
```

Note that the MPLAB C18 compiler uses RC0 as bit 0 of PortC, RC1 as bit 1 of Port C, and so on. The same thing applies to other ports (PA0 as bit 0 of Port A and so on).

Now, a '1' can be output to bit 2 of PORT C using the following statement:

```
portbit = 1;
```

Similarly, the statement, `portbit = 0;` will output a '0' to bit 2 of PORT C.

TABLE 8.2　　PIC18F4321 I/O Ports And TRISx Registers

Port Name	Size	Comment
Port A	8-bit	Port A
TRISA	8-bit	Data Direction Register for Port A
Port B	8-bit	Port B
TRISB	8-bit	Data Direction Register for Port B
Port C	8-bit	Port C
TRISC	8-bit	Data Direction Register for Port C
Port D	8-bit	Port D
TRISD	8-bit	Data Direction Register for Port D
Port E	4-bit	Port E
TRISE	4-bit	Data Direction Register for Port E

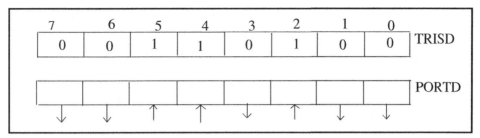

FIGURE 8.5   PORT D along with TRISD.

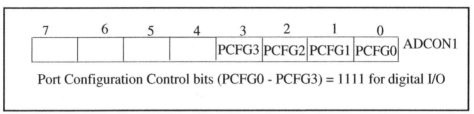

FIGURE 8.6   ADCON1 register for digital I/O.

Next, bit 3 of PORT D can be configured as an input by writing a '1' at bit 3 of TRISD as follows:

```
# define portbit PORTDbits.RD3 // Declare a bit (bit 3) of PORT D
TRISDbits. TRISD3 = 1 ; // Configure bit 3 of PORT D as an input
```

Note that configuring Port A, Port B, and Port E is different than configuring Port C and Port D. This is because, certain bits of Port A, Port B, and Port E are multiplexed with analog inputs (Figure 8.1). Upon reset, all bits of ports A, B, and E multiplexed with AN0–AN12 are configured as analog inputs. However, writing 0x0F to bits 0-3 of the ADCON1 register (Figure 8.6) will configure these multiplexed port bits as digital I/O. Also, upon reset, all TRISx registers are loaded with 0xFF and the associated port bits are configured as inputs. Hence, upon loading 0x0F into ADCON1 register will make ports A, B and E as inputs. Note that TRISA, TRISB, and TRISE are not required to configure Ports A, B, and E as inputs.

However, it is recommended to use the following C statements to configure Ports A, B, and E as digital inputs:

```
ADCON1 = 0x0F; // Configure of ports A, B, and E as digital I/O
TRISA = 0xFF;
TRISB = 0xFF;
TRISE = 0x0F;
```

Note that TRISx registers associated with ports A, B, and E must be used to configure these ports as outputs.

For example, the following C statements will configure Ports A, B and E as outputs:

```
ADCON1 = 0x0F:
 TRISA = 0;
 TRISB = 0;
 TRISE = 0;
```

8.2.2   Interfacing LEDs (Light Emitting Diodes) and Seven-segment Displays

PIC18F sources and sinks adequate currents so that LEDs and seven-segment displays can be interfaced to the PIC18F without buffers (current amplifiers) such as 74HC244. An LED can be connected in two ways. Figure 8.7 shows these configurations.

In Figure 8.7(a), the PIC18F will output a HIGH to turn the LED ON; the PIC18F will output a 'LOW' will turn it OFF. In Figure 8.7 (b), the PIC18F will output a LOW to turn the LED ON; the PIC18F will output a 'HIGH' will turn it OFF. Also, when an LED is turned on, a typical current of 10 mA flows through the LED with a voltage drop of 1.7 V. Hence,

$$R = \frac{5 - 1.7}{10 \text{ mA}} = 330 \ \Omega$$

As discussed in Chapter 6, a seven-segment display can be used to display, for example, decimal numbers from 0 to 9. The name "seven segment" is based on the fact that there are seven LEDs — one in each segment of the display. Figure 8.8 shows a typical seven-segment display.

All decimal numbers from 0 to 9 can be displayed by turning the appropriate segment "ON" or "OFF". For example, a zero can be displayed by turning the LED in segment *g* "OFF" and turning the other six LEDs in segments *a* through *f* "ON." There are two types of seven-segment displays. These are common cathode and common anode. Figure 8.9 shows two different seven-segment display configurations, namely, common cathode and common anode.

In a common cathode arrangement, the microcontroller can be programmed to send a HIGH to light a segment and a LOW to turn it off. In a common anode configuration, on the other hand, the microcontroller can send a LOW to light a segment and a HIGH to turn it off. In both configurations, R = 330 ohms can be used.

Figure 8.10 shows a typical interface between the PIC18F4321 and a common cathode seven-segment display via Port C. Each bit of Port C is connected to a segment of the seven-segment display via 330 ohm resistor. Note that the seven resistors are not shown in the figure. A common anode seven-segment display can similarly be interfaced to the PIC18F4321.

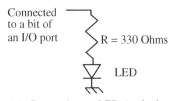
(a) Connecting an LED (cathode grounded) to an I/O port bit

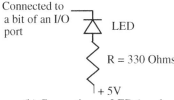
(b) Connecting an LED (anode tied to 5V) to an I/O port bit

FIGURE 8.7   Interfacing LED to PIC18F.

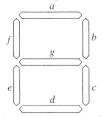

FIGURE 8.8   A seven-segment display.

8.2.3   Microchip MPLAB C18 compiler and the PICkit3 interface

As mentioned in the preface, a basic background in C is required to understand the
C-programs covered in this chapter and chapter 9. These C-programs are associated with
the PIC18F I/O and interfacing techniques. In order to obtain hands-on experience, all
hardware and C-programs can be developed as follows:

1. First, the PIC18F-based hardware for a particular application can be designed
    and built on a breadboard.
2. Next, a C-program can be written to perform the specified task. In this book,
    these C-programs are compiled using Microchip's C18 compiler. Appendix C
    provides a tutorial showing step-by-step procedure to download the C18 compiler
    from Microchip website (www.microchip.com), and compile a simple C-program
    using the MPLAB C18 compiler v3.47. Microchip introduces newer versions of
    the C18 compiler frequently. One may download the C18 v3.47 from the archive
    or a newer version from Microchip website.
3. The compiled C-program can then be downloaded into a PIC18F chip on the
    breadboard by connecting a Personal computer or a Laptop via USB to the
    PIC18F chip using Microchip's PICkit3 interface. A step-by-step procedure for
    accomplishing this is provided in Appendix D.
4. Finally, after successful downloading of the desired program, the PIC18F-based
    breadboard can be disconnected from the Personal computer or Laptop. The PIC18F
    chip on the breadboard can then execute the programs downloaded into its memory
    upon activating the RESET pushbutton. Thus, the desired task can be accomplished.

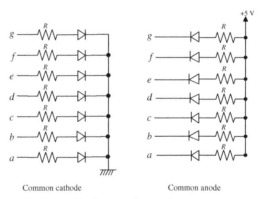

FIGURE 8.9   Seven-segment display configurations.

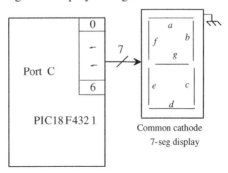

FIGURE 8.10   PIC18F4321 interface to a common cathode seven-segment
display via PORT C.

8.2.4   Configuration commands

For C programs with I/O to work, the configuration commands are necessary. The following configuration commands should be inserted after #include <p18f4321.h>:

```
#pragma config OSC = INTIO2 // Select internal oscillator
#pragma config WDT = OFF // Watch Dog Timer OFF
#pragma config LVP = OFF // Low Voltage Programming OFF
#pragma config BOR = OFF // Brown On Reset OFF
```

Example 8.1

Assume PIC18F4321. Suppose that three switches are connected to bits 0-2 of Port C and an LED to bit 6 of Port D. If the number of HIGH switches is even, turn the LED on; otherwise, turn the LED off. Write a C language program to accomplish this.

Solution

The C language program is shown below:

```
#include <p18f4321.h>
#define portc0 PORTCbits.RC0
#define portc1 PORTCbits.RC1
#define portc2 PORTCbits.RC2
void main (void)
{
      unsigned char mask = 0x07; // Data for masking off
                                 // upper 5 bits of Port C
      unsigned char masked_in;
      unsigned char xor_bit;
      TRISC = 0xFF; // Configure Port C as an input port
      TRISD = 0; // Configure Port D as an output port
      while(1)
      {
        masked_in = PORTC & mask; // Mask input bits
        if (masked_in == 0)
         PORTD = 0x40; //For all low switches (even), turn led on
        else
         xor_bit = portc0 ^ portc1 ^ portc2; // Xor input bits
        if (xor_bit == 0) // For even # of high switches
         PORTD = 0x40; // turn led on
        else
         PORTD = 0; // For odd # of high switches, turn led off
      }
}
```

Example 8.2

Assume PIC18F4321. Suppose that it is desired to input a switch connected to bit 4 of Port C, and then output it to an LED connected to bit 2 of Port D. Write a C language program to accomplish this.

Solution

The following C code will accomplish this:

```
# include <P18F4321.h>
# define portc_bitin PORTCbits.RC4 // Declare a bit (bit 4) of Port C
# define portd_bitout PORTDbits.RD2 // Declare a bit (bit 2) of Port D
     void main (void)
     {
        TRISCbits.TRISC4 = 1; //Configure bit 4 of Port C as an input bit
        TRISDbits.TRISD2 = 0; //Configure bit 2 of Port D as an output bit
        while (1) // Halt
        {
        portd_bitout = portc_bitin; // Output switch to LED
        }
     }
```

Example 8.3

The PIC18F4321 microcontroller shown in Figure 8.11 is required to output a BCD digit (0 to 9) to a common-anode seven-segment display connected to bits 0 through 6 of Port D. The PIC18F4321 inputs the BCD number via four switches connected to bits 0 through 3 of Port C. Write a C language program that will display a BCD digit (0 to 9) on the seven-segment display based on the switch inputs.

Solution

The C code is provided below:

```
#include <p18f4321.h>
  void main ()
  {
  unsigned char input;
  unsigned char code[10] = {0x40, 0x79, 0x24, 0x30, 0x19, 0x12, 0x03, 0x78, 0x00, 0x18};
  TRISD = 0; //Configure PortD as Output
  TRISC = 0xFF; //Configure PortC as Input
  while (1) {
       input = PORTC & 0x0F;
       PORTD = code [input];
             }
  }
```

In the above program, first the PORTB is set as an output port and PORTC is set as an input port. A variable 'input' is then declared. The program moves to an infinite 'while' loop where it will first take the input from the four switches via PORTC, and mask the first four bits. An unsigned char array code is set up in order to contain the seven-segment code for each decimal digit from 0 through 9. The input is used as the index to the array, and the corresponding LED code is sent to PORTD. The code then repeats this process and displays the proper digit on the seven-segment display based on switch inputs.

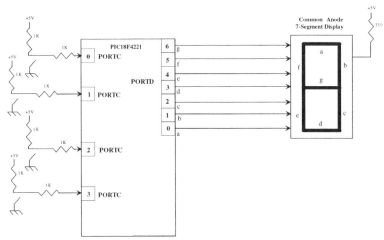

FIGURE 8.11   Figure for Example 8.3.

Example 8.4

Assume that the PIC18F4321 micrococontroller shown in Figure 8.12 is required to perform the following:
If Vx > Vy , turn the LED ON if the switch is open; otherwise, turn the LED OFF. Write a C program to accomplish the above by inputting the comparator output via bit 0 of port B.

Solution

The C program is provided below:

```c
#include <p18f4321.h>
void main (void)
{
  TRISD = 0; //PORTD is output
  ADCON1 = 0x0F; //Configure PORTB
  TRISB = 0xFF; // as digital inputs
  PORTD = 0; // Turn LED OFF
  while(1)
  {
   PORTD = 0; //Turn LED OFF
   while(PORTBbits.RB0 ==1) //While Vx > Vy
   {
     if (PORTBbits.RB1==1)
     PORTD = 1; //Turn LED ON
     else if (PORTBbits.RB1== 0)
     PORTD = 0; //Turn LED OFF
   }
  }
}
```

In the above code, the register ADCON1 is used to configure Port B. Within the infinite while loop, the code checks to see when the comparator output is one indicating Vx > Vy. The LED is then turned ON or OFF based on the state of the switch.

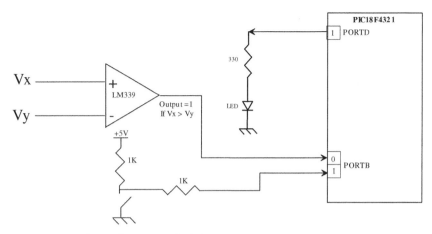

FIGURE 8.12   Figure for Example 8.4.

8.3   PIC18F Interrupts

The concept of interrupt is discussed in detail in Chapter 6. The PIC18F interrupts will be covered in this section.

8.3.1   PIC18F Interrupt Types

The PIC18F4321 interrupts can be of two types. These are external interrupts initiated via PIC18F4321's interrupt pins, and internal interrupts initiated by internal peripheral devices such as on-chip A/D converter and on-chip timers. The PIC18F4321 is provided with three external maskable interrupt pins. These interrupts are INT0, INT1, and INT2.

These interrupts are edge-triggered. Upon power-on reset, each of these interrupts (INT0, INT1, INT2) is activated by a rising edge pulse (LOW to HIGH). This triggering level can be changed to a falling edge via programming. This topic will be discussed later.

The PIC18F4321 on-chip peripherals can generate internal interrupts. These interrupts are also maskable. These peripheral interrupts will be covered in Chapter 9.

The PIC18F4321 does not have any nonmaskable interrupts. Note that the nonmaskable interrupt cannot be enabled or disabled by instructions, whereas a microcontroller's instruction set typically contains instructions to enable or disable maskable interrupt.

8.3.2   PIC18F External Interrupts in Default Mode

The concept of PIC18F external interrupts (default mode) described in this section can be used to perform simple and meaningful experiments in laboratories.

Upon power-on reset, the PIC18F handles the three external interrupts (INT0, INT1, INT2) in default mode. The interrupt address vector for all three interrupts is 0x000008 in the program memory in this mode. The INT0 interrupt has an individual interrupt enable bit along with a corresponding flag bit located in the INTCON register.

Each of the two other external interrupts (INT1 and INT2), on the other hand, has an individual interrupt enable bit along with the corresponding flag bit located in the INTCON3 register. Figure 8.13 shows the INTCON and INTCON3 registers along with the associated interrupt bits. The other bits in these registers are specified for other functions such as timers, and will be discussed in Chapter 9.

(a) <u>INTCON Register</u>

bit 7 **GIE/GIEH:** Global Interrupt Enable bit
When IPEN = 0: 1 = Enables all unmasked interrupts   0 = Disables all interrupts
When IPEN = 1: 1 = Enables all high priority interrupts 0 = Disables all interrupts

bit 6 **PEIE/GIEL:** Peripheral Interrupt Enable bit
When IPEN = 0: 1 = Enables all unmasked peripheral interrupts   0 = Disables all peripheral interrupts
When IPEN = 1: 1 = Enables all low priority peripheral interrupts 0 = Disables all low priority peripheral interrupts

bit 5 **TMR0IE:** TMR0 Overflow Interrupt Enable bit,
1 = Enables the TMR0 overflow interrupt 0 = Disables the TMR0 overflow interrupt

bit 4 **INT0IE:** INT0 External Interrupt Enable bit
1 = Enables the INT0 external interrupt, 0 = Disables the INT0 external interrupt

bit 3 **RBIE:** RB Port Change Interrupt Enable bit
1 = Enables the RB port change interrupt, 0 = Disables the RB port change interrupt

bit 2 **TMR0IF:** TMR0 Overflow Interrupt Flag bit
1 = TMR0 register has overflowed (must be cleared in software), 0 = TMR0 register did not overflow

bit 1 **INT0IF:** INT0 External Interrupt Flag bit
1 = The INT0 external interrupt occurred (must be cleared in software), 0 = The INT0 external interrupt did not occur

bit 0 **RBIF:** RB Port Change Interrupt Flag bit
1 = At least one of the RB7:RB4 pins changed state (must be cleared in software)
0 = None of the RB7:RB4 pins have changed state

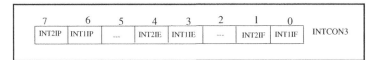

(b) <u>INTCON3 Register</u>

bit 7 **INT2IP:** INT2 External Interrupt Priority bit, 1 = High priority 0 = Low priority
bit 6 **INT1IP:** INT1 External Interrupt Priority bit, 1 = High priority 0 = Low priority
bit 5 **Unimplemented:** Read as '0'
bit 4 **INT2IE:** INT2 External Interrupt Enable bit, 1 = Enables the INT2 interrupt   0 = Disables the INT2 interrupt
bit 3 **INT1IE:** INT1 External Interrupt Enable bit, 1 = Enables the INT1 interrupt   0 = Disables the INT1 interrupt
bit 2 **Unimplemented:** Read as '0'
bit 1 **INT2IF:** INT2 External Interrupt Flag bit, 1 = The INT2 interrupt occurred (must be cleared in software)
0 = The INT2 external interrupt did not occur
bit 0 **INT1IF:** INT1 External Interrupt Flag bit, 1 = The INT1 external interrupt occurred (must be cleared in software)
0 = The INT1 external interrupt did not occur

FIGURE 8.13   INTCON (interrupt control) and INTCON3 (interrupt control 3).

All PIC18F interrupts are disabled upon reset. However, these interrupts can be enabled via software by setting the GIE bit (bit 7 in INTCON register) to one. Next, the respective Interrupt Enable (IE) for each one of the three external interrupts must be set to 1 in order to enable a particular interrupt. Finally, since the external interrupts (INT2, INT1, and INT0) are multiplexed with analog inputs (AN8, AN10, and AN12), the ADCON1 register must be configured as digital input. The PIC18F will now recognize any interrupt via the INT0-INT2 pins.

Once an interrupt is recognized by the PIC18F, the corresponding flag bits (INT0IF for INT0 in INTCON, INT1IF for INT1 in INTCON3, and INT2IF for INT2 in INTCON3) are set to one. Figure 8.14 shows a simplified schematic for the PIC18F external interrupts for power-on reset. The PIC18F external interrupts must first be enabled using instructions by setting the GIE and INTxIE bits to one. The PIC18F will automatically set the corresponding interrupt flag bit to one after occurrence of each interrupt (INT0 through INT2).

In Figure 8.14, consider the occurance of INT0 interrupt. Note that the GIE input of AND gate #5 and the INT0IE input of AND gate #1 can be set to one via software. Now, as soon as the interrupting device connected to the PIC18F4321 INT0 pin interrupts the microcontroller, the PIC18F4321 saves the current program counter contents internally, and sets the INT0IF flag to 1, indicating that the INT0 interrupt has occurred. This will make the INT0IF input of AND gate #1 to one. Hence, the output of AND gate #5 in Figure 8.14 will be one. This will enable the appropriate hardware inside the PIC18F4321, and will load the program counter with 0x000008. The user can write a service routine at this address.

8.3.3 Interrupt Registers and Priorities

The PIC18F4321 contains ten registers which are used to control interrupt operation. These registers are

- INTCON (Figure 8.13)
- INTCON3 (Figure 8.13)
- RCON (Figure 8.15)
- INTCON2 (Figure 8.16)
- PIR1, PIR2 (to be discussed in Chapter 9)
- PIE1, PIE2 (to be discussed in Chapter 9)
- IPR1, IPR2 (discussed in Microchip's PIC18F4321 manual)

The PIC18F4321 interrupts can be classified into two groups: high-priority interrupt levels and low-priority interrupt levels. The high-priority interrupt vector

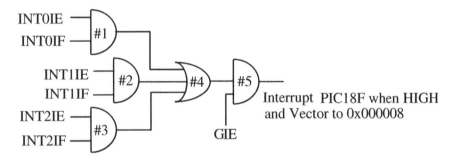

FIGURE 8.14   Simplified schematic for the PIC18F external interrupts for power-on reset.

7	6	5	4	3	2	1	0
IPEN							

Bit 7 = IPEN (interrupt pending, 1 = enable, 0 = disable)

FIGURE 8.15   RCON (RESET CONTROL) register.

is at address 0x000008 and the low-priority interrupt vector is at address 0x000018 in the program memory. High-priority interrupt events will interrupt any low-priority interrupts that may be in progress.

When the IPEN bit (Figure 8.15) is cleared (default state), the interrupt priority feature is disabled; on the other hand, when the IPEN bit in the RCON register is set to one via programming, interrupt priority is enabled. Bit 6 of the INTCON register is the PEIE bit, which enables/disables all peripheral interrupt sources. Bit 7 of the INTCON register is the GIE bit, which enables/disables all interrupt sources. All interrupts branch to address 0x000008 upon power-on reset (default). When an interrupt is responded to, the global interrupt enable bit is cleared to disable further interrupts. If interrupt priority levels are used, high-priority interrupt sources can interrupt a low-priority interrupt. Low-priority interrupts are not processed while high-priority interrupts are in progress. The return address is saved internally and the program counter is loaded with the interrupt vector address (0x000008 or 0x000018). Once in the interrupt service routine, the source(s) of the interrupt can be determined by polling the interrupt flag bits. The interrupt flag bits must be cleared in software before re-enabling interrupts to avoid recursive interrupts. Upon exiting the interrupt service routine, the PIC18F sets the GIE bit (GIEH or GIEL if priority levels are used), which re-enables interrupts.

Next, interrupt priorities associated with the PIC18F4321 external interrupts (INT0, INT1, and INT2) will be discussed. Setting up interrupt priorities for peripherals such as the hardware timers will be discussed in Chapter 9.

Table 8.3 shows the three external interrupts of the PIC18F4321 along with the corresponding IP bit. As mentioned in the table, since an IP is not assigned to INT0, it always has the high-priority level. However, INT1 and INT2 can be programmed as high- or low-level priorities.

8.3.4   Setting the Triggering Levels of INTn Pin Interrupts

As mentioned before, external interrupts on the INT0, INT1, and INT2 pins are edge-triggered. Upon power-on reset, each of these external interrupts (INT0, INT1, INT2) is activated by a rising edge pulse (LOW to HIGH). The PIC18F has the flexibility of changing the triggering levels for these interrupts to falling edge pulses (HIGH to LOW). This can be accomplished by programming bits 4 through 6 of the INTCON2 register (Figure 8.16).

bit 7 **RBPU:** PORTB Pull-up Enable bit
1 = All PORTB pull-ups are disabled
0 = PORTB pull-ups are enabled by individual port latch values
bit 6 **INTEDG0:** External Interrupt 0 Edge Select bit, 1 = Interrupt on rising edge, 0 = Interrupt on falling edge
bit 5 **INTEDG1:** External Interrupt 1 Edge Select bit, 1 = Interrupt on rising edge, 0 = Interrupt on falling edge
bit 4 **INTEDG2:** External Interrupt 2 Edge Select bit, 1 = Interrupt on rising edge, 0 = Interrupt on falling edge
bit 3 **Unimplemented:** Read as '0'
bit 2 **TMR0IP:** TMR0 Overflow Interrupt Priority bit, 1 = High priority, 0 = Low priority
bit 1 **Unimplemented:** Read as '0'
bit 0 **RBIP:** RB Port Change Interrupt Priority bit, 1 = High priority, 0 = Low priority

FIGURE 8.16   INTCON2 register.

TABLE 8.3   PIC18F4321 external interrupts along with interrupt priority (IP) bits

Interrupt name	Interrupt priority (IP) bit	Comment
INT0	Unassigned	Since no interrupt priority bit is assigned, INT0 always has the high priority level.
INT1	INT1IP	Bit 6 of INTCON3 register (Figure 8.13); 1 = high priority, 0 = low priority
INT2	INT2IP	Bit 7 of INTCON3 register (Figure 8.13); 1 = high priority, 0 = low priority

If the corresponding INTEDGx bit in the INTCON2 register is set (= 1), the interrupt is triggered by a rising edge; if the bit is clear, the trigger is on the falling edge. When a valid edge appears on the INTx pin, the corresponding flag bit, INTxF, is set. This interrupt can be disabled by clearing the corresponding enable bit, INTxE. Flag bit INTxF must be cleared in software in the interrupt service routine before re-enabling the interrupt.

8.3.5   Programming the PIC18 interrupts using C

As mentioned before, the PIC18F4321 interrupts can be classified into two groups: high priority interrupt levels and low priority interrupt levels. INT0 always has the high priority level while INT1 and INT2 can be programmed using C as high- or low-level priorities.

The Microchip C18 compiler uses the keywords `interrupt` and `interruptlow` to specify high and low priority interrupt levels respectively. The high priority interrupt vector is at address 0x000008 and the low priority interrupt vector is at address 0x000018 in the program memory. High priority interrupt events will interrupt any low priority interrupts that may be in progress. Also, using the C18 compiler, the programmer can use the directive `#pragma code begin` to specify an address to a program or data at address `begin`. For example, the C statement `#pragma code int_vect = 0x000008` will assign the address 0x000008 to label `int_vect`. Note that `pragma` and `code` are keywords of the C18 compiler.

As mentioned before, upon power-on reset, the interrupt address vector is 0x000008 (default), and no interrupt priorities are available. The IPEN bit (bit 7 of the RCON register) of the RCON register (Figure 8.15) can be programmed to assign interrupt priorities. Upon power-on reset, IPEN is automatically cleared to 0, and the PIC18F operates as a high-priority interrupt (single interrupt) system. Hence, the interrupt vector address is 0x000008 for all interrupts. During normal operation, the IPEN bit can be set to one by using the following C-code to assign priorities in the system:

```
RCONbits.IPEN = 1;
```

When interrupt priority is enabled (IPEN = 1), there are two bits which enable interrupts globally. Setting the GIEH bit (bit 7 of INTCON register of Figure 8.13) enables all interrupts that have the priority bit set (high priority). Setting the GIEL bit (bit 6 of INTCON register of Figure 8.13) enables all interrupts that have the priority bit cleared (low priority). When the interrupt enable bit, and appropriate global interrupt enable bit are set, and the interrupt flag bit is cleared, the interrupt will vector to address 0x000008 or 0x000018, depending on the priority bit setting. Individual interrupts can be disabled through their corresponding enable bits.

Based on the above-mentioned discussions, the steps for programming INT0 to recognize interrupts are as follows:

1. Configure the interrupts, INT0-INT2 as digital inputs using the ADCON1 register.
2. Enable INT0 by setting the INT0IE to 1 in the INTCON register (Figure 8.13).
3. Clear the interrupt flag bit for INT0 by clearing the INT0IF in the INTCON register.
4. Enable global interrupts by setting the GIE bit in the INTCON register.

The following C-code will accomplish this:

```
TRISB = 0xFF;
ADCON1=0x0F; // Configure INT0-INT2 as digital inputs
INTCONbits.INT0IE=1; // Enable INT0 interrupt
INTCONbits.INT0IF=0; // Clear the INT0 interrupt flag
INTCONbits.GIE=1; // Enable global interrupts
```

As an example, the following steps will configure INT0 as high priority interrupt and INT1 as low priority interrupt:

1. Configure INT0 and INT1 multiplexed with Port B as digital inputs using the ADCON1 register.
2. Set INT0IE (INTCON register) and INT1IE (INTCON3 register) to 1's.
3. Clear interrupt flag bits INT0IF (INTCON register) and INT1IF (INTCON3 register) to 0's.
4. Set INT1 as low priority interrupt.
5. Enable IPEN in the RCON register (Figure 8.15).
6. Enable global HIGH and LOW interrupts.
    The following C-code will accomplish this:

```
TRISB = 0xFF;
ADCON1=0x0F; // Configure INT0 and INT1 as digital inputs
INTCONbits.INT0IE=1; // Enable external interrupt INT0
INTCON3bits.INT1IE=1; // Enable external interrupt INT1
INTCONbits.INT0IF=0; // Clear INT0 interrupt flag
INTCON3bits.INT1IF=0; // Clear INT1 interrupt flag
INTCON3bits.INT1IP=0; // Set INT1 to low priority interrupt, INT0
                      // always has high priority
RCONbits.IPEN=1; // Enable priority interrupts
INTCONbits.GIEH=1; // Enable global high priority interrupts
INTCONbits.GIEL=1; // Enable global low priority interrupts
```

Note that _asm and _endasm MPLAB C18 directives allow the insertion of the PIC18F assembly language code into the C code. For example, the following PIC18F assembly code GOTO ISR included within the _asm and _endasm directives can be used to branch to the service routine called *ISR* (written in C):

```
{
_asm // Using assembly language to branch to interrupt
   // service routine
GOTO ISR
_endasm
}
```

Example 8.5

Assume that the PIC18F4321 micrococontroller shown in Figure 8.17 is required to perform the following:

If Vx > Vy , turn the LED ON if the switch is open; otherwise, turn the LED OFF. Write a C program to accomplish the above by interrupting the PIC18F4321 by the comparator output via INT0. Also, write the main program in C which will initialize Port B and Port D, and then wait for interrupt in an infinite loop.

Solution

In this example, an LM339 comparator is interfaced to the PIC18F4321 via the INT0 Pin. An external interrupt allows the microcontroller to trigger an interrupt from a source outside the PIC18F4321 such as the comparator. The code starts with the `#pragma` command which will place code fragments at specific locations in memory, and when the interrupt is triggered, the microcontroller will automatically jump to memory location 0x000008, and then to COMP_ISR.

In the main program, PORTD is configured as an output, and PORTB is configured as a digital input. The external interrupt flag is cleared to 0, and the global interrupt is enabled. The main program then waits in an infinite 'while' loop that turns the LED OFF until the comparator output is HIGH, interrupting the microcontroller. After recognizing the interrupt, the code will automatically jump to address 0x000008, and then jump to the service routine at COMP_ISR via the code at COMP_int. Within the service routine, the code will continue to take the switch data from PORTB and output the data to the LED via PORTD. It will continue to do this as long as the comparator output stays HIGH. When Vx is lower than Vy, the comparator will output 0V and the code will return to the infinite 'while' loop and turn the LED OFF.

The PIC18F program using C is provided below:

```c
#include <P18F4321.h>
void COMP_ISR (void);
#pragma code COMP_Int=0x08 //At interrupt code jumps here
void COMP_Int(void)
{
  _asm //Using assembly language
  GOTO COMP_ISR
  _endasm
}
#pragma code
void main( ) //Start of the main program
{
  TRISD=0x00; //PORTD is output
  TRISB = 0xFF;
  ADCON1=0x0F; //Configure INT0 to be digital input
  INTCONbits.INT0IE=1; //Enable external interrupt
  INTCONbits.INT0IF=0; //Clear the external interrupt flag
  INTCONbits.GIE=1; //Enable global interrupts
  PORTD=0; //LED is off
  while(1){ //Wait in an infinite loop for the interrupt to occur
   PORTD=0; //LED is off
  }
}
```

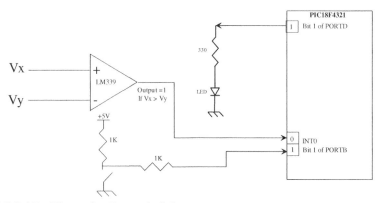

FIGURE 8.17   Figure for Example 8.5.

```
  # pragma interrupt COMP_ISR
void COMP_ISR(void) //Start of the Comparator interrupt
                    //service routine
{

  INTCONbits.INT0IF=0; //Clear external interrupt flag
  while(PORTBbits.RB0==1){ //Check if comparator is high
  PORTD = PORTB; //Move PORTB into PORTD
  }
}
```

Example 8.6

In Figure 8.18, if Vx > Vy, the PIC18F4321 is interrupted via INT0. On the other hand, opening the switch will interrupt the microcontroller via INT1. Note that in the PIC18F4321, INT0 has the higher priority than INT1. Write the main program in C that will perform the following:

-Configure interrupt inputs.
-Clear interrupt flag bits of INT0 and INT1.
-Set INT1 as low priority interrupt.
-Enable IPEN in RCON register.
-Enable global HIGH and LOW interrupts.
-Turn both LEDs at PORTD OFF.
-Wait in an infinite loop for one or both interrupts to occur.
Also, write a service routine for the high priority interrupt (INT0) in C that will perform the following:
-Turn LED on at bit 0 of PORTD.

Finally, write a service routine for the low priority interrupt (INT1) in C that will perform the following:
-Turn LED on at bit 1 of PORTD.

Solution

This example will demonstrate the interrupt priority system of the PIC18F microcontroller. Using interrupt priority, the user has the option to have various interrupts assigned as either low-priority or high-priority interrupts. If a low-priority

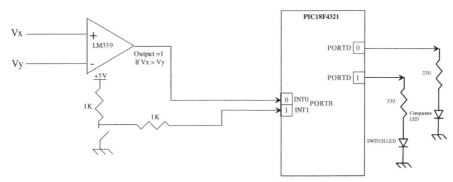

FIGURE 8.18   Figure for Example 8.6.

interrupt and a high-priority interrupt occur at the same time, the PIC18F will always service the high priority interrupt first.

In the above example, the high priority is assigned to the comparator while the switch is assigned with the low priority. Hence, if both interrupts were triggered at the same time, the LED associated with the comparator would be turned ON first, and then the LED associated with the switch will be turned ON. Note that the external interrupt INT0 can only be a high-priority interrupt. Hence, INT0 is connected to the comparator output while the switch is connected to INT1 since it has the low priority. At the end of the code provided below, it can be seen that there are two interrupt service routines, HP_COMP_ISR and LP_SWITCH_ISR, which are the high-priority and low-priority service routines respectively.

The following code implements priority interrupts on the PIC18F using C:

```c
#include <P18F4321.h>
void HP_COMP_ISR (void);
void LP_SWITCH_ISR(void);
#pragma code High_Priority_COMP_Int=0x08 //High interrupt
                                         //code jumps here
void COMP_Int(void)
{
_asm //Using assembly language
GOTO HP_COMP_ISR
_endasm
}
#pragma code Low_Priority_SWITCH_Int=0x018 //Low interrupt
                                           //code jumps here
void Switch_Int(void)
{
_asm //Using assembly language
GOTO LP_SWITCH_ISR
_endasm
}
#pragma code
void main( )
{
  TRISBbits.TRISB0=1; //Set pin 0 of PORTB (INT0) as input
```

```
   TRISBbits.TRISB1=1; //Set pin 1 of PORTB (INT1) as input
   TRISD=0x00; //PORTD is output
   ADCON1=0x0F; //Configure INT0 - INT1 to be digital inputs
   INTCONbits.INT0IE=1; //Enable external interrupt INT0
   INTCON3bits.INT1IE=1; //Enable external interrupt INT1
   INTCONbits.INT0IF=0; //Clear INT0 external interrupt flag
   INTCON3bits.INT1IF=0; //Clear INT1 external interrupt flag
   INTCON3bits.INT1IP=0; //Set INT1 to low priority interrupt
   RCONbits.IPEN=1; //Enable priority interrupts
   INTCONbits.GIEH=1; //Enable global high priority interrupts
   INTCONbits.GIEL=1; //Enable global low priority interrupts
   PORTD=0; //LEDs are off
   while(1){
    PORTD=0; //Turn off LEDs
   }
} // end of main

#pragma interrupt HP_COMP_ISR
void HP_COMP_ISR(void){ //High-priority interrupt service
INTCONbits.INT0IF=0; //Clear external interrupt flag
   while(PORTBbits.RB0)==1 //Check if comparator is high
   {
    PORTD=0x01; //Turn on LED
   }
} // end of HP_COMP_ISR

#pragma interrupt low LP_SWITCH_ISR
void LP_SWITCH_ISR(void){ //Low-priority interrupt service
INTCON3bits.INT1IF=0; //Clear external interrupt flag
   while(PORTBbits.RB1)==1 //Check if switch is still on
   {
    PORTD=0x02; //Turn on LED
   }
} // end of LP_SWITCH_ISR
```

Questions and Problems

8.1   What is the default clock period of the PIC18F4321?

8.2   What is the minimum reset timing requirement for the PIC18F manual reset?

8.3   Name all the I/O ports included with the PIC18F4321? What are their sizes?

8.4   Write a C language statement to configure

    (a)   all bits of Port C as inputs
    (b)   all bits of Port D as outputs
    (c)   bits 0 through 4 of Port B as input
    (d)   all bits of Port A as outputs

8.5   The PIC18F4321 microcontroller is required to drive the LEDs connected to bit 0 of Ports C and D based on the input conditions set by switches connected to bit 1 of Ports C and D. The I/O conditions are as follows:

    • If the input at bit 1 of Port C is HIGH and the input at bit 1 of Port D is LOW, the LED at Port C will be ON and the LED at Port D will be OFF.
    • If the input at bit 1 of Port C is LOW and the input at bit 1 of Port D is HIGH, the LED at Port C will be OFF and the LED at Port D will be ON.
    • If the inputs at both Ports C and D are HIGH, turn both LEDs ON; on the other hand, if the inputs at both Ports C and D are LOW, turn both LEDs OFF. Assume a '1' will turn the LED ON while a '0' will turn it OFF.

    Write a C language program to accomplish this.

8.6   The PIC18F4321 microcontroller is required to test a NAND gate. Figure P8.6 shows the I/O hardware needed to test the NAND gate. The PIC18F4321 is to be programmed to generate the various logic conditions for the NAND inputs, input the NAND output, and turn the LED ON connected to bit 3 of Port D if the NAND gate chip is found to be faulty. Otherwise, turn the LED ON connected to bit 4 of Port D. Write a C language program to accomplish this.

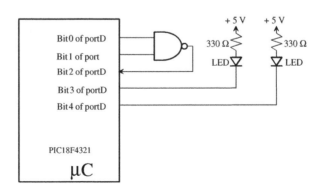

FIGURE P8.6.

8.7   The PIC18F4321 microcontroller (Figure P8.7) is required to add two 3-bit
        numbers entered via DIP switches connected at bits 0-2 and bits 3-5 of Port
        D and output the sum (not to exceed 9) to a common-cathode seven-segment
        display connected to Port C as shown in Figure P8.7. Write a C language
        program to accomplish this by using a lookup table.

8.8   The PIC18F4321 microcontroller is required to input a number from 0 to 9
        from an ASCII keyboard interfaced to it and output to an EBCDIC printer.
        Assume that the keyboard is connected to Port C and the printer is connected
        to Port D. Write a C language program using a lookup table for EBCDIC
        codes from 0 to 9 to accomplish this. Note that decimal numbers 0 through 9
        are represented by F0H through F9H in EBCDIC code, and by 30H through
        39H in ASCII code as mentioned in Chapter 2.

8.9   In Figure P8.9, the PIC18F4321 is required to turn on an LED connected to
        bit 1 of Port D if the comparator voltage Vx > Vy; otherwise, the LED will be
        turned off. Write a C language program to accomplish this using conditional
        or polled I/O.

8.10   Repeat Problem 8.9 using Interrupt I/O by connecting the comparator output
        to INT1. Note that RB1 is also multiplexed with INT1. Write main program
        and interrupt service routine in C language. The main program will configure
        the I/O ports, enable interrupt INT1, turn the LED OFF, and then wait for
        interrupt. The interrupt service routine will turn the LED ON and return to
        the main program at the appropriate location so that the LED is turned ON
        continuously until the next interrupt.

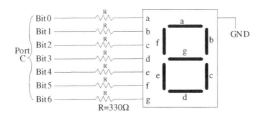

FIGURE P8.7.

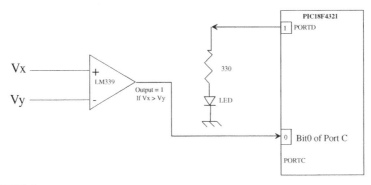

FIGURE P8.9.

8.11   In Figure P8.11, if $V_M > 12$ V, turn an LED ON connected to bit 3 of port A. If V_M < 11 V, turn the LED OFF. Using ports, registers, and memory locations as needed and INT0 interrupt:

(a)   Draw a neat block diagram showing the PIC18F4321 microcontroller and the connections to ports in the diagram in Figure P8.11.

(b)   Write the main program and the service routine in C language. The main program will initialize the ports and wait for an interrupt. The service routine will accomplish the task and stop.

8.12   Write C language program to set interrupt priority of INT1 as the high level, and interrupt priority for INT2 level as low level.

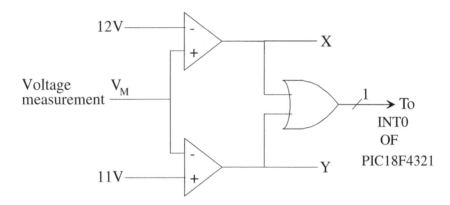

FIGURE P8.11.

PIC18F HARDWARE AND INTERFACING USING C: PART 2

In this chapter we describe the second part of hardware aspects of the PIC18F4321. Topics include PIC18F4321's on-chip timers, LCD displays, analog interfaces (ADC and DAC), serial I/O, and CCP (Capture/Compare/Pulse Width Modulation). Several C-programs are written to illustrate these concepts.

9.1 PIC18F Timers

The PIC18F4321 microcontroller includes four timers, namely, Timer0 (8-bit or 16-bit), Timer1 (16-bit), Timer2 (8-bit), and Timer3 (16-bit). These timers can be used to generate time delays using on-chip hardware. Because these timers are on-chip hardware devices, they accurately keep time in the background while other functions are performed by the PIC18F4321.

The basic hardware inside each of these timers is a register that can be incremented or decremented at the rising or falling edge of a clock. The register can be loaded with a count for a specific time delay. Time delay is computed by subtracting the initial starting count from the final count in the register, and then multiplying the subtraction result by the clock frequency.

These timers can also be used as event counters. Note that an event counter is basically a register with the clock replaced by an event such as a switch. The counter is incremented or decremented whenever the switch is activated. Thus the number of times the switch is activated (occurrence of the event) can be determined.

Timer0 can operate as an 8-bit timer or a 16-bit timer. Timer0 utilizes the internal clock when used as a timer, and an external clock when used as a counter. Timer1 can only be used as a 16-bit timer or a 16-bit counter; it cannot be used as an 8-bit timer or 8-bit counter.

Like Timer0, Timer1 uses the internal clock when used as a timer, and the external clock when used as a counter. Timer2 can be used as an 8-bit timer using the internal clock. It cannot be used as a counter. Finally, Timer3 can operate as a 16-bit timer or a 16-bit counter. It cannot be used as an 8-bit timer or an 8-bit counter.

Timer1, Timer2, and Timer3 are used for the CCP (Capture/Compare/Pulse Width Modulation) operation. For example, the PIC18F4321 utilizes Timer1 or Timer3 for the capture and compare modes, and Timer2 for the PWM mode. Note that the Timer0 is not used by the CCP module. These topics will be discussed later in this chapter.

Polling Timer flags Timer0 counts up to a maximum value of 0xFF when it operates as an 8-bit timer and then overflows to 0. Also, Timer0 or Timer1 or Timer3 counts up to a maximum value of 0xFFFF while operating as a 16-bit timer, and then overflows to 0.

The PIC18F4321 Timer2 contains an 8-bit timer register (TMR2) and an 8-bit period register (PR2). Timer2 overflows to 0 for TMR2-to-PR2 match. The overflow is indicated automatically by the PIC18F4321 by setting a flag bit (TMRxIF bit where $x = 0, 1, 2, 3$) associated with each of these timers. This flag bit can be polled in software to provide time-related functions. All four timers include the flag bit. Examples 9.1 through 9.4 illustrate how to write C-programs for these timers by polling each timer's individual flag bits.

Timer interrupts The problem with polling the timer flag bit is that the PIC18F4321 has to wait in a loop checking for the TMRxIF bit to be HIGH. In this case, the PIC18F4321 will not be able to do any other task. Hence, timer interrupts can be used. This concept is similar to the advantages of using interrupt I/O over polled I/O. In order to accomplish interrupts with timers, each timer is provided with an interrupt enable bit. This enable bit for a particular timer, when set to 1 in software, automatically interrupts the PIC18F4321. The interrupt address vector is 0x00008 upon power-on reset. A service routine can be written at this address to perform timer-related functions. Example 9.5 illustrates how to write C-programs for interrupt-driven timers using Timer0 through Timer3.

7	6	5	4	3	2	1	0	
TMR0ON	T08BIT	T0CS	T0SE	PSA	TOPS2	TOPS1	TOPS0	T0CON

bit 7 **TMR0ON**: Timer0 On/Off Control bit
1 = Enables Timer0
0 = Stops Timer0

bit 6 **T08BIT**: Timer0 8-Bit/16-Bit Control bit
1 = Timer0 is configured as an 8-bit timer/counter
0 = Timer0 is configured as a 16-bit timer/counter

bit 5 **T0CS**: Timer0 Clock Source Select bit
1 = External clock connected to RA4/T0CKI pin (pin 6; Timer0 external clock input)
0 = Internal clock from crystal oscillator (divide by 4; crystal frequency can vary from 4 to 25 MHz)

bit 4 **T0SE**: Timer0 Source Edge Select bit
1 = Increment on high-to-low transition on T0CKI pin
0 = Increment on low-to-high transition on T0CKI pin

bit 3 **PSA**: Timer0 Prescaler Assignment bit
1 = TImer0 prescaler is NOT assigned. Timer0 clock input bypasses prescaler.
0 = Timer0 prescaler is assigned. Timer0 clock input comes from prescaler output.

bit 2-0 **TOPS2:TOPS0**: Timer0 Prescaler Select bits (Prescaler is used to obtain more delays)
111 = 1:256 prescale value
110 = 1:128 prescale value
101 = 1:64 prescale value
100 = 1:32 prescale value
011 = 1:16 prescale value
010 = 1:8 prescale value
001 = 1:4 prescale value
000 = 1:2 prescale value

FIGURE 9.1   T0CON (Timer0 Control) Register.

9.1.1 Timer0

As mentioned before, Timer0 can operate as a timer or as a counter in 8-bit or 16-bit mode. The Timer0 uses the internal clock when used as a timer, and external clock (T0CK1) when used as a counter.

Timer0 as a timer Timer0 can be used as a timer by setting the TMR0ON (bit 7 of T0CON register of Figure 9.1) to 1 via programming. After Timer0 is started, it counts up by incrementing the contents of the register (TMR0L for an 8-bit timer mode or TMR0H:TMR0L for a 16-bit timer mode) by 1 at each instruction cycle. The TMR0L counts up until the TMR0L reaches 0xFF in the 8-bit mode. The TMR0IF (interrupt on overflow) flag bit in the INTCON register is set to 1 when the TMR0L rolls over from 0xFF to 0x00. In the 8-bit mode, only the TMR0L register is used; the TMR0H register is not used, and contains a value of 0.

Similarly, in the 16-bit mode, after the Timer0 is started, the TMR0H:TMR0L register pair counts up until the TMR0H:TMR0L reaches 0xFFFF in the 16-bit timer mode. The TMR0IF (interrupt on overflow) flag bit in the INTCON register is set to 1 when the TMR0H:TMR0L rolls over from 0xFFFF to 0x0000.

The timer can be stopped in either 8-bit or 16-bit mode by clearing the TMR0ON (bit 7 of T0CON register of Figure 9.1) to 0 via programming. Note that the T0CON register shown in Figure 9.1 controls all aspects of the Timer0 operation, including the prescale selection. Timer0 is both readable and writable.

Timer0 as a counter The Timer0 can be configured as a counter by setting the T0CS bit (bit 5 in the T0CON register of Figure 9.1) to 1. This will enable the PIC18F4321 to

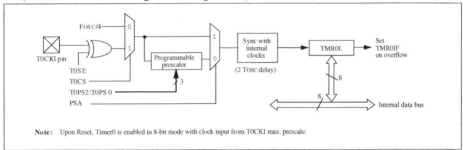

FIGURE 9.2 TIMER0 block diagram (8-bit mode).

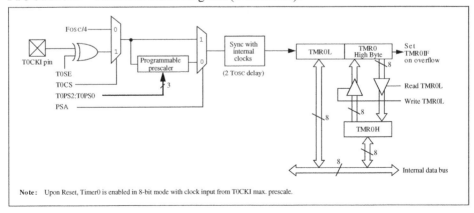

FIGURE 9.3 TIMER0 block diagram (16-bit mode).

use an external clock connected to the T0CK1 pin. The T0SE bit (bit 4 of the T0CON register can then be cleared to 0 to increment the Timer0 register on the rising edge of the clock or set to one to increment the Timer0 register on the falling edge of the clock.

Timer0 block diagrams   Figure 9.2 and Figure 9.3 show the simplified block diagrams of the Timer0 module in 8-bit and 16-bit modes, respectively. In Figure 9.2, TOSE (bit 4 of T0CON register of Figure 9.1) is exclusive-ORed with the T0CK1 clock (pin 6 of the PIC18F4321, Figure 8.1). The output of the Exclusive-OR gate is selected when TOCS (bit5 of T0CON of Figure 9.1) is one. When TOCS is 0, the internal oscillator (Fosc/4) is used and the Timer0 operates as a timer; otherwise, the external clock is selected and the Timer0 operates as a counter. The Timer0 bypasses the prescaler if the PSA bit (bit 3 of T0CON register of Figure 9.1) is 1; otherwise, the prescaler is selected (PSA = 0), and is specified by TOPS2:TOPS0 (T0CON of Figure 9.1). The next block provides a two-cycle (Tosc) delay. This is because when the register is written to, the increment operation is inhibited for the following two instruction cycles. This may be negligible for large time delays. However, if desired, the user can load an adjusted value in the TMR0 register for accuracy.

An interrupt on overflow indicated by TMR0IF (bit 2 of INTCON register in Figure 9.4) is set to one if the TMR0L rolls over from 0xFF to 0x00. The block diagram for the Timer0 16-bit mode of Figure 9.3 can be similarly explained.

Timer0 Read/Write in 16-bit Mode Two 8-bit registers (TMR0H and TMR0L) are used to hold the 16-bit value in this mode. The register pair TMR0H:TMR0L is used as a 16-bit register in the 16-bit mode. The 8-bit high byte TMR0H is latched (buffered). It is not the actual high byte of Timer0 in 16-bit mode, and is not directly readable or writable (refer to Figure 9.3). Since TMR0H is not the actual high byte register of Timer0, one should initialize TMR0H before TMR0L to avoid any errors. Note that the upper 8-bit value of the timer is stored in the latched register, and loaded into actual TMR0H when TMR0L is loaded.

Similarly, a write to the high byte of Timer0 must also take place through the TMR0H Buffer register. The high byte is updated with the contents of TMR0H when a write occurs to TMR0L. This allows all 16 bits of Timer0 to be updated at once.

Prescaler An 8-bit counter is available as a prescaler for the Timer0. The value of the prescaler is set by the PSA (bit 3 of T0CON) and T0PS2:T0PS0 bits (bits 0 through 2 of T0CON) which determine the prescaler assignment and prescale ratio. Clearing the PSA bit assigns the prescaler to the Timer0. When it is assigned, prescale values from 1:4 through 1:256 in power-of-2 increments are selectable. This is shown in Figure 9.1.

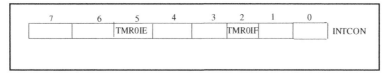

Bit 5 **TMR0IE** (TMR0 Overflow Interrupt Enable bit):   1 = Enables the TMR0 overflow interrupt
                                                            0 = Disables the TMR0 overflow interrupt.
Bit 2 **TMR0IF** (TMR0 Overflow Interrupt Flag bit):   1 = TMR0 register has overflowed (must be cleared in
                                                              software),
                                                            0 = TMR0 register did not overflow

FIGURE 9.4   INTCON register with the TMR0IE and TMR0IF bits.

Timer0 Interrupt As mentioned before, the Timer0 counts up in increments of one from 0x00 to 0xFF for 8-bit mode and from 0x0000 to 0xFFFF for 16-bit mode. Hence, any value between 0x00 and 0xFF can be loaded into TMR0L register for 8-bit mode. For 16-bit mode, any value between 0x0000 and 0xFFFF can be loaded into the 16-bit register TMR0H:TMR0L. An extra clock cycle is required when Timer0 rolls over from 0xFF to 0x00 in 8-bit mode or from 0xFFFF to 0x0000 in 16-bit mode.

The INTCON register (Figure 9.4) shows the TMR0IE and TMR0IF bits. The TMR0 interrupt is generated (if enabled by setting TMR0IE to 1) using

```
INTCONbits.TMR0IE = 1;
```

when the TMR0 register overflows from 0xFF to 0x00 in 8-bit mode, or from 0xFFFF to 0x0000 in 16-bit mode. This overflow sets the TMR0IF flag bit in the INTCON register to 1 (Figure 9.4). The interrupt can be disabled by clearing the TMR0IE bit to 0. Before re-enabling the interrupt, the TMR0IF bit must be cleared in software by the Interrupt Service Routine. Note that Timer0 interrupt flag bit can be cleared using the C-code as follows:

```
INTCONbits.TMR0IF = 0;
```

The following C-code will clear TMR0IF to 0, set TMR0IE to 1, GIE to 1, and then start Timer0:

```
INTCONbits.TMR0IF = 0;
INTCONbits.TMR0IE = 1; // Set TMR0IE to 1
INTCONbits.GIE = 1; // Enable global interrupt
T0CONbits.TMR0ON = 1; // Start TMR0
```

Once TMR0 overflows (counts from 0xFF to 0x00 for 8-bit or from 0xFFFF to 0x0000 for 16-bit), the PIC18F4321 sets the TMR0IF bit to one in the INTCON register. The PIC18F4321 is then interrupted (See Figure 9.5). The microcontroller completes execution of the current instruction, saves the program counter (PC) internally, and automatically loads PC with 0x00008. A branch instruction can be placed at this interrupt address vector to branch to another address so that adequate memory locations are available to write the service routine to perform the desired task.

Interrupt priority for Timer0 is determined by the value contained in the interrupt priority bit, TMR0IP (bit 2 of INTCON2, Figure 8.16). The PIC18F4321 can use the Timer0 interrupt as a high priority interrupt (vector address 0x000008) by setting TMR0IP to 1 using the C-code:

```
INTCON2bits.TMR0IP = 1;
```

The PIC18F4321 can use the Timer0 interrupt as a low priority interrupt (vector address 0x000018) by clearing TMR0IP to 0 using the C-code:

```
INTCON2bits.TMR0IP = 0;
```

FIGURE 9.5   Schematic for TMR0 interrupt.

Example 9.1

Using Timer0 in 16-bit mode, write a C language program to obtain a time delay of 1 ms. Assume 8-MHz crystal, leading edge clock, and a prescale value of 1:128.

Solution

Since the timer works with a divide-by-4 crystal, timer frequency = (8 MHz)/4 = 2 MHz.
Instruction cycle clock period = (1/2 MHz) = 0.5 µ sec.
The bits in register T0CON of Figure 9.1 are as follows:
TMR0ON(bit 7) = 0, T08bit (bit 6) = 0, T0CS (bit 5) = 0, T0SE (bit 4) = 0, and PSA (bit 3) = 0.
TOPS2 TOPS1 TOPS0 = 110 for a prescale value of 1:128. Hence, the T0CON register will be initialized with 0x06.
Time delay = Instruction cycle × Prescale value × Count
Hence, Count = (1 ms) / (0. 5 µ sec × 128) = 15.625 which can be approximated to an integer value of 16 (0x0010). The timer counts up from an initialized value to 0xFFFF, and then rolls over (increments) to 0000H. The number of counts for rollover is (0xFFFF -0x0010) = 0xFFEF.
Note that an extra cycle is needed for the rollover from 0xFFFF to 0x0000, and the TMR0IF flag is then set to 1. Because of this extra cycle, the total number of counts for rollover will be 0xFFF0 (0xFFEF + 1).
The following C language program will provide a time delay of 1 ms:

```
#include<p18f4321.h>
void main(void)
{
  T0CON=0x06; // Initialize T0CON
  TMR0H=0xFF; // Initialize TMR0H first with 0xFF
  TMR0L=0xF0; // Initialize TMR0L next
  INTCONbits.TMR0IF=0; // Clear Timer0 flag bit
  T0CONbits.TMR0ON=1; // Start Timer0
  while(INTCONbits.TMR0IF==0); // Wait for Timer0 flag bit to be 1
  T0CONbits.TMR0ON=0; // Stop Timer0
  while(1); // Halt
}
```

9.1.2 Timer1

The Timer1 can be used as a 16-bit timer or a counter. It consists of two 8-bit registers, namely, TMR1H and TMR1L. The Timer1 overflows when it counts from 0xFFFF to 0x0000. An extra cycle is required when the Timer1 rolls over from 0xFFFF to 0x0000.
Timer1 is controlled through the T1CON Control register (Figure 9.6). It also contains the Timer1 Oscillator Enable bit (T1OSCEN). Timer1 can be enabled or disabled by setting or clearing the TMR1ON (bit 0 of T1CON) control bit.

Timer1 Operation   Timer1 can operate as a timer, a synchronous counter, or an asynchronous counter. The operating mode is determined by the clock select bit, TMR1CS (bit 1 of T1CON). When TMR1CS is cleared to 0, Timer1 operates as a timer using the internal clock, and increments on every internal instruction cycle

(Fosc/4). When the TMR1CS bit is set to 1, Timer1 increments on every rising edge of the Timer1 external clock input or the Timer1 oscillator, if enabled. Note that the on-chip crystal oscillator circuit can be enabled by setting the Timer1 Oscillator Enable bit, T1OSCEN (bit 3 of T1CON). Finally, the timer1 is enabled by setting the TMR1ON (bit 0 of T1CON register) to 1.

Timer1 interrupts The TMR1 register pair (TMR1H:TMR1L) increments from 0x0000 to 0xFFFF and rolls over to 0x0000. The Timer1 interrupt, if enabled, is generated on overflow, which is latched as the interrupt flag bit, TMR1IF of PIR1 register shown in Figure 9.7. The TMR1 overflow interrupt bit can be enabled or disabled by setting or clearing the Timer1 Interrupt Enable bit, TMR1IE of the PIE1 register shown in Figure 9.8. The other bits in the PIR1 and PIE1 registers contain the individual flag and enable bits for the peripheral interrupts.

Timer1 is a peripheral interrupt, and hence, the PEIE, peripheral interrupt enable in the INTCON register (Figure 8.13) must be set to high using the following C-code:

```
INTCONbits.PEIE=1; // Enable peripheral interrupts
```

7	6	5	4	3	2	1	0	
RD16	T1RUN	T1CKPS1	T1CKPS0	T1OSCEN	T1SYNC	TMR1CS	TMR1ON	T1CON

bit 7 **RD16:** 16-Bit Read/Write Mode Enable bit
1 = Enables register read/write of TImer1 in one 16-bit operation
0 = Enables register read/write of Timer1 in two 8-bit operations

bit 6 **T1RUN:** Timer1 System Clock Status bit
1 = Device clock is derived from Timer1 oscillator
0 = Device clock is derived from another source

bit 5-4 **T1CKPS1:T1CKPS0:** Timer1 Input Clock Prescale Select bits
11 = 1:8 prescale value
10 = 1:4 prescale value
01 = 1:2 prescale value
00 = 1:1 prescale value

bit 3 **T1OSCEN:** Timer1 Oscillator Enable bit
1 = Timer1 oscillator is enabled
0 = Timer1 oscillator is shut off

bit 2 **T1SYNC:** Timer1 External Clock Input Synchronization Select bit
When TMR1CS = 1:
1 = Do not synchronize external clock input
0 = Synchronize external clock input
When TMR1CS = 0:
This bit is ignored. Timer1 uses the internal clock when TMR1CS = 0.

bit 1 **TMR1CS:** Timer1 Clock Source Select bit
1 = External clock from pin RC0/T1OSO/T13CKI (on the rising edge)
0 = Internal clock (Fosc/4)

bit 0 **TMR1ON:** Timer1 On bit
1 = Enables Timer1
0 = Stops Timer1

FIGURE 9.6 T1CON (Timer1 Control) Register.

The following C-code must be used in order for the PIC18F4321 to be interrupted as soon as the Timer1 overflow occurs:

```
PIR1bits.TMR1IF = 0; // Clear TMR1IF to 0
PIE1bits.TMR1IE = 1; // Set TMR1IE to 1
INTCONbits.GIE = 1;  // Enable global interrupt
INTCONbits.PEIE = 1; // Enable peripheral interrupts
T1CONbits.TMR1ON = 1; // Start TMR1
```

Example 9.2

Write a C language program to provide a delay of 1 msec using Timer1 with an internal clock of 4 MHz. Use 16-bit mode of Timer1 and the prescale value of 1:4.

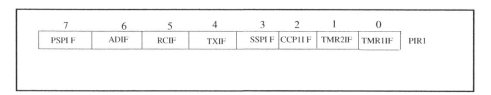

7	6	5	4	3	2	1	0	
PSPI F	ADIF	RCIF	TXIF	SSPI F	CCP1I F	TMR2IF	TMR1IF	PIR1

bit 7 **PSPIF:** Parallel Slave Port Read/Write Interrupt Flag bit
1 = A read or a write operation has taken place (must be cleared in software)
0 = No read or write has occurred

bit 6 **ADIF:** A/D Converter Interrupt Flag bit
1 = An A/D conversion completed (must be cleared in software)
0 = The A/D conversion is not complete

bit 5 **RCIF:** EUSART Receive Interrupt Flag bit
1 = The EUSART receive buffer, RCREG, is full (cleared when RCREG is read)
0 = The EUSART receive buffer is empty

bit 4 **TXIF:** EUSART Transmit Interrupt Flag bit
1 = The EUSART transmit buffer, TXREG, is empty (cleared when TXREG is written)
0 = The EUSART transmit buffer is full

bit 3 **SSPIF:** Master Synchronous Serial Port Interrupt Flag bit
1 = The transmission/reception is complete (must be cleared in software)
0 = Waiting to transmit/receive

bit 2 **CCP1IF:** CCP1 Interrupt Flag bit
Capture mode:
1 = A TMR1 (or TMR3) register capture occurred (must be cleared in software)
0 = No TMR1 (or TMR3) register capture occurred
Compare mode:
1 = A TMR1 (or TMR3) register compare match occurred (must be cleared in software)
0 = No TMR1 (or TMR3) register compare match occurred
PWM mode:
Unused in this mode.

bit 1 **TMR2IF:** TMR2-to-PR2 Match Interrupt Flag bit
1 = TMR2-to-PR2 match occurred (must be cleared in software), 0 = No TMR2-to-PR2 match occurred

bit 0 **TMR1IF:** TMR1 Overflow Interrupt Flag bit,
1 = TMR1 register overflowed (must be cleared in software)
0 = TMR1 register did not overflow

FIGURE 9.7   PIR1 (Peripheral Interrupt Request) Register 1.

Solution

For 4 MHz clock, each instruction cycle = $4 \times (1/4 \text{ MHz}) = 1$ µsec
Total instruction cycles for 1 msec delay = $(1 \times 10^{-3} / 10^{-6}) = 1000$
With the prescale value of 1:4, instruction cycles = $1000 / 4 = 250$
Counter value = $65536_{10} - 250_{10} = 65286_{10} = 0xFF06$
Because of the extra rollover cycle, Counter value = 0xFF07
Hence, TMR1H must be loaded with 0xFF, and TMR1L with 0x07
The C language program for one msec delay is provided below:

```
#include<p18f4321.h>
void main(void)
{
  T1CON=0xC0; //16-bit mode, 1:4 prescaler, Timer1 OFF
  TMR1H=0xFF; //Initialize TMR1H with 0xFF
  TMR1L=0x07; //Initialize TMR1L with 0x07
```

7	6	5	4	3	2	1	0	
PSPI E	ADIE	RCIE	TXIE	SSPI E	CCP1I E	TMR2IE	TMR1IE	PIE1

bit 7 **PSPIE:** Parallel Slave Port Read/Write Interrupt Enable bit
1 = Enables the PSP read/write interrupt
0 = Disables the PSP read/write interrupt

bit 6 **ADIE:** A/D Converter Interrupt Enable bit
1 = Enables the A/D interrupt
0 = Disables the A/D interrupt

bit 5 **RCIE:** EUSART Receive Interrupt Enable bit
1 = Enables the EUSART receive interrupt
0 = Disables the EUSART receive interrupt

bit 4 **TXIE:** EUSART Transmit Interrupt Enable bit
1 = Enables the EUSART transmit interrupt
0 = Disables the EUSART transmit interrupt

bit 3 **SSPIE:** Master Synchronous Serial Port Interrupt Enable bit
1 = Enables the MSSP interrupt
0 = Disables the MSSP interrupt

bit 2 **CCP1IE:** CCP1 Interrupt Enable bit
1 = Enables the CCP1 interrupt
0 = Disables the CCP1 interrupt

bit 1 **TMR2IE:** TMR2-to-PR2 Match Interrupt Enable bit
1 = Enables the TMR2-to-PR2 match interrupt
0 = Disables the TMR2-to-PR2 match interrupt

bit 0 **TMR1IE:** TMR1 Overflow Interrupt Enable bit
1 = Enables the TMR1 overflow interrupt
0 = Disables the TMR1 overflow interrupt

FIGURE 9.8   PIE1 (Peripheral Interrupt Enable) Register 1.

```
    PIR1bits.TMR1IF=0; //Clear Timer1 overflow flag in PIR1
    T1CONbits.TMR1ON=1; //Start Timer1
    while(PIR1bits.TMR1IF==0); //Wait for Timer1 Flag to be 1
    T1CONbits.TMR1ON=0; //Stop Timer1
    while(1); //Halt
}
```

9.1.3   Timer2

Timer2 contains an 8-bit timer register (TMR2) and an 8-bit period register (PR2). The Timer2 can be programmed only as a timer (not as a counter) with prescale values of 1:1, 1:4, and 1:16, and postscale values of 1:1 through 1:16.

The module is controlled through the T2CON register shown in Figure 9.9. The T2CON register enables or disables the timer and configures the prescaler and postscaler.

Timer2 Operation   In normal operation, the PR2 register is initialized to a specific value, and the 8-bit timer register (TMR2) is incremented from 0x00 on each internal clock (FOSC/4). A 4-bit counter/prescaler on the clock input gives direct input, divide-by-4 and divide-by-16 prescale options. These are selected by the prescaler control bits, T2CKPS1:T2CKPS0 (bits 1, 0 of T2CON). The value of TMR2 is compared to that of the 8-bit period register, PR2, on each clock cycle. When the two values match, the Timer2 outputs a HIGH on the TMR2IF flag in the PIR1 register, and also resets the value of TMR2 to 0x00 on the next cycle. The output frequency is divided by a counter/postscale value (1:1 to 1:16) as specified in the T2CON register. Note that the interrupt is generated and the TMR2IF flag bit in the PIR1 register (Figure 9.7) is set to 1, indicating the match between TMR2 and PR2 registers. The TMR2IF bit must be cleared to 0 using software. Note that an additional clock cycle is needed when Timer2 rolls over from 0xFF to 0x00.

7	6	5	4	3	2	1	0	
--------------	T2OUTPS3	T2OUTPS2	T2OUTPS1	T2OUTPS0	TMR2ON	T2CKPS1	T2CKPS0	T2CON

bit 7 **Unimplemented:** Read as '0'

bit 6-3 **T2OUTPS3:T2OUTPS0:** Timer2 Output Postscale Select bits (Postscaler is used to obtain more delays).
0000 = 1:1 postscale
0001 = 1:2 postscale
•
•
•
1111 = 1:16 postscale

bit 2 **TMR2ON:** Timer2 On bit
1 = Timer2 is on
0 = Timer2 is off

bit 1-0 **T2CKPS1:T2CKPS0:** Timer2 Clock Prescale Select bits (Prescaler is used to obtain more delay).
00 = Prescaler is 1
01 = Prescaler is 4
1x = Prescaler is 16

FIGURE 9.9   T2CON (Timer2 Control) Register.

Timer2 Interrupt Timer2 can also generate an optional device interrupt. The Timer2 output signal (TMR2-to-PR2 match) provides the input for the 4-bit output counter/postscaler. This counter generates the TMR2 match interrupt flag which is latched in TMR2IF (bit 1 of PIR1, Figure 9.7). The interrupt is enabled by setting the TMR2 Match Interrupt Enable bit, TMR2IE (bit 1 of PIE1, Figure 9.8). A range of 16 postscale options (from 1:1 through 1:16 inclusive) can be selected with the postscaler control bits, T2OUTPS3:T2OUTPS0 (bits 6-3 of T2CON, Figure 9.9).

Example 9.3

Write a C language program using Timer2 to turn an LED connected at bit 0 of PORT D after 10 sec. Assume an internal clock of 4 MHz, a prescale value of 1:16, and a postscale value of 1:16.

Solution

For 4 MHz clock, each instruction cycle = $4 \times 1/(4\text{MHz})$ = 1 μ sec. TMR2 is incremented every 1 μ sec. When the TMR2 value matches with the value in PR2, the value in TMR2 is cleared to 0 in one instruction cycle. Since the PR2 is 8-bit wide, one can have a maximum PR2 value of 255. Let us calculate the delay with this PR2 value.

Delay = (Instruction cycle) x (Prescale value) x (Postscale value) x (PR2 value + 1)
= $(1 \mu \sec) \times (16) \times (16) (255 + 1)$
= 65.536 msec

Note that, in the above, one is added to the PR2 value since an additional clock cycle is needed when it rolls over from 0xFF to 0x00, and sets the TMR2IF to 1. External counter value for 10 sec delay using 65.536 msec as the inner loop = (10 sec)/(65.536 msec), which is approximately 153 in decimal.

The C-code is provided below:

```
#include<p18f4321.h>
void main(void)
{
  unsigned char i;
  TRISDbits.TRISD0=0; // Configure bit 0 of Port D as an
                      //output
  PORTDbits.RD0=0;  // Turn LED OFF
  T2CON=0x7A; //1:16 prescaler, 1:16 postscaler Timer2 off
  TMR2=0x00; // Initialize TMR2 with 0x00
  for(i=0;i<153;i++)
  {
      PR2=255; // Load PR2 with 255
      PIR1bits.TMR2IF=0;// Clear Timer2 interrupt flag in PIR1
      T2CONbits.TMR2ON=1;// Set TMR2ON bit in T2CON to start timer
      while(PIR1bits.TMR2IF==0); // Wait for Timer2
                                 // Flag to be 1
  }
  PORTDbits.RD0=1; // Turn LED ON
  T2CONbits.TMR2ON=0; // Turn off Timer2
  while(1); // Halt
}
```

9.1.4   Timer3

Timer3 can be used as a 16-bit timer or a 16-bit counter. Although Timer3 consists of two 8-bit registers, namely, TMR3H (high byte) and TMR3L (low byte), it can only be programmed in 16-bit mode. The Timer3 module is controlled through the T3CON register (Figure 9.10). Some of the bits of the T3CON register are associated with the CCP module. This topic will be discussed later in this chapter.

Timer3 Operation Timer3 can operate in one of three modes, namely, timer, synchronous counter, and asynchronous counter. The operating mode is determined by the clock select bit, TMR3CS (bit 1 of T3CON, Figure 9.10). When TMR3CS is cleared to 0, Timer3 increments on every internal instruction cycle (FOSC/4). When the bit is set to 1, Timer3 increments on every rising edge of the Timer1 external clock input or the Timer1 oscillator, if enabled.

7	6	5	4	3	2	1	0	
RD16	T3CCP2	T3CKPS1	T3CKPS0	T3CCP1	T3SYNC	TMR3CS	TMR3ON	T3CON

bit 7 **RD16:** 16-Bit Read/Write Mode Enable bit
1 = Enables register read/write of Timer3 in one 16-bit operation
0 = Enables register read/write of Timer3 in two 8-bit operations

bit 6,3 **T3CCP2:T3CCP1:** Timer3 and Timer1 to CCPx Enable bits
1x = Timer3 is the capture/compare clock source for the CCP modules
01 = Timer3 is the capture/compare clock source for CCP2;
Timer1 is the capture/compare clock source for CCP1
00 = Timer1 is the capture/compare clock source for the CCP modules

bit 5-4 **T3CKPS1:T3CKPS0**: Timer3 Input Clock Prescale Select bits
11 = 1:8 prescale value
10 = 1:4 prescale value
01 = 1:2 prescale value
00 = 1:1 prescale value

bit 2 **T3SYNC:** Timer3 External Clock Input Synchronization Control bit
(not usable if the device clock comes from Timer1/Timer3)
When TMR3CS = 1:
1 = Do not synchronize external clock input
0 = Synchronize external clock input
When TMR3CS = 0:
This bit is ignored. Timer3 uses the internal clock when TMR3CS = 0.

bit 1 **TMR3CS:** Timer3 Clock Source Select bit
1 = External clock input from Timer1 oscillator or T13CKI (on the rising edge after the first falling edge)
0 = Internal clock (FOSC/4)

bit 0 **TMR3ON:** Timer3 On bit
1 = Enables Timer3
0 = Stops Timer3

FIGURE 9.10   T3CON (Timer3 Control) Register.

Timer3 Interrupt The TMR3 register pair (TMR3H:TMR3L) increments from 0x0000 to 0xFFFF and overflows to 0x0000. The Timer3 interrupt, if enabled, is generated on overflow and is latched in interrupt flag bit TMR3IF (bit 1 of PIR2, Figure 9.11). This interrupt can be enabled or disabled by setting or clearing the Timer3 Interrupt Enable bit, TMR3IE (bit 1 of PIE2, Figure 9.12). An additional clock cycle is required when Timer3 rolls over from 0xFFFF to 0x0000.

7	6	5	4	3	2	1	0	
OSCFIF	CMIF	------------	EEIF	BCLIF	HLVDIF	TMR3IF	CCP2IF	PIR2

bit 7 **OSCFIF:** Oscillator Fail Interrupt Flag bit
1 = Device oscillator failed, clock input has changed to INTOSC (must be cleared in software)
0 = Device clock operating

bit 6 **CMIF:** Comparator Interrupt Flag bit
1 = Comparator input has changed (must be cleared in software)
0 = Comparator input has not changed
bit 5 **Unimplemented:** Read as '0'

bit 4 **EEIF:** Data EEPROM/Flash Write Operation Interrupt Flag bit
1 = The write operation is complete (must be cleared in software)
0 = The write operation is not complete or has not been started

bit 3 **BCLIF:** Bus Collision Interrupt Flag bit
1 = A bus collision occurred (must be cleared in software)
0 = No bus collision occurred

bit 2 **HLVDIF:** High/Low-Voltage Detect Interrupt Flag bit
1 = A high/low-voltage condition occurred; direction determined by VDIRMAG bit (HLVDCON<7>)
0 = A high/low-voltage condition has not occurred

bit 1 **TMR3IF:** TMR3 Overflow Interrupt Flag bit
1 = TMR3 register overflowed (must be cleared in software)
0 = TMR3 register did not overflow

bit 0 **CCP2IF:** CCP2 Interrupt Flag bit
Capture mode:
1 = A TMR1 register capture occurred (must be cleared in software)
0 = No TMR1 register capture occurred
Compare mode:
1 = A TMR1 register compare match occurred (must be cleared in software)
0 = No TMR1 register compare match occurred
PWM mode:
Unused in this mode.

FIGURE 9.11   PIR2 (Peripheral Interrupt Request) Register 2.

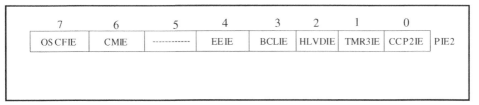

7	6	5	4	3	2	1	0
OSCFIE	CMIE	------------	EEIE	BCLIE	HLVDIE	TMR3IE	CCP2IE	PIE2

bit 7 **OSCFIE:** Oscillator Fail Interrupt Enable bit
1 = Enabled
0 = Disabled

bit 6 **CMIE:** Comparator Interrupt Enable bit
1 = Enabled
0 = Disabled

bit 5 **Unimplemented:** Read as '0'

bit 4 **EEIE:** Data EEPROM/Flash Write Operation Interrupt Enable bit
1 = Enabled
0 = Disabled

bit 3 **BCLIE:** Bus Collision Interrupt Enable bit
1 = Enabled
0 = Disabled

bit 2 **HLVDIE:** High/Low-Voltage Detect Interrupt Enable bit
1 = Enabled
0 = Disabled

bit 1 **TMR3IE:** TMR3 Overflow Interrupt Enable bit
1 = Enabled
0 = Disabled

bit 0 **CCP2IE:** CCP 2 Interrupt Enable bit
1 = Enabled

FIGURE 9.12   PIE2 (Peripheral Interrupt Enable) Register 2.

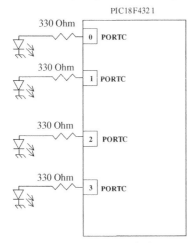

FIGURE 9.13   Figure for Example 9.4.

Example 9.4

Four LEDs are connected at bits 0-3 of PORTC of the PIC18F4321 in Figure 9.13. Write a C-program to flash the LEDs ON or OFF based on time delay provided by Timer3 by polling the TMR3IF flag. Use maximum delay provided by Timer3, and a TMR3 prescaler value of 1:4.

Solution

The program is required to do the following:
1. Configure PORTC as an output, output 0's to LEDs to turn them OFF.
2. Configure T3CON for one 16-bit operation, 1:4 prescaler, and then turn Timer3 OFF.
3. Initialize TIMR3H:TMR3L to 0:0.
4. Clear TMR3IF to 0
5. Start Timer3, and then wait in a 'while' loop for the TMR3IF to become HIGH.
6. As soon as the Timer3 counts up to 0xFFFF (providing maximum time delay), and overflows to 0x0000, the TMR3IF becomes 1.
7. Stop Timer3.
8. Re-initialize TMR3H:TMR3L to 0:0 for the next round.
9. Invert PORTC so that all four LEDs will be turned ON the first time through the loop.
10. Clear TMR3IF to 0.
11. The process continues.

The following C-code will accomplish the above:

```c
#include <p18f4321.h>
#pragma config OSC = INTIO2
#pragma config WDT=OFF
#pragma config LVP=OFF
#pragma config BOR =OFF
void main(void)
{
  TRISC=0x00;        // PortC is output
  PORTC=0x00;
  T3CON=0xA0;        // Timer3 OFF with prescaler of 4
  TMR3H=0x00;        // TMR3H initialized to 0
  TMR3L=0x00;                // TMR3L initialized to 0
  PIR2bits.TMR3IF=0;         // Clear TMR3IF to 0
  while(1)
    {
      T3CONbits.TMR3ON=1;           // Turn on TMR3
      while(PIR2bits.TMR3IF==0);// Wait for TMR3IF to be 1
      T3CONbits.TMR3ON=0;      // Stop TMR3
      TMR3H=0x00;             // Re-initialize TMR3H to 0
      TMR3L=0x00;             // Re-initialize TMR3L to 0
      PORTC=~PORTC;               // Invert PORTC output
      PIR2bits.TMR3IF=0;      // Clear TMR3IF to 0
    }
}
```

Example 9.5

Four LEDs are connected at bits 0-3 of PORTC of the PIC18F4321 in Figure 9.14. Write a C-program to do the following:
1. Turn all LEDs OFF.
2. Turn LED0 ON using time delay provided by interrupt-driven Timer0. Use maximum time delay with a prescaler value of 1:4.
3. Turn LED1 ON using time delay provided by interrupt-driven Timer1. Use maximum time delay with a prescaler value of 1:4.
4. Turn LED2 ON using time delay provided by interrupt-driven Timer2. Use maximum value for PR2 with a prescaler value of 1:4 and a postscaler value of 1:16.
5. Turn LED3 ON using time delay provided by interrupt-driven Timer3. Use maximum time delay with a prescaler value of 1:4.
6. Repeat the process.

Solution

The C-program is provided below:

```
#include <p18f4321.h>
#pragma config OSC = INTIO2
#pragma config WDT=OFF
#pragma config LVP=OFF
#pragma config BOR =OFF
void TMR0_ISR (void);
void TMR1_ISR (void);
void TMR2_ISR (void);
void TMR3_ISR (void);
#pragma interrupt check_int
void check_int(void) // Check what interrupt flag triggered
{
  if(INTCONbits.TMR0IF==1)
   TMR0_ISR();
  if(PIR1bits.TMR1IF==1)
```

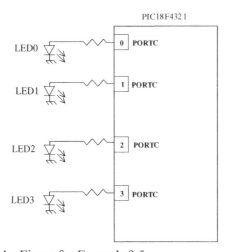

FIGURE 9.14　Figure for Example 9.5.

```
    TMR1_ISR();
  if(PIR1bits.TMR2IF==1)
    TMR2_ISR();
  if(PIR2bits.TMR3IF==1)
    TMR3_ISR();
}
#pragma code TMR_Int=0x08 // At interrupt, code jumps here
void TMR_Int(void)
{
  _asm // Using assembly language
  GOTO check_int
  _endasm
}
#pragma code // End of code
void main(void)
{
      TRISC=0x00;                  // PortC is output
      PORTC=0x00;                  // All LEDs OFF
      T0CON=0x21;                  // Timer0 OFF, 8-bit with
                              // prescaler of 4
      TMR0L=0x00;              // Value placed in lower 8 bits
                                   // of TMR0
      INTCONbits.TMR0IF=0;     // Clear TMR0 interrupt flag
      INTCONbits.TMR0IE=1;     // Enable TMR0 interrupt
      T1CON=0xA0;              // Timer1OFF,16-bit with
                                   // prescaler of 4
      TMR1H=0x00;              // TMR1H = 0
      TMR1L=0x00;              // TMR1L = 0
      PIR1bits.TMR1IF=0;       // Clear TMR1 interrupt flag
      PIE1bits.TMR1IE=1;       // Enable TMR1 interrupt
      T2CON=0x79;              // Timer2 OFF, prescaler of 4
                                   // and postscaler of 16
      TMR2=0x00;               // Value placed in Timer2
      PR2=0xFF;                // Value to compare with Timer2
      PIR1bits.TMR2IF=0;       // Clear TMR2 interrupt flag
      PIE1bits.TMR2IE=1;       // Enable TMR2 interrupt
      T3CON=0xA0;              // Timer3 OFF, 16-bit width
                                   // prescaler of 4
      TMR3H=0x00;              // TMR3H = 0
      TMR3L=0x00;              // TMR3L = 0
      PIR2bits.TMR3IF=0;       // Clear TMR3 interrupt flag
      PIE2bits.TMR3IE=1;       // Enable TMR3 interrupt
      INTCONbits.GIE=1;        // Enable global interrupts
      INTCONbits.PEIE=1;       // Enable peripheral interrupts
      T0CONbits.TMR0ON=1;      // Turn on TMR0
      while(1);                // Do nothing while timers
                                   // are counting
}
```

```
void TMR0_ISR(void)
{
      T0CONbits.TMR0ON=0;       // Turn off TMR0
      PORTC=0x01;               // Light ON LED connected to
                                // PORTC bit 0
      TMR0H=0x00;               // TMR0H = 0
      TMR0L=0x00;               // TMR0L = 0
      INTCONbits.TMR0IF=0;      // Clear TMR0 interrupt flag
      T1CONbits.TMR1ON=1;       // Turn on TMR1
}
void TMR1_ISR(void)
{
      T1CONbits.TMR1ON=0;       // Turn off TMR1
      PORTC=0x02;               // Light LED connected to
                                // PORTC bit 1
      TMR1H=0x00;               // TMR1H = 0
      TMR1L=0x00;               // TMR1L = 0
      PIR1bits.TMR1IF=0;        // Clear TMR1 interrupt flag
      T2CONbits.TMR2ON=1;       // Turn on TMR2
}
void TMR2_ISR(void)
{
      T2CONbits.TMR2ON=0;       // Turn off TMR2
      PORTC=0x04;               // Light LED connected to
                                // PORTC bit 2
      TMR2=0x00;                // Value placed in Timer2
      PIR1bits.TMR2IF=0;        // Clear TMR2 interrupt flag
      T3CONbits.TMR3ON=1;       // Turn on TMR3
}
void TMR3_ISR(void)
{
      T3CONbits.TMR3ON=0;       // Turn off TMR3
      PORTC=0x08;               // Light LED connected to
                                // PORTC bit 3
      TMR3H=0x00;               // TMR3H = 0
      TMR3L=0x00;               // TMR3 = 0
      PIR2bits.TMR3IF=0;        // Clear TMR3 interrupt flag
      T0CONbits.TMR0ON=1;       // Turn on TMR0
}
```

9.2 PIC18F Interface to an LCD (Liquid Crystal Display)

Seven-segment LEDs are easy to use, and can display only numbers and limited characters. LCDs are very useful for displaying numbers, and several ASCII characters along with graphics. Furthermore, LCDs consume low power. Because of inexpensive price of LCDs these days, they have been becoming popular. LCDs are widely used in notebook computers.

Figure 9.15 shows the PIC18F4321's interface to a typical LCD display such as the Optrex DMC16249 LCD with a 2-line x 16-character display screen. A C-program is written to display the phrase "Switch Value:" along with the numeric BCD value (0 through 9) of the four switch inputs.

The Optrex DMC16249 LCD shown in Figure 9.15 contains 14 pins. The VCC pin is connected to +5 V and the VSS pin is connected to ground the VEE pin is the contrast control for brightness of the display. VEE is connected to a potentiometer with a value between 10k and 20k. The eight data pins (D0-D7) are used to input data and commands to display desired the message on the screen.

The three control pins, EN, R/W̄, and RS, allow the user to let the display know what kind of data is sent. The EN pin latches the data from the D0-D7 pins into the LCD display. Data on D0-D7 pins will be latched on the trailing edge (high-to-low) of the EN pulse. The EN pulse must be at least 450 ns wide. The R/W̄ (read/write) pin, allows the user to either read from the LCD or write to the LCD. In this example, the R/W̄ pin will always be zero since only a string of ASCII data is written to the LCD. The R/W̄ pin is set to one for reading data from the LCD.

The command or data can be output to the LCD in two ways. One way is to provide time delays of a few milliseconds before outputting the next command or data.

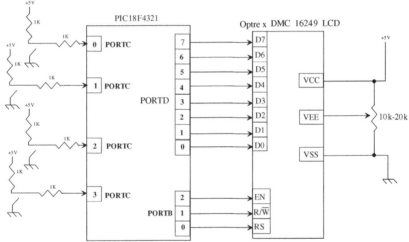

FIGURE 9.15   PIC18F4321 interface to Optrex DMC 16249 LCD.

TABLE 9.1   Typical LCD commands along with 8-bit codes in hex

Hex	Command
0x01	Clear the screen
0x02	Return home
0x04	Shift cursor to left
0x05	Shift display to right
0x06	Shift cursor to right
0x07	Shift display to left
0x08	Display off, cursor off
0x0A	Display off, cursor on
0x0C	Display on, cursor off
0x0E	Display on, cursor blinking
0x10	Shift cursor position to the left
0x14	Shift cursor position to the right
0x80	Move cursor to the start of the first line

The second approach utilizes a busy flag to determine whether the LCD is free for the next data or command. For example, in order to display ASCII characters one at a time, the LCD must be read by outputting a HIGH on the R/W̄ pin. The busy flag can be checked to ensure whether the LCD is busy or not before outputting another string of data. Note that the busy flag can thus be used instead of time delays.

Finally, the RS (Register Select) pin is used to determine whether the user is sending command or data. The LCD contains two 8-bit internal registers. They are command register and data register. When RS = 0, the command register is accessed, and typical LCD commands such as shift cursor left (hex code 0x04) can be used. Table 9.1 shows a list of some of the commands. Note that the busy flag is bit 7 of the LCD's command register. The busy bit can be read by outputting 0 to RS pin, 1 to R/W̄ pin, and a leading edge (LOW to HIGH) pulse to the EN pin.

When attempting to send data or commands to the LCD, the user must make sure that the values of EN, R/W̄, and RS are correct, along with appropriate timing. A PIC18F C- program can be written to output appropriate values to these pins via I/O ports. For example, in order to send the 8-bit command code to the LCD, write a C- program to perform the following steps:

- Output the command value to the PIC18F4321 I/O port that is connected to the LCD's D0-D7 pins.

- Send 0 to RS pin and 0 to R/W̄ pin.

- Send a '1' and then a '0' to the EN pin to latch the LCD's D0-D7 code.

As mentioned earlier, the example in Figure 9.15 will display the phrase "Switch Value:" along the BCD value of the four switch inputs. Four switches are connected to bits 0 through 3 of PORTC. The D0-D7 pins of the LCD are connected to bits 0 through 7 of PORTD. The RS, R/W̄, and EN pins of the LCD are connected to bits 0, 1, and 2 of PORTB of PIC18F4321.

The complete LCD program in C is shown in this section. Note that time delay rather than the busy bit is used before outputting the next character to the LCD. Two functions are used: one for outputting command code, and the other for the delay. PORTB and PORTD are configured as input ports, and PORTC is set up as an input port. Also, assume 1MHz default crystal frequency for the PIC18F4321.

As an example, let us consider the code for outputting a command code such as the command "move cursor to the beginning of the first line" (Start at line 1 position 0) to the LCD. From Table 9.1, the command code for this is 0x80. From the complete LCD program shown below, the statement cmd(0x80); will execute the following C code:

```
void cmd(unsigned char value)
{
      PORTD=value; //Command is sent to PORTD
      PORTB=0x04; //rs=0 rw=0 en=1
      delay(10); //20msec delay
      PORTB=0x00; //rs=0 rw=0 en=0
}
```

The above CMD function first outputs the value = 0x80 to PORTD. Since PORTD is connected to LCD's D0-D7 pins, these data will be available to be latched by the LCD. The following few lines of the above code of the CMD function are for outputting 0's to RS and R/W̄ pins, and a trailing edge (1 to 0) pulse to EN pin along with a delay of 20 msec. Hence, the LCD will latch 0x80, and the cursor will move to the start of the first line.

The following C loop obtained from the complete C-program will provide 2 msec delay:

```c
void delay(unsigned int itime) //2 msec delay
{
        unsigned int i,j;
        for(i=0; i<itime; i++)
                for(j=0; j<255;j++);
}
```

The above C code along with the statement delay(10); will provide 20 msec delay. Similarly, the program logic (shown below) for outputting other ASCII characters and switch input data can be explained.

The complete LCD program using C is provided below:

```c
#include <P18F4321.h>
void cmd(unsigned char);
void data(unsigned char);
void delay(unsigned int);
void main(void)
{
  unsigned char input,output,i;
  unsigned char tstr [13]={'S', 'w', 'i', 't', 'c', 'h', ' ', 'V', 'a', 'l', 'u', 'e', ':'};
    TRISD=0; //PORTD is output
       ADCON1 = 0x0F; // Configure PORTB as digital I/O
       TRISB=0; //PORTB is output
       TRISC=0xFF; //PORTC is input
       PORTB=0x00; //rs=0 rw=0 en=0
       delay(10); //20msec delay
       cmd(0x0C); //Display On, Cursor Off
       delay(10); //20msec delay
       cmd(0x01); //Clear Display
       delay(10); //20msec delay
       cmd(0x06); //Shift cursor to the right
       delay(10); //20msec delay
       cmd(0x80); //Start at line 1 position 0
       delay(10); //20msec delay
       for (i = 0; i<13; i++)
       data (tstr [i]);
       while(1)
       {
       input= PORTC&0x0F; //Mask switch value
       output=0x30 | input; //Logically OR switch inputs with
                            //0x30 to obtain the ASCII code
       data(output); //Display switch value on screen
       delay(10); //20msec delay
       cmd(0x10); //Shift cursor to the left
   }
}     //end main
```

```
void cmd(unsigned char value)
{
      PORTD=value; //Command is sent to PORTD
      PORTB=0x04; //rs=0 rw=0 en=1
      delay(10); //20msec delay
      PORTB=0x00; //rs=0 rw=0 en=0
}
void data(unsigned char value)
{
      PORTD=value; //Data sent to PORTD
      PORTB=0x05; //rs=1 rw=0 en=1
      delay(10); //20msec delay
      PORTB=0x01; //rs=1 rw=0 en=0
}
void delay(unsigned int itime) //2 msec delay
{
      unsigned int i,j;
      for(i=0; i<itime; i++)
            for(j=0; j<255;j++);
}
```

9.3   Analog Interface

A/D (Analog-to-Digital) and D/A (Digital-to-Analog) converters are widely used these days for performing data acquisition and control. Separate A/D and D/A converter chips are commercially available. These chips are interfaced externally in microprocessor-based applications. The PIC18F4321 includes an on-chip A/D. However, an external D/A chip needs to be interfaced if digital-to-analog conversion is desired. Figure 1.9 (Chapter 1) depicts a typical control application where A/D and D/A converters are used.

Suppose that it is necessary to maintain the temperature of a furnace (Figure 1.9) to a desired level to maintain the quality of a product. Assume that the designer has decided to control this temperature by adjusting the fuel. This can be accomplished using a typical microcontroller such as the PIC18F4321 along with the interfacing components as follows. Temperature is an analog (continuous) signal. It can be measured by a temperature-sensing (measuring) device such as a thermocouple. The thermocouple provides the measurement in millivolts (mV) equivalent to the temperature.

Since microcontrollers only understand binary numbers (0's and 1's), each analog mV signal must be converted to a binary number using the microcontroller's on-chip A/D converter. Note that the PIC18F contains an on-chip A/D converter. The PIC18F does not include an on-chip D/A converter. However, the D/A converter chip can be interfaced to the PIC18F externally.

First, the millivolt signal is amplified by a mV/V amplifier to make the signal compatible for A/D conversion. A microcontroller such as the PIC18F4321 can be programmed to solve an equation with the furnace temperature as an input. This equation compares the temperature measured with the temperature desired which can be entered into the microcontroller using the keyboard. The output of this equation will provide the appropriate opening and closing of the fuel valve to maintain the appropriate temperature. Since this output is computed by the microcontroller, it is a binary number. This binary output must be converted into an analog current or voltage signal.

The D/A converter chip inputs this binary number and converts it into an analog current (I). This signal is then input into the current/pneumatic (I/P) transducer for opening or closing the fuel input valve by air pressure to adjust the fuel to the furnace. The furnace temperature desired can thus be achieved. Note that a transducer converts one form of energy (electrical current, in this case) to another form (air pressure, in this example).

9.3.1 On-chip A/D Converter

The PIC 18F4321 contains an on-chip A/D converter (or sometimes called ADC) module with 13 channels (AN0-AN12). An analog input can be selected as an input on one of these 13 channels, and can be converted to a corresponding 10-bit digital number. Three control registers, namely, ADCON0 through ADCON2, are used to perform the conversion.

The ADCON0 register, shown in Figure 9.16, controls the operation of the A/D module. The ADCON0 register can be programmed to select one of 13 channels using bits CHS3 through CHS0 (bits 5 through 2). The conversion can be started by setting the GO/$\overline{DONE}$ (bit 1) to 1. Once the conversion is completed, this bit is automatically cleared to 0 by the PIC18F4321.

7	6	5	4	3	2	1	0	
------------	-----------	CHS3	CHS2	CHS1	CHS0	GO/$\overline{DONE}$	ADON	ADCON0

bit 7-6 **Unimplemented:** Read as '0'
bit 5-2 **CHS3:CHS0:** Analog Channel Select bits
0000 = Channel 0 (AN0)
0001 = Channel 1 (AN1)
0010 = Channel 2 (AN2)
0011 = Channel 3 (AN3)
0100 = Channel 4 (AN4)
0101 = Channel 5 (AN5)
0110 = Channel 6 (AN6)
0111 = Channel 7 (AN7)
1000 = Channel 8 (AN8)
1001 = Channel 9 (AN9)
1010 = Channel 10 (AN10)
1011 = Channel 11 (AN11)
1100 = Channel 12 (AN12)
1101 = Unimplemented
1110 = Unimplemented
1111 = Unimplemented

bit 1 **GO/$\overline{DONE}$:** A/D Conversion Status bit

When GO/$\overline{DONE}$ = 1:
1 = A/D conversion in progress
0 = A/D idle

bit 0 **ADON:** A/D On bit
1 = A/D converter module is enabled

FIGURE 9.16   ADCON0 (A/D Control Register 0).

The ADCON1 register, shown in Figure 9.17, configures the functions of the port pins as Analog (A) input or Digital (D) I/O. The table shown in Figure 9.17 shows how the port bits are defined as analog or digital signals by programming the PCFG3 through PCFG0 bits (bits 3 through 0) of the ADCON1 register. This register can also be programmed to select the reference voltages for the A/D.

The ADCON2 register, shown in Figure 9.18, configures the A/D clock source, programmed acquisition time and justification. The A/D conversion time per bit is defined as TAD. The A/D conversion requires 11 TAD per 10-bit conversion. The source of the A/D conversion clock is software selectable. For correct A/D conversions, the A/D conversion clock (TAD) must be as short as possible, but greater than the minimum requirement of 0.7 μ sec for a Tosc-based clock with Vref $\geq$3V After conversion, the 10-bit binary output of the A/D is placed in a 16-bit register (two 8-bit register pair) ADRESH:ADRESL. Since six bits of the 16-bit register will not be used, the ADFM bit (bit 7) of the ADCON2 can be set to 1 or cleared to 0 to provide the conversion reading, respectively, as right or left justified with unused bits as 0's.

The PIC18F4321 contains a 10-bit on-chip A/D with 13 channels. Figure 9.19 shows a block diagram of the PIC18F4321 A/D. The ADRESH and ADRESL registers contain the result of the A/D conversion. When the A/D conversion is complete,

7	6	5	4	3	2	1	0	
------------	-----------	VCFG1	VCFG0	PCFG3	PCFG2	PCFG1	PCFG0	ADCON1

bit 7-6 **Unimplemented:** Read as '0'
bit 5 **VCFG1:** Voltage Reference Configuration bit (V$_{REF}$ - source)
1 = V$_{REF}$- (AN2)
0 = V$_{SS}$
bit 4 **VCFG0:** Voltage Reference Configuration bit (V$_{REF}$+ source)
1 = V$_{REF}$+ (AN3)
0 = V$_{DD}$

bit 3-0 **PCFG3:PCFG0:** A/D Port Configuration Control bits

bit 7-6	**Unimplemented**: Read as '0'
bit 5	**VCFG1:** Voltage Regerence Configuration bit (VREF-source)
	1 = VREF- (AN2)
	0 = VSS
bit 4	**VCFG0:** Voltage Reference Configuration bit (VREF + source)
	1 = VREF +(AN3)
	0 = VDD

bit 3-0 (decimal) in Column 1: **PCFG3:PCFG0:**A/D Port Configuration Control bits

	AN12	AN11	AN10	AN9	AN8	AN7	AN6	AN5	AN4	AN3	AN2	AN1	AN0
0	A	A	A	A	A	A	A	A	A	A	A	A	A
1	A	A	A	A	A	A	A	A	A	A	A	A	A
2	A	A	A	A	A	A	A	A	A	A	A	A	A
3	D	A	A	A	A	A	A	A	A	A	A	A	A
4	D	D	A	A	A	A	A	A	A	A	A	A	A
5	D	D	D	A	A	A	A	A	A	A	A	A	A
6	D	D	D	D	A	A	A	A	A	A	A	A	A
7	D	D	D	D	D	A	A	A	A	A	A	A	A
8	D	D	D	D	D	D	A	A	A	A	A	A	A
9	D	D	D	D	D	D	D	A	A	A	A	A	A
10	D	D	D	D	D	D	D	D	A	A	A	A	A
11	D	D	D	D	D	D	D	D	D	A	A	A	A
1 2	D	D	D	D	D	D	D	D	D	D	A	A	A
13	D	D	D	D	D	D	D	D	D	D	D	A	A
14	D	D	D	D	D	D	D	D	D	D	D	D	A
15	D	D	D	D	D	D	D	D	D	D	D	D	D

A = Analog input   D = Digital I/O

FIGURE 9.17   ADCON1 (A/D Control Register 1).

the result is loaded into the ADRESH:ADRESL register pair, the GO/DONE bit (ADCON0 register) is cleared, and the A/D Interrupt Flag bit, ADIF, is set.

The following steps should be followed to perform an A/D conversion:

1. Configure the A/D module:
   - Configure analog pins, voltage reference, and digital I/O (ADCON1)
   - Select A/D input channel (ADCON0)
   - Select A/D acquisition time (ADCON2)
   - Select A/D conversion clock (ADCON2)
   - Turn on A/D module (ADCON0)

2. Configure A/D interrupt (if desired):
   - Clear ADIF bit (bit 6 of PIR1, Figure 9.7)
   - Set ADIE bit (bit 6 of PIE1, Figure 9.8)
   - Set GIE bit (bit 7) and PEIE (bit 6) of INTCON register, Figure 8.17(a)

   - All interrupts including A/D Converter interrupt, branch to address 0x000008 (default) upon power-on reset. However, the A/D Converter interrupt can be configured as low priority by setting the ADIP bit (bit 6) of the IPR1 register (See Microchip manual) to branch to address 0x000018. The instruction, BSF IPR1, ADIP can be used for this purpose.

3. Wait for the required acquisition time (if required).

4. Start conversion:

7	6	5	4	3	2	1	0	
ADFM	------------	ACQT2	ACQT1	ACQT0	ADCS2	ADCS1	ADCS0	ADCON2

bit 7 **ADFM:** A/D Result Format Select bit
1 = Right justified; 10-bits in lower 2 bits of ADRESH with upper 6 bits as 0's and in 8 bits of ADRESL
0 = Left justified; 8-bit result in ADRESH and the contents of ADRESL are ignored. Used for 8-bit conversion.
bit 6 **Unimplemented:** Read as '0'
bit 5-3 **ACQT2:ACQT0:** A/D Acquisition Time Select bits
$111 = 20\ T_{AD}$
$110 = 16\ T_{AD}$
$101 = 12\ T_{AD}$
$100 = 8\ T_{AD}$
$011 = 6\ T_{AD}$
$010 = 4\ T_{AD}$
$001 = 2\ T_{AD}$
$000 = 0\ T_{AD(1)}$
bit 2-0 **ADCS2:ADCS0:** A/D Conversion Clock Select bits
$111 = F_{RC}$ (clock derived from A/D RC oscillator)(1)
$110 = F_{OSC}/64$
$101 = F_{OSC}/16$
$100 = F_{OSC}/4$
$011 = F_{RC}$ (clock derived from A/D RC oscillator)(1)
$010 = F_{OSC}/32$
$001 = F_{OSC}/8$
$000 = F_{OSC}/2$
Note 1: If the A/D F_{RC} clock source is selected, a delay of one T_{CY} (instruction cycle) is
added before the A/D clock starts. This allows the SLEEP instruction to be executed before starting a conversion.

FIGURE 9.18   ADCON2 (A/D Control Register 2).

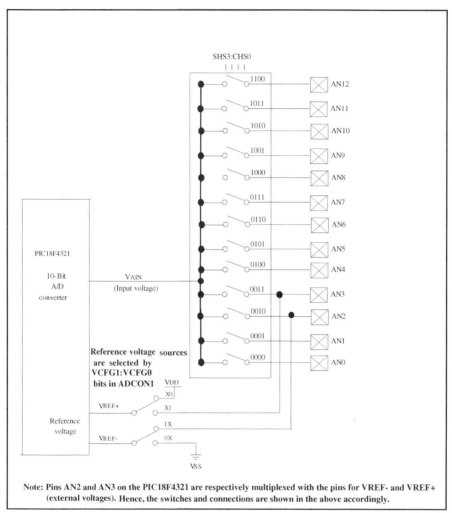

FIGURE 9.19   Block diagram of the PIC18F4321 A/D.

• Set GO/$\overline{\text{DONE}}$ bit (ADCON0 register) to 1.

5.   Wait for A/D conversion to complete, by either

• Polling for the GO/$\overline{\text{DONE}}$ bit to be cleared to 0 (conversion completed) or waiting for the A/D interrupt.

6.   Read A/D Result registers (ADRESH:ADRESL); clear bit ADIF, if required.
7.   For next conversion, go to step 1 or step 2, as required. The A/D conversion time per bit is defined as TAD. A minimum wait of 2 TAD is required before the next acquisition starts.

Three control registers namely, ADCON0 through ADCON2 are used to perform the conversion. The ADCON0 register (Figure 9.16) controls the operation of the ADC module. The ADCON0 register can be programmed to select one of 13 channels using bits CHS3 through CHS0 (bits 5 through 2). For example, the C-code
```
ADCON0 = 0x01;
```

will select AN0, keep the ADC idle (conversion not started), and also enable the ADC module.

The conversion can be started by setting the GO/$\overline{\text{DONE}}$ bit (bit 1 of ADCON0) to 1 using the following C-code:

```
ADCON0bits.Go = 1;
```

Once the conversion is started, the GO/$\overline{\text{DONE}}$ bit can be checked in the program by waiting in a loop until the GO/$\overline{\text{DONE}}$ bit becomes LOW indicating completion of the A/D conversion (polled I/O) using the C-code:

```
while(ADCON0bits.DONE == 1);
```

The GO/$\overline{\text{DONE}}$ bit can also be checked using peripheral interrupt I/O to be discussed next. The following steps can be used by the PIC18F4321 for programming the ADC using interrupt I/O:

1. Set the ADIE bit to 1 in the PIE1 register (Figure 9.8). This will enable the A/D interrupt.
2. Clear the ADIF bit in the PIR1 register (Figure 9.7) to 0.
3. Set the PEIE bit in the INTCON register (Figure 9.4) to 1 to enable peripheral interrupt.
4. Set the GIE bit in the INTCON register to 1 to enable global interrupt.
5. Start A/D conversion by setting the GO/DONE bit in the ADCON0 register (Figure 9.14). As soon as the A/D conversion is completed, the ADIF bit in the PIR1 register will automatically be set to 1. The PIC18F4321 will complete execution of the current instruction, and then branch to the interrupt address vector 0x00008 (high priority interrupt address upon power-on reset). Note that low priority can be assigned to ADC interrupt by clearing the ADIP (A/D converter interrupt priority bit) in the IPR1 register (Page 100 of the PIC18F4321 manual) to 0. In that case, the interrupt address vector will be 0x00018.

The following C-code will accomplish the interrupt-driven ADC upon power-on reset:

```
PIE1bits.ADIE = 1;   // set ADIE bit in PIR1 to 1
PIR1bits.ADIF = 0;   // Clear ADIF to 0
INTCONbits.PEIE = 1; // Enable peripheral interrupt
INTCONbits.GIE = 1;  // Enable global interrupt
ADCON0bits.GO = 1;   // Start A/D conversion
```

The ADCON1 register, shown in Figure 9.17, configures the functions of the port pins as Analog (A) input or Digital (D) I/O. The table in Figure 9.17 shows how the port bits are defined as analog or digital signals by programming the PCFG3 through PCFG0 (bits 3 through 0) of the ADCON1 register. This register can also be programmed to select the reference voltages for the A/D. As an example, the C-code

```
ADCON1 = 0x0E;
```

will configure bit 0 of Port A as AN0 with VDD and VSS as reference voltages.

The ADCON2 register (Figure 9.18) configures the A/D clock source, programmed acquisition time, and justification. Using ADCS2 through ADCS0 bits in the ADCON2 register, the A/D conversion clock source can be selected. For example, with ADCS2 ADCS1 ADCS0 = 001_2, a conversion clock source of Fosc/8 is selected. Note that the Fosc is the frequency of the crystal of the PIC18F.

Using ACQT2 through ACQT0 of ADCON2 register, the A/D acquisition time can be selected. For example, with ACQT2 ACQT1 ACQT0 = 101_2, an A/D acquisition time of 12 TAD is selected. The A/D acquisition time is a function of the A/D conversion time per bit. Note that the A/D conversion time per bit is defined as TAD. The A/D conversion time is 11 times the TAD per 10-bit conversion. For correct A/D conversion, the A/D conversion clock (TAD) must be as short as possible, but greater than the minimum requirement of 0.7 microsecond for Tosc based clock with Vref $\geq$ 3V.

With ACQT2 ACQT1 ACQT0 = 101_2, and ADCS2 ADCS1 ADCS0 = 001_2, the A/D conversion clock of Fosc/8 and an acquisition time of 12 TAD will be selected. Hence, using the default clock of 1 MHz, the value of Fosc/8 will be 0.125 MHz. TAD = (1/0.125 MHz) = 8 microseconds which satisfies the minimum requirement of 0.7 microsecond. The conversion time = 12 TAD = 12 x 8 microseconds = 96 microseconds.

After conversion, the 10-bit binary output of the A/D is placed in a 16-bit register (two 8-bit register pair) ADRESH:ADRESL. Since six bits of the 16-bit register pair will not be used, the ADFM bit in the ADCON2 register is used to specify the result as left justified or right justified. By clearing the ADFM bit to 0 in ADCON2 will format the result as left justified. The left justified format provides 8-bit result in ADRESH. This means that the result will be placed in the ADRESH register as 8-bit; the contents of ADRESL are discarded. By setting the ADFM bit to 1 in ADCON2, on the other hand, will format the result as right justified. The right justified format provides the 10-bit result in ADRESH:ADRESL with the upper two bits in bits 0-1 of ADRESH and lower eight bits in ADRESL; the upper six bits of ADRESH are discarded.

Example 9.6

A PIC18F4321 microcontroller shown in Figure 9.20 is used to implement a voltmeter to measure voltage in the range 0 to 5 V and display the result in two decimal digits: one integer part and one fractional part. Write a C language program to accomplish this using:

(a)   Polled I/O,
(b)   Interrupt I/O

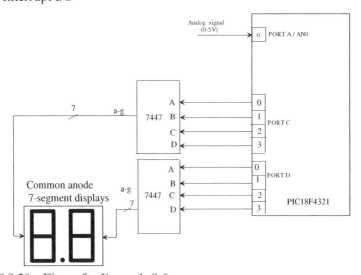

FIGURE 9.20   Figure for Example 9.6.

Solution

(a)   Using polled I/O:

In order to design the voltmeter, the PIC18F4321 on-chip A/D converter will be used. Three registers (ADCON0, ADCON1, ADCON2) need to be configured. In ADCON0, bit 0 of PORT A (RA0/AN0) is designated as the analog signal to be converted. Hence, CHS3-CHS0 bits (bits 5-2) are programmed as 0000 to select channel 0 (AN0). The ADCON0 register is also used to enable the A/D, start the A/D, and then check the "end of conversion" bit.

In order to design the voltmeter, the PIC18F4321 on-chip A/D converter available will be used. Three registers, ADCON0 through ADCON2, need to be configured. In ADCON0, bit 0 of Port A (AN0) is designated as the analog signal to be converted. Hence, CHS3CHS0 bits (bits 5-2) are programmed as 0000 to select channel 0 (AN0). The ADCON0 register is also used to enable the A/D, start the A/D, and then check the "End of conversion" bit.

The reference voltages are chosen by programming the ADCON1 register. In this example, V_{DD} (by clearing bit 4 of of ADCON1 to 0) and V_{ss} (by clearing bit 5 of ADCON1 to 0) will be used. Note that V_{DD} and V_{ss} are already connected to the PIC18F4321. The ADCON1 register is also used to configure AN0 (bit 0 of Port A) as an analog input by writing 1101 at PCFG3-PCFG0 bits (bits 3-0 of ADCON1). Note that there are several choices to configure AN0 as an analog input.

The ADCON2 is used to set up the acquisition time, conversion clock, and, also, if the result is to be left or right justified. In this example, an 8-bit result is assumed. The A/D result is configured as left justified, and, therefore, the 8-bit register ADRESH will contain the result. The contents of ADRESL are ignored. When 8 bits are obtained from 10 bits using left justified, lower two bits in ADRESL are discarded.

Because the maximum decimal value that can be accommodated in 8 bits of ADRESH is 255_{10} (FF_{16}), the maximum voltage of 5 V will be equivalent to 255_{10}. This means the display in decimal is given by

$$D = 5 \times \ (\text{input} / 255)$$

$$= \text{input} /51$$

$$= \underbrace{\text{quotient}}_{\text{Integer part}} + \text{remainder}$$

This gives the integer part. The fractional part in decimal is

$$F = (\ \text{remainder}/51\ ) \times 10$$

$$\simeq \text{remainder})/5$$

For example, suppose that the decimal equivalent of the 8-bit output of A/D is 200.
$$D = 200/51 \Rightarrow \text{quotient} = 3, \text{remainder} = 47$$
Hence, integer part = 3 and fractional part, F = 47/5 = 9
Therefore, the display will show 3.9 V.

From these equations, the final result will be in BCD, which can then be sent to the 7447 decoders.

The integer value is placed in D1 and the remainder part is placed in D0. Finally, the result is displayed on the seven-segment displays.

The C language program for the voltmeter using polled I/O is provided below:

```
#include <P18F4321.h>
unsigned int FINAL,ADCONRESULT; //Initialize variables
unsigned char D1,D0;
void CONVERT(void);
void main ( )
 {
      TRISD = 0; // Port D is Output
      TRISC = 0; // Port C is Output
      ADCON0 = 0x01; // Configure the ADC registers
      ADCON1 = 0x0D;
      ADCON2 = 0x29;
      D0=0; // Data to display '0' on integer 7seg display
      D1=0;
      while(1) // Data to display '0' on fractional 7seg display
      {
         ADCON0bits.GO = 1; // Start the ADC
    while(ADCON0bits.GO == 1); //Delay until conversion complete

         PORTC = D1; // Output D1 to integer 7segment display
         PORTD = D0; // Output D0 to fractional 7segment
    ADCONRESULT = ADRESH; // Move the ADC result into
                             //ADCONRESULT
    FINAL = (ADCONRESULT*10)/51; // Conversion factor
    CONVERT( );
      }
}   //end main
void CONVERT( )
{
 D1 = FINAL/10;
 D0 = FINAL%10; // D0 is remainder of FINAL divided by 10
}
```

(b) Using interrupt I/O: The basics are already described in the last section and also in part (a).

The C language program for the voltmeter using PIC18F interrupt I/O is provided below:

```
#include <p18f4321.h>
#pragma config WDT=OFF
#pragma config LVP=OFF
#pragma config BOR =OFF
#pragma config OSC = INTIO2
unsigned int result;
void convert (void);
#pragma code ADCINT=0x08       // At interrupt, code jumps here
void ADCINT (void)
{ _asm
```

```
GOTO convert
_endasm
}
#pragma code // End code
void main()
{
  TRISC = 0;              // Configure Port C as output
  TRISD = 0;              // Configure Port D as output
  PIE1bits.ADIE = 1;      // enable ADC interrupt
  PIR1bits.ADIF = 0;      // clear ADC interrupt flag
  INTCONbits.PEIE = 1;    // enable peripheral interrupt
  INTCONbits.GIE = 1;     // enable global interrupt
  ADCON0=0x01;            // Select AN0
  ADCON1=0x00;            // Enable AN0 and select
                          // reference voltages
  ADCON2=0x29;            // Left justified, 12 TAD, Fosc/8
  ADCON0bits.GO = 1;      // Start A/D conversion
  while(1);               // Wait for interrupt
}
#pragma interrupt convert
void convert (void)
{
  PIR1bits.ADIF = 0;      // Clear ADIF to 0
  result = ADRESH;
  PORTC = result/51;
  PORTD = result%51;
  ADCON0bits.GO = 1;      // Start A/D conversion again
}
```

9.3.2 Interfacing an External D/A (Digital-to-Analog) Converter to the PIC18F4321

Several microcontrollers such as the PIC18F4321 do not have any on-chip D/A converter (or sometimes called DAC). Hence, external D/A converter chip is interfaced to the PIC18F4321 to accomplish this function. Some microcontrollers such as the Intel/Analog Devices 8051 include an on-chip D/A converter. In order to illustrate the basic concepts associated with interfacing a typical D/A converter such as the Maxim MAX5102 to the PIC18F4321, consider Figure 9.21.

The MAX5102 is a 16-pin chip. In this example, the PIC18F4321 microcontroller is interfaced with the MAX5102 chip to convert an 8-bit binary input to an analog voltage from 0 to 5 V. Eight switches connected to PORTD of the PIC18F4321 will provide a value between 0 and 255 that will be converted by the MAX5102 into a DC voltage between 0 V and 5 V. This analog voltage will then appear on the OUTA or OUTB pin of the MAX5102 D/A converter.

The MAX5102 contains two independent D/A converters, namely, DAC A and DAC B. These D/A converters are selected by the A0 input pin on the MAX5102. The two converters share the same 8-bit input pins, D0-D7. The $\overline{WR}$ input pin on the MAX5102 when HIGH latches the 8-bit input data for DAC A and DAC B for conversion to analog voltage. For example, A0 = 0 and $\overline{WR}$ = 0, the 8-bit data on DAC A input is transparent while A0 = 1 and $\overline{WR}$ = 0, the 8-bit data on DAC B input is transparent. The analog voltage output for either DAC A or DAC B will be available when $\overline{WR}$ = 1. One must make sure that 8-bit input data are valid before $\overline{WR}$ goes to 0 to get rid of any glitches.

The manufacturer recommends that for proper operation of the MAX5102, the VSS should be connected to ground via a 0.1 µF capacitor. Note that the programmer must ensure that the timing requirements for $\overline{WR}$, A0, and D0-D7 are met according to the manufacturer's specification. Hence, each time the PIC18F4321 outputs 8-bit new data on the data pins of the MAX5102, a delay of a few milliseconds (for example, 2 msec) may be required so that the data will be valid before outputting a LOW on the $\overline{WR}$ pin.

Example 9.7

Assume the block diagram of Figure 9.21. Write a C language program that will input eight switches via PORTD of the PIC18F4321, and output the byte to D0-D7 input pins of the MAX5102 D/A converter. The microcontroller will send appropriate signals to the $\overline{WR}$ and A0 pins so that the D/A converter will convert the input byte to an analog voltage between 0 and 5 V, and output the converted voltage on its OUTA pin.

Solution

The steps for writing a PIC18F C language program for the D/A converter interface of Figure 9.20 is provided in the following:

1. Configure PORTB and PORTC as outputs, and PORTD as input.
2. Output a LOW to A0 Pin of the D/A via bit 1 of PORTB to select OUTA.
3. Output a LOW to $\overline{WR}$ pin of the D/A via bit 0 of PORTB.
4. Input the switches via PORTD, and output to PORTC.
5. Output a HIGH to $\overline{WR}$ pin of the A/D via bit 0 of PORTB to latch 8-bit input data for converting to analog voltage. No delay is needed since the program will be written to input one byte of data from the switches.

The C language program is provided below:

```c
#include <p18f4321.h>
void main (void)
{
        TRISB=0x00;    // Configure PORTB as output
        TRISC=0x00;    // Configure PORTC as output
```

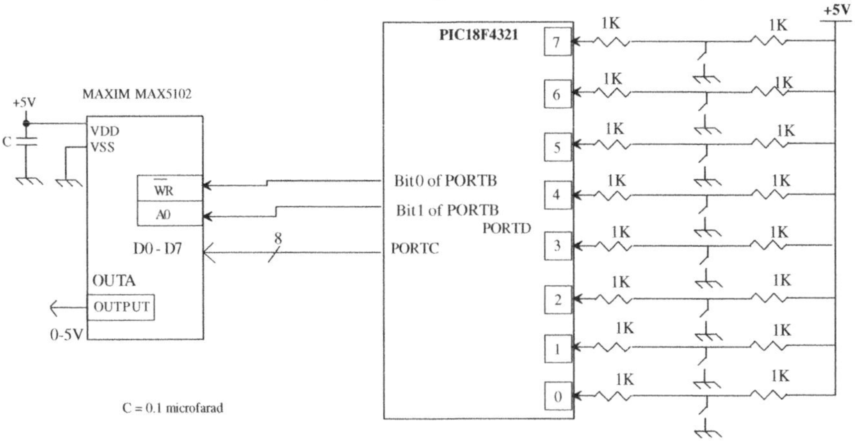

FIGURE 9.21   Figure for Example 9.7.

```
        TRISD=0xFF;    // Configure PORTD as input
        PORTBbits.RB1=0; // Clear A0 to 0 to select OUTA
        PORTBbits.RB0=0; // Output LOW on bit 0 of PORTB
        PORTC=PORTD;   // Input switches, output to PORTD
        PORTBbits.RB0=1; // Latch data for conversion
        while(1);     // Halt
}
```

9.4 Serial Interface

In various instances, it is desirable to transmit binary data from one microcontroller to another. In such situations, data can be transmitted using either parallel or serial transmission techniques. In parallel transmission, each bit of the binary data is transmitted over a separate wire or line.

In serial transmission, only one line is used to transmit the complete binary data bit by bit. Hence, the transmitting device such as a microcontroller must convert parallel data into a string of serial bits. The receiving device such as another microcontroller must convert data from serial to parallel. Data are usually sent starting with the least significant bit. In order to differentiate among various bits, a clock signal is used. Serial data transmission can be divided into two types: synchronous and asynchronous. We now briefly describe them.

9.4.1 Synchronous Serial Data Transmission

The basic feature of synchronous serial data transmission is that data are transmitted or received based on a clock signal. After deciding on a specific rate of data transmission, commonly known as "baud rate" (bits per second), the transmitting device sends a data bit at each clock pulse. In order to interpret data correctly, the receiving device must know the start and end of each data unit. Therefore, in synchronous serial data transmission, the receiver must know the number of data units to be transferred. Also, the receiver must be synchronized with data boundaries. Usually, one or two SYNC characters (a string of bits) are used to indicate the start of each synchronous data stream.

The data unit normally contains error bits such as parity. In some transmissions, the least significant bit is used as a parity bit. The synchronous receiver usually waits in a "hunt" mode while looking for data. As soon as it matches one or more SYNC characters based on the number of SYNC characters used, the receiver starts interpreting the data. In synchronous serial transmission, the transmitting device needs to send data continuously to the receiving device. However, if data are not ready to be transmitted, the transmitter will pad with SYNC characters until data are available.

As mentioned before, in synchronous serial transfer, the receiver must know the number of SYNC characters used, and the number of data units to be transferred. Once the receiver matches the SYNC characters, it receives the specified number of data units, and then goes into a "hunt" mode for matching the SYNC pattern for next data.

9.4.2 Asynchronous Serial Data Transmission

In this type of data transfer, the transmitting device does not need to be synchronized to the receiving device. The transmitting device can send one or more data units when it has data ready to be sent. Each data unit must be formatted. In other words, each data unit must contain "start" and "stop" bits, indicating the beginning and the end of each data unit. The interface circuits between the transmitting device and the receiving device must perform the following functions:

1.   Converts an 8-bit parallel data unit from the transmitting device into serial data for transmitting them to the receiving device.
2.   Converts serial data from the receiving device into parallel data for sending them back to the transmitting device assuming two-way (full duplex) transmission.

Each data unit can be divided into equal time intervals, called "bit intervals." A data bit can be either HIGH or LOW during each bit interval. For example, 8-bit data will have eight bit intervals. Each data bit will correspond to one of the eight bit intervals.

The format for asynchronous serial data typically contains the following information:

1.   A LOW START bit.
2.   5-8 bits, denoting the actual data being transferred.
3.   An optional parity bit for either odd or even parity.
4.   1, 1½, or 2 STOP bits having HIGH levels. Note that 1½ STOP bits mean a HIGH level with a duration of 1.5 times the bit interval.

9.4.3   PIC18F Serial I/O

Serial I/O is typically fabricated as an on-chip module with microcontrollers. This will facilitate interfacing microcontrollers with other microcontrollers or peripheral devices. Several protocol (rules) standards for serial data transmission have been introduced over the years. Two such standards implemented include SPI (Serial Peripheral Interface) developed by Motorola and I²C (Inter-Integrated Circuit) developed by Philips. Both protocols are based on synchronous serial data transmission.

The SPI is a protocol established for data transfer between a master and a slave device. The master device can be a microcontroller while the slave device can be devices such as another microcontroller, EEPROMs, and A/D converters. The I²C protocol, on the other hand, is widely used for transferring data among the ICs (Integrated Circuits) on PCBs (Printed Circuit Boards).

The PIC18F4321 contains an on-chip Master Synchronous Serial Port (MSSP) module which is a serial interface, useful for communicating with other peripheral or microcontroller devices. The MSSP module can operate in either SPI or I²C mode. We will cover the PIC18F SPI in this section.

PIC18F4321 pins and signals for the SPI mode The PIC18F SPI primarily uses three pins of the PIC18F4321. They are SCK (Serial Clock, pin 18), SDI (SPI Data In, pin 23), and SDO (SPI Data Out, pin 24). A fourth pin, namely, $\overline{SS}$ (SPI Slave select input, pin 7), is provided for applications requiring multiple slave devices.

PIC18F4321 registers in SPI mode The MSSP module uses four registers for SPI mode operation. These are:

•   MSSP Control Register 1 (SSPCON1)

•   MSSP Status Register (SSPSTAT)

•   Serial Receive/Transmit Buffer Register (SSPBUF)

•   MSSP Shift Register (SSPSR) – Not directly accessible

Figures 9.22 and 9.23 show the SSPCON1 and SSPSTAT registers, respectively. The SSPCON1 and SSPSAT are the control and status registers in SPI mode operation. The SSPCON1 can be used to enable and configure the serial port pins by setting the SSPEN bit (bit 5) of the SSPCON1 register. The SSPCON1 can also be used to select the master or slave mode using the SSPM3-SSPM0 bits (bits 3-0), and the clock polarity using the CKP bit (bit 4).

Both SPI and I²C modes use the SSPSTAT register. This register can be used to select the SPI mode (master or slave using bit 7), SPI clock (bit 6), and the buffer full status of the buffer register of the receiver (BF bit, bit 0). Bits 1-5 of the SSPSTAT are used in the I²C mode.

Operation When initializing the SPI, several options need to be specified. This is done by programming the appropriate control bits (bits 0-5 of SSPCON1 and bits 7-6 of SSPSTAT). These control bits allow the following to be specified:

- Master mode (SCK is the clock output)
- Slave mode (SCK is the clock input)
- Clock Polarity (idle state of SCK)
- Data Input Sample Phase (middle or end of data output time)
- Clock Edge (output data on rising/falling edge of SCK)
- Clock Rate (master mode only)
- Slave Select mode (slave mode only)

The MSSP consists of a transmit/receive shift register (SSPSR) and a buffer register (SSPBUF). The SSPSR shifts the data in and out of the device, the most significant bit (MSB) first. The SSPBUF holds the data that were written to the SSPSR until the received data are ready. Once the 8 bits of data have been received, that byte is moved to the SSPBUF register. Then, the Buffer Full detect bit, BF (bit 0 of SSPSTAT), and the interrupt flag bit, SSPIF, are set to 1. This double-buffering of the received data (SSPBUF) allows the next byte to start reception before reading the data that were just received. Any write to the SSPBUF register during transmission/ reception of data will be ignored and the write collision detect bit, WCOL (bit 7 of SSPCON1), will be set. User software must clear the WCOL bit so that it can be determined if the following write(s) to the SSPBUF register completed successfully. When the application software is expecting to receive valid data, the SSPBUF should be read before the next byte of data to transfer is written to the SSPBUF. The BF indicates when SSPBUF has been loaded with the received data (transmission is complete). When the SSPBUF is read, the BF bit is cleared. These data may be irrelevant if the SPI is only a transmitter. Generally, the MSSP interrupt is used to determine when the transmission/reception has completed. The SSPBUF must be read and/or written. If the interrupt method is not going to be used, then software polling can be done to ensure that a write collision does not occur.

Enabling SPI I/O To enable the serial port, MSSP Enable bit, SSPEN (bit 5 of SSPCON1) must be set. To reset or reconfigure SPI mode, clear the SSPEN bit, reinitialize the SSPCON registers, and then set the SSPEN bit. This configures the SDI, SDO, SCK, and S̄S̄_pins as serial port pins. For the pins to behave as the serial port function, the data direction bits (in the TRIS register) must be appropriately programmed as follows:

- SDI is automatically controlled by the SPI module.
- SDO must have TRISC (bit 5) bit cleared to 0.
- SCK (Master mode) must have TRISC (bit 3) bit cleared to 0.
- SCK (Slave mode) must have TRISC (bit 3) bit set to 1.
- $\overline{SS}$ must have TRISA (bit 5) bit set to 1 for multiple slaves. Note that RA5 is multiplexed with $\overline{SS}$.

Any serial port function that is not desired may be overridden by programming the corresponding data direction (TRIS) register to the opposite value.

Figure 9.24 shows a simplified block diagram of SPI Master/Slave connection between two PIC18F4321s. In Figure 9.24, the master PIC18F4321 initiates the data transfer by sending the SCK signal. The master can initiate the data transfer at any time because it controls the SCK. The master determines when the slave is to broadcast data by the software protocol. Data are shifted out of both shift registers on their programmed clock

7	6	5	4	3	2	1	0	
WCOL	SSPOV	SSPEN	CKP	SSPM3	SSPM2	SSPM1	SSPM0	SSPCON1

bit 7 **WCOL:** Write Collision Detect bit (Transmit mode only)
1 = The SSPBUF register is written while it is still transmitting the previous word (must be cleared in software)
0 = No collision

bit 6 **SSPOV:** Receive Overflow Indicator bit
SPI Slave mode:
1 = A new byte is received while the SSPBUF register is still holding the previous data. In case of overflow, the data in SSPSR is lost. Overflow can occur only in Slave mode. The user must read the SSPBUF, even if only transmitting data, to avoid setting overflow (must be cleared in software).
0 = No overflow
Note: In Master mode, the overflow bit is not set since each new reception (and transmission) is initiated by writing to the SSPBUF register.

bit 5 **SSPEN:** Synchronous Serial Port Enable bit
1 = Enables serial port and configures SCK, SDO, SDI, and SS as serial port pins
0 = Disables serial port and configures these pins as I/O port pins
Note: When enabled, these pins must be properly configured as input or output.

bit 4 **CKP:** Clock Polarity Select bit
1 = Idle state for clock is a high level
0 = Idle state for clock is a low level

bit 3-0 **SSPM3:SSPM0:** Synchronous Serial Port Mode Select bits
0101 = SPI Slave mode, clock = SCK pin, SS pin control disabled, SS can be used as I/O pin
0100 = SPI Slave mode, clock = SCK pin, SS pin control enabled
0011 = SPI Master mode, clock = TMR2 output/2
0010 = SPI Master mode, clock = FOSC/64
0001 = SPI Master mode, clock = FOSC/16
0000 = SPI Master mode, clock = FOSC/4
Note: Bit combinations not specifically listed here are either reserved or implemented in I2C™ mode only.

FIGURE 9.22   SSPCON1 (MSSP CONTROL) Register 1 in SPI mode.

edge and latched on the opposite edge of the clock. Both processors should be programmed to the same Clock Polarity (CKP); then both controllers would send and receive data at the same time.

Since read and write operations must be performed on each data byte, some data may not be useful.

Note that whether the data are meaningful (or dummy data) depends on the application software. This leads to three types of data transmission:

• Master sends data, and Slave sends dummy data

• Master sends data, and Slave sends data

• Master sends dummy data, and Slave sends data

Example 9.7

Figure 9.25 shows a block diagram for interfacing two PIC18F4321s in SPI mode. One of the microcontrollers is the master while the other is the slave. The master PIC18F4321 will input four switches via bits 0-3 of PORTB, and then transmit the

7	6	5	4	3	2	1	0	
SMP	CKE	D/$\overline{\text{A}}$	P	S	R/$\overline{\text{W}}$	UA	BF	SSPSTAT

bit 7 **SMP:** Sample bit
SPI Master mode:
1 = Input data sampled at end of data output time
0 = Input data sampled at middle of data output time
SPI Slave mode:
SMP must be cleared when SPI is used in Slave mode.

bit 6 **CKE:** SPI Clock Select bit
1 = Transmit occurs on transition from active to Idle clock state
0 = Transmit occurs on transition from Idle to active clock state
Note: Polarity of clock state is set by the CKP bit (SSPCON1<4>).

bit 5 **D/A:** Data/Address bit
Used in I^{2}C™ mode only.

bit 4 **P:** Stop bit
Used in I^{2}C mode only. This bit is cleared when the MSSP module is disabled, SSPEN is cleared.

bit 3 **S:** Start bit
Used in I^{2}C mode only.

bit 2 **R/W:** Read/Write Information bit
Used in I^{2}C mode only.

bit 1 **UA:** Update Address bit
Used in I^{2}C mode only.

bit 0 **BF:** Buffer Full Status bit (Receive mode only)
1 = Receive complete, SSPBUF is full
0 = Receive not complete, SSPBUF is empty

FIGURE 9.23   SSPSTAT (MSSP Status Register) in SPI mode.

4-bit data using its SDO pin to the slave's SDI pin. The slave PIC18F4321 will output these data to four LEDs, and turn them ON or OFF based on the switch inputs.

Write a C language for the master PIC18F4321 that will configure PORTB and PORTC, initialize SSPSTAT and SSPCON1, input switches, and place these data into its SSPBUF register.

Also, write a C language program for the slave PIC18F4321 that will configure PORTC and PORTD, initialize SSPSTAT and SSPCON1 registers, input data from its SDI pin, place the data in the slave's SSPBUF, and then output to the LEDs.

Solution

The PIC18F C-programs for the master PIC18F4321 and the slave PIC18F4321 in Figure 9.25 are written using the following steps as the guidelines:

Master PIC18F4321
1. Configure PORTB as input, and SD0 and SCK as outputs.
2. Select CKE (SPI clock select bit) using the master's SSPSTAT register.
3. Enable serial functions, select master mode with clock such as fosc/4 using the SSPCON1 register.
4. Input switches and moves switch data into the master's Serial Buffer register (SSPBUF).
5. Wait in a loop, and check whether BF bit in the master's SSPSTAT register is 1, indicating completion of transmission.
6. As soon as BF = 1, the program returns to Step 4.

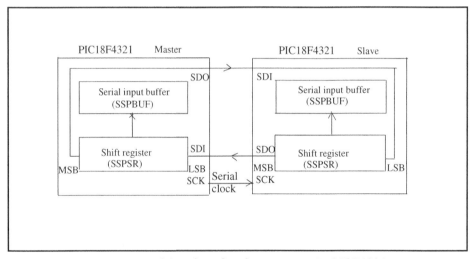

FIGURE 9.24   SPI Master/Slave interface between two PIC18F4321s.

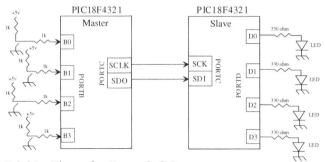

FIGURE 9.25   Figure for Example 9.7.

Slave PIC18F4321

1. Initialize SDI and SCK pins as inputs, and PORTD as output. Note that the SCK is controlled by the master, and, therefore, it is configured as an input by the slave.
2. Select CKE same as the master CKE (high to low clock in this example) using the slave's SSPSTAT register.
3. Enable serial functions, disable the $\overline{SS}$ pin, and select slave mode using the slave's SSPCON1 register. Note that the $\overline{SS}$ pin is used by multiple slaves.
4. Wait in a loop, and check whether BF = 1 in the slave's SSPSTAT register.
5. If BF = 0; wait. However, if BF = 1, output the contents of the slave's Serial Buffer register (SSPBUF) to slave's PORTD.
6. Go to Step 5.

```c
//The following C-code is used to program the master PIC18F4321 device:

#include <p18f4321.h>
void SPI_out(unsigned char);
void main (void)
{
 unsigned char output;
 TRISCbits.TRISC5 = 0; // RC5 is output
 TRISCbits.TRISC3 =0; // RC3 is output
 TRISB = 0xFF;
 ADCON1=0x0F; // Configure PORTB to be digital input
 SSPSTAT= 0x40; // Transmission occurs on high to low clock
 SSPCON1 = 0x20; // Enable serial functions and set as master device
 while(1){
    output = PORTB; // Move switch value to output
    SPI_out(output); // Send variable 'output' to SPI_out
    }
}
void SPI_out(unsigned char SPI_data)
{
    SSPBUF = SPI_data; // Place switch value into the serial buffer
    while (SSPSTATbits.BF == 0); // Wait for transmission to finish
}

//The following C-code is used on the slave PIC18F4321 device:
#include <p18f4321.h>
void main (void)
{
 TRISCbits.TRISC4 = 1; // RC4 is input
 TRISCbits.TRISC3 =1; // RC3 is input
 TRISD=0x00; // PORTD is output
 SSPSTAT = 0x40; // Transmission occurs on high to low clock
 SSPCON1 = 0x25; // Enable serial functions and disable the slave device
 while(1){
  while (SSPSTATbits.BF == 0); // Wait for transmission to finish
  PORTD=SSPBUF; // Move serial buffer to PORTD
  }
}
```

Let us now explain the above C-programs. First, consider the master PIC18F4321. The CKE bit (bit 6) in the SSPSTAT is set to one so that data transmission will occur from an active to an idle (HIGH to LOW) clock. Next, the register SSPCON1 is configured in order to set up the parameters for serial transmission. The bit SSPEN (bit 5) in the SSPCON1 is set to HIGH in order to enable the three pins, namely, SCK, SDO, and SDI. Writing 0000 to bits 3-0 of the SSPCON1 register define the master mode operation with a clock of Fosc/4.

The following C-code will accomplish this:

```
SSPCON1= 0x20;    Enable serial functions and set as master device
```

Next, consider the C-program for the slave; the four bits (bits 3-0) of the slave's SSPCON1 are initialized with 0101. This will place the microcontroller in the slave mode, and, also, the $\overline{SS}$ pin will be disabled since there is only one serial device in this example. Note that the $\overline{SS}$ pin is required if multiple slave devices are used. Also, the SCK pin will be used as the clock.

Let us now briefly explain the program logic. In the C- program for the master, the PIC18F4321 will first perform all initializations, input the switches, and place the switch values to SPI_out. The switch inputs are placed into the SSPBUF register using the C-code:

```
SSPBUF = SPI_data; //Place switch value into the serial buffer
```

As soon as the serial parameters for the master such as the SCK clock pin is set up, data are automatically transmitted to the slave device. Once all the data have been written, the BF bit (bit 0) in the SSPSTAT register of the master microcontroller will go to HIGH, indicating completion of transmission.

The program for the slave microcontroller waits in a loop until the BF flag in its SSPSTAT register goes to HIGH, indicating that the transmission is completed. The switch values from the slave's SSPBUF register are output to the LEDs connected at PORTD using the following C-code:

```
PORTD = SSPBUF;    //Move serial buffer value to PORTD
```

Upon hardware reset of the master PIC18F4321 and the slave PIC18F4321, both the master and the slave start executing the respective programs. Note that the processors do not need to be reset at the same time. The master microcontroller performs initalizations, moves switch data input continuously, and waits in a loop. The slave microcontroller also performs initializations, and then waits in a loop until BF = 1. As soon as the serial communication is established between the master and the slave, the master transmits the contents of SSPBUF via its SDO pin to the slave's SDI pin using the SCK clock. The switch data are transferred to the slave's SSPBUF register. After completion of the transfer, the slave's BF bit in the SSPSTAT register becomes 1. The slave then outputs these data to the LEDs via PORTD.

This example has been successfully implemented in the laboratory. This example can also be implemented using the $\overline{SS}$ pin. In that case, the slave's $\overline{SS}$ pin should be connected to the master's one of the I/O port bits. The I/O port bit must be configured as an output by the master via programming. Also, the $\overline{SS}$ pin must be configured as an input pin. The SSPCON1 should be loaded with 0x24, which will initialize the slave PIC18F4321 in slave mode, and enable its $\overline{SS}$ pin.

9.5   PIC18F4321 Capture/Compare/PWM (CCP) Modules

The CCP module is implemented in the PIC18F4321 as an on-chip feature to provide measurement and control of time-based pulse signals. Timer1 through Timer3 are used for this purpose. Note that Timer0 is not used by the CCP module.

Capture mode causes the contents of a 16-bit timer (Timer1 or Timer3) to be written in special function registers upon detecting an *nth* rising or falling edge of a pulse. Compare mode generates an interrupt or change on output pin, when Timer1 or Timer3 matches a preset comparison value. PWM mode creates a re-configurable square wave duty cycle output at a user set frequency; the PWM uses Timer2. The application software can change the duty cycle or period by modifying the value written to specific special function registers.

The PIC18F4321 contains two CCP modules, namely, CCP1 and CCP2. The CCP1 module of the PIC18F4321 is implemented as a standard CCP with enhanced PWM capabilities for better DC motor control. Hence, the CCP1 module in the PIC18F4321 is also called ECCP (Enhanced CCP). Note that the CCP2 module is provided with standard capture, compare, and PWM features. The CCP1 and CCP2 modules will be referred to as CCPx in the following discussion.

9.5.1   CCP Registers

Each CCP module is associated with an 8-bit control register (CCPxCON) shown in Figure 9.26. The CCPxCON can be used to select one of the three modes, namely, Compare, Capture, or PWM.

Each CCP module also contains a 16-bit data register (CCPRx). The 16-bit data register, in turn, is comprised of two 8-bit registers: CCPRxL (low byte) and CCPRxH (high byte). This 16-bit data register can operate as a 16-bit Capture register, a 16-bit Compare register, or an 8-bit PWM register holding the 8-bit decimal part of the duty cycle.

9.5.2   CCP Modules and Associated Timers

The CCP modules utilize Timer1, Timer2, or Timer3, depending on the mode selected. Timer1 and Timer3 are available to modules in Capture or Compare modes, while Timer2 is available for modules in PWM mode. The assignment of a particular timer to a module is determined by the Timer to CCP enable bits in the T3CON register (Figure 9.10). Both modules may be active at any given time and may share the same timer resource if they are configured to operate in the same mode (Capture, Compare, or PWM) at the same time. The assignment of the timers is summarized in Table 9.2.

9.5.3   PIC18F4321 Capture Mode

In Capture mode, the CCPRxH:CCPRxL register pair captures the 16-bit value of the TMR1 or TMR3 registers when an event (such as every rising or falling edge) occurs on the corresponding CCPx pin. The event is selected by the mode select bits, CCPxM3:CCPxM0 (bits 3-0 of CCPxCON, Figure 9.26). When a capture is made, the interrupt request flag bit, CCPxIF (PIR1 register in Figure 9.7), is set; it must be cleared in software. If another capture occurs before the value in register CCPRx is read, the old captured value is overwritten by the new captured value.

In Capture mode, the appropriate CCPx pin (CCP1 pin or CCP2 pin) of the PIC18F4321 should be configured as an input by setting the corresponding TRIS direction bit. Also, the timers that are to be used with the capture feature (Timer1 or Timer3) must be running in Timer mode or synchronized counter mode. In asynchronous counter mode, the capture operation will not work. The timer to be used with each CCP module is selected in the T3CON register (Figure 9.10).

When the Capture mode is changed, a false capture interrupt may be generated. The user should keep the CCPxIE interrupt enable bit clear to avoid false interrupts. The interrupt flag bit, CCPxIF, should also be cleared following any such change in operating mode.

In summary, the following steps can be used to program the PIC18F4321 in capture mode to determine the period of a waveform (assume CCP1; similar procedure for CCP2):

1. Load the CCP1CON register (Figure 9.26) with appropriate data for capture mode.
2. Configure RC2/CCP1/P1A as an input pin using the TRISC register.
3. Select Timer1 or Timer3 by loading appropriate data respectively into T1CON register (Figure 9.6) and/or T3CON register (Figure 9.10).

7	6	5	4	3	2	1	0	
-------------	-------------	DCxB1	DCxB0	CCP xM3	CCPxM2	CCPxM1	CCPxM0	CCPxCON Register

bit 7-6 **Unimplemented:** Read as '0'

bit 5-4 **DCxB1:DCxB0**: PWM Duty Cycle bit 1 and bit 0 for CCP Module x
Capture mode:
Unused.
Compare mode:
Unused.
PWM mode:
These bits are the lower two bits (bit 1 and bit 0) of the 10-bit PWM duty cycle. The higher eight bits (DCx9:DCx2) of the duty cycle are found in CCPRxL.

bit 3-0 **CCPxM3:CCPxM0**: CCPx Module Mode Select bits
0000 = Capture/Compare/PWM disabled (resets CCP module)
0001 = Reserved
0010 = Compare mode, toggle output on match (CCPxIF bit is set)
0011 = Reserved
0100 = Capture mode, every falling edge
0101 = Capture mode, every rising edge
0110 = Capture mode, every 4th rising edge
0111 = Capture mode, every 16th rising edge
1000 = Compare mode: initialize CCP pin low; on compare match, force CCP pin high
(CCPxIF bit is set)
1001 = Compare mode: initialize CCP pin high; on compare match, force CCP pin low
(CCPxIF bit is set)
1010 = Compare mode: generate software interrupt on compare match (CCPxIF bit is set,
CCP pin reflects I/O state)
1011 = Compare mode: trigger special event, reset timer, start A/D conversion on
CCPx match (CCPxIF bit is set)
11xx = PWM mode

FIGURE 9.26   CCPxCON register.

TABLE 9.2   Assignment of timers for the PIC18F4321 CCP mode

CCP mode selected	Timer
Capture mode	Timer1 or Timer3
Compare mode	Timer1 or Timer3
PWM mode	Timer2

4. Clear the interrupt request flag, CCP1IF for CCP1 (Register PIR1 of Figure 9.7) or CCP2IF for CCP2 (Register PIR2 of Figure 9.11), after a capture so that the next capture can be made.

5. Clear the interrupt enable bit, CCP1IE for CCP1 (Register PIE1 of Figure 9.8) or CCP2IE for CCP2 (Register PIE2 of Figure 9.12), to avoid false interrupts.

6. Clear CCPR1H and CCPR1L to 0.

7. Check CCP1IF flag in PIR1 and wait in a loop until CCP1IF is 1 for the first rising edge. As soon as the first rising edge is detected, start Timer1 (or Timer3).

8. Save CCPR1H and CCPR1L in data memory such as REGX and REGY.

9. Clear CCP1IF to 0.

10. Check CCP1IF flag in PIR1 and wait in a loop until CCP1IF is 1 for the second rising edge. As soon as the second rising edge is detected, stop Timer1 (or Timer3).

11. Disable capture by clearing CCP1CON register.

12. Perform 16-bit subtraction: [CCPR1H:CCPR1L] -[REGX:REGY].

13. 16-bit result in register pair [REGX:REGY] will contain the period of the incoming waveform in terms of the number of clock cycles.

Typical applications of the capture mode include:

- measurement of the pulse width of an unknown periodic signal by capturing the subsequent leading (rising) and trailing (falling) edges of a pulse.

- measurement of the period of a signal by capturing two subsequent leading or trailing edges.

- measurement of duty cycle. Note that the duty cycle is defined as (t1/T) x 100 where t1 is the fraction of the time the signal is HIGH in a period T.

Example 9.8

Assume PIC18F4321. Write a C language program to measure the period (in terms of the number of clock cycles) of an incoming periodic waveform connected at the CCP1 pin. Use Timer3, and capture mode of CCP1.

Solution

```
#include<p18f4321.h>
void main(void)
{
unsigned char FIRST_CCPR1L, FIRST_CCPR1H, HIGH_BYTE, LOW_BYTE;
CCP1CON=0x05; // Select capture mode rising edge
TRISCbits.TRISC2=1; // Configure RC2/CCP1/P1A pin as input
T3CON=0x40; // Select TIMER3 as clock source for capture
PIE1bits.CCP1IE=0; // Disable CCP1IE to avoid false interrupt
CCPR1H=0x00; // Clear CCPR1H to 0
CCPR1L=0x00; // Clear CCPR1L to 0
PIR1bits.CCP1IF=0; // Clear CCP1IF
while(PIR1bits.CCP1IF==0); // Wait for the first rising edge
T3CONbits.TMR3ON=1;  // Turn Timer3 ON
FIRST_CCPR1L=CCPR1L;  // Save CCPR1L in FIRST_CCPR1L at
                      // 1st rising edge
FIRST_CCPR1H=CCPR1H;  // Save CCPR1H in FIRST_ CCPR1H
                      // at 1st rising edge
```

```
PIR1bits.CCP1IF=0;    // Clear CCP1IF
while(PIR1bits.CCP1IF==0); // Wait for next rising edge
T3CONbits.TMR3ON=0;   // Turn OFF Timer3
CCP1CON=0x00;         // Disable capture
LOW_BYTE=CCPR1L-FIRST_CCPR1L; // Low byte of result
HIGH_BYTE=CCPR1H-FIRST_CCPR1L; // High byte of result
while(1);                      // Halt
}
```

9.5.4 PIC18F4321 Compare Mode

In Compare mode, the 16-bit CCPRx (CCPR1H:CCPR1L for CCP1 or CCPR2H: CCPR2L) register value is constantly compared against the value in either the TMR1 or the TMR3 register. When a match occurs, the CCPx pin (CCP1 pin or /CCP2 pin) can be:

* driven high

* driven low

* toggled (high-to-low or low-to-high)

* remain unchanged (that is, reflects the state of the I/O latch)

The action on the pin is based on the value of the mode select bits (CCPxM3:CCPxM0) in CCPxCON register (Figure 9.26). As soon as a match occurs, the interrupt flag bit, CCPxIF, is set to one. The user must configure the CCPx pin as an output by clearing the appropriate TRIS bit. Timer1 or Timer3 must be running in Timer mode or synchronized counter mode if the CCP module is using the compare feature. In asynchronous counter mode, the compare operation may not work.

When the Generate Software Interrupt mode is chosen (CCPxM3:CCPxM0 = 1010), the corresponding CCPx pin is not affected. Only a CCP interrupt is generated, if enabled and the CCPxIE bit is set. Both CCP modules are equipped with a Special Event Trigger. This is an internal hardware signal generated in Compare mode to trigger actions by other modules. The Special Event Trigger is enabled by selecting the Compare Special Event Trigger mode (CCPxM3:CCPxM0 = 1011). For either CCP module, the Special Event Trigger resets the Timer register pair for whichever timer resource is currently assigned as the module's time base. This allows the CCPRx registers to serve as a programmable period register for either timer. The Special Event Trigger for CCP2 can also start an A/D conversion. In order to do this, the A/D converter must already be enabled.

Typical applications of the compare mode include generation of a certain time delay, a pulse train, or a waveform with a specific duty cycle.

The following steps can be used to program the PIC18F4321 in compare mode to provide time delay or determine the period of a waveform:

1. Load the CCP1CON (or CCP2CON) register (Figure 9.26) with appropriate data for compare mode.

2. Configure the CCP1 pin (or CCP2 pin) as an output.

3. Load the CCPR1H:CCPR1L (or CCPR2H:CCPR2L) register pair with appropriate values.

4. Load Timer1 (or Timer3) in the timer mode or synchronized counter mode by loading appropriate data into T1CON (or T3CON) register.

5. Initialize Timer1H:Timer1L (or Timer3H:Timer3L) to 0.

6. Clear CCP1IF in PIR1 (or CCP2IF in PIR2).

7. Start Timer1 (or Timer3).

8. Wait in a loop until the CCP1IF (or CCP2IF) is HIGH.

9. As soon as match occurs (CCP1IF or CCP2IF is HIGH), stop Timer1 (or Timer3).

Example 9.9

Assume PIC18F4321 with an internal crystal clock of 20 MHz. Write a C language program that will toggle the CCP1 pin after a time delay of 10 msec. Use Timer3, and compare mode of CCP1.

Solution

With 20 MHz internal crystal, Fosc = 20 MHz. Since Timer3 uses Fosc/4.
Timer clock frequency = Fosc/4 = 5 MHz. Hence, clock period of Timer3 = 0.2 μ sec
Counter value = (10 msec)/(0.2 μ sec) = 500_{10} = 0x01F4. Hence, CCPR1H :CCPR1L should be loaded with 0x01F4 for the PIC18F4321 compare mode.
The C language program is provided below:

```
#include<p18f4321.h>
void main(void)
{
CCP1CON=0x02; // Select compare mode, toggle CCP1 pin on match
TRISCbits.TRISC2=0; // Configure CCP1 pin as output
T3CON=0x40; // Select TIMER3 as clock for compare, 1:1 prescale
CCPR1H=0x01; // Load CCPR1H with 0x01
CCPR1L=0xF4; // Load CCPR1L with 0xF4
TMR3H=0; // Initialize TMR3H to 0
TMR3L=0; // Initialize TMR3L to 0
PIR1bits.CCP1IF=0; // Clear CCP1IF
T3CONbits.TMR3ON=1; // Start Timer3
while(PIR1bits.CCP1IF==0); // Wait in a loop until CCP1IF is 1
T3CONbits.TMR3ON=0; // Stop Timer3
while(1); // Halt
}
```

9.5.5 PIC18F4321 PWM (Pulse Width Modulation) Mode

In PWM mode, the CCPx pin can be configured as an output to generate a periodic waveform with a specified frequency, and a 10-bit duty cycle. Timer2 is used for the PWM mode. The PWM period is specified by writing to the 8-bit PR2 register in the CCP module. Note that PWM waveform can be generated using timers. However, it is easier to produce PWM waveform using the CCPx module.

The PWM period is specified by writing to the PR2 register. From Microchip PIC18F4321 data sheet, the PWM period can be calculated using the following formula:
PWM Period = [(PR2) + 1] × 4 x Tosc × (TMR2 Prescale Value)

where Tosc = (1/Fosc), Fosc is the crystal frequency, and TMR2 Prescale Value can be initialized as 1, 4, or 16 using the T2CON register.

Hence, PR2 = [(Fosc)/(4 × Fpwm × TMR2 Prescale Value)] -1

Note that PWM frequency (Fpwm) is defined as 1/[PWM period].

The PWM duty cycle is specified by writing to the CCPRxL register and to the CCPxCON<5:4> bits. Up to 10-bit resolution is available. The CCPRxL contains the eight most significant bits, and the CCPxCON (bits 5 and 4) contains the two least significant bits. This 10-bit value is represented by CCPRxL:CCPxCON (bits 5 and 4). The following equation is used to calculate the PWM duty cycle in time:

PWM Duty Cycle = (CCPRxL:CCPxCON<5:4>) × Tosc × (TMR2 Prescale Value).

As mentioned before, the duty cycle is defined as the percentage of the time the pulse is high in a clock period. Note that the upper eight bits in the CCPRxL are the decimal part of the duty cycle while bits 5 and 4 of the CCPxCON register contain the fractional part of the duty cycle. For example, consider 25% duty cycle. Since duty cycle is a fraction of the PR2 register value, decimal value for the duty cycle with a PR2 value of 30 is 7.5 (0.25 x 30). Hence, the 8-bit binary number 00000111_2 must be loaded into CCPRxL, and $10_2 (0.5_{10})$ must be loaded for DCxB1 and DCxB0 bits in the CCPxCON register (Figure 9.26).

The CCP1 PWM output waveform with the specified duty cycle in CCPR1L:CCPICON register is generated as follows: CCPR1L is copied to CCPR1H. Two bits of CCP1CON <5:4> are latched internally to provide 10-bit duty cycle. Also, 8-bit TMR2 value is concatenated with two-bit internal latch to create 10-bit duty cycle. After TMR2 is started from 0, the CCP1 pin goes to HIGH to indicate the start of a cycle. The CCPR1H and 2-bit latch values are compared with TMR2 and 2-bit latch values for a match. As soon as the match occurs, the CCP1 pin goes to LOW. At this point, the waveform will be HIGH for the duration specified by the duty cycle. TMR2 keeps incrementing. As soon as the contents of TMR2 and PR2 match, CCP1 pin is driven to HIGH, and TMR2 is cleared to 0. This competes a cycle, and another cycle is then started. The same process continues for subsequent cycles.

The following procedure should be followed when configuring the CCP module for PWM operation:

1. The PR2 register should be initialized with the PWM period.
2. Load the PWM duty cycle by writing to the CCPRxL register for higher eight bits, and bits 5, 4 of CCPxCON (Figure 9.26) for lower two bits.
3. Make the CCPx pin an output by clearing the appropriate TRIS bit.
4. Set the TMR2 prescale value, then enable Timer2 by writing to T2CON.
5. Initialize TMR2 register to 0.
6. Set up the CCPx module for PWM operation, and turn Timer2 ON.

Example 9.10

Write a C language program to generate a 4 KHz PWM with a 50% duty cycle on the CCP1 pin of the PIC18F4321. Assume 4 MHz crystal. Note that PWM uses Timer2.

Solution

PR2 = [(Fosc)/(4 x Fpwm x TMR2 Prescale Value)] -1

PR2 = [(4 MHz)/(4 x 4 KHz x 1)] -1 assuming Prescale value of 1

PR2 = 249. With 50% duty cycle, count = 0.5 x 249 = 124.5. Hence, the CCPR1L register will be loaded with 124, and bits DC1B1:DC0B0 (CCP1CON register) with 10 (binary).

The C language program is provided below:

```
#include<p18f4321.h>
void main(void)
{
  PR2=249; // Initialize PR2 register
  CCPR1L=124; // Initialize CCPR1L
  CCP1CON=0x20; // PWM mode disabled, CCP1 OFF, DC1B1:DC0B0=10
  TRISCbits.TRISC2=0; // Configure CCP1 pin as output
  CCP1CON=0x2C; // PWM mode enabled
  TMR2=0; // Clear Timer2 to 0
  while(1)
  {
   PIR1bits.TMR2IF=0; // Clear TMR2IF to 0
   T2CONbits.TMR2ON=1; // Turn Timer2 ON
   while(PIR1bits.TMR2IF==0); // Wait until TMR2IF is HIGH
  }
}
```

9.6 DC Motor Control

Typical applications of the PWM mode include DC motor control. The speed of a DC motor is directly proportional to the driving voltage. The speed of a motor increases as the voltage is increased. In earlier days, voltage regulator circuits were used to control the speed of a DC motor. But voltage regulators dissipate lots of power. Hence, the PIC18F in the PWM mode is used to control the speed of a DC motor. In this scheme, power dissipation is significantly reduced by turning the driving voltage to the motor ON and OFF. The speed of the motor is a direct function of the ON time divided by the OFF time.

Sometimes, it is desirable to change direction of rotation of the DC motor. This can be accomplished by reversing the direction of the motor via software by interfacing a device called an H-Bridge to an I/O port of the PIC18F. Note that the speed of the motor, on the other hand, can be controlled using the PWM mode, and by connecting the DC motor to a PWM pin such as the PIC18F CCP1. The basic concepts associated with the DC motor control using the PIC18F4321's PWM mode will be illustrated in Example 9.11.

Microcontrollers such as the PIC18F4321 are not capable of outputting the required large current and voltage to control a typical DC motor. Hence, a driver such as the CNY17F Optocoupler is needed to amplify the current and voltage provided by the PIC18F's output, and provide appropriate levels for the DC motor. One of the many useful applications for employing a PWM signal is its ability to control a mechanical device, such as a motor.

Note that the motor will run faster or slower based on the duty cycle of the PWM signal. The motor runs faster as the duty cycle of the PWM signal at the CCPx pin is increased. To illustrate this concept, two different duty cycles will be used in the following example (Example 9.11).

In Pulse-Width Modulation (PWM) mode, the CCPx pin can be configured as an output to generate a periodic waveform with a specified frequency, and a 10-bit duty cycle. TMR2 is used for the PWM mode. The PWM period is specified by writing to an 8-bit internal register called PR2 in the CCP module. Note that PWM waveform can be generated using timers. However, it is easier to produce PWM waveform using the CCP module.

The PWM period is specified by writing to the PR2 register. From PIC18F4321 data sheet, the PWM period can be calculated using the following formula:

PWM Period = [(PR2) + 1] x 4 x Tosc x (TMR2 Prescale Value)

where Tosc = (1/Fosc), Fosc is the crystal frequency, TMR2 Prescale value can be initialized as 1, 4, or 16 using the T2CON register.

Hence, PR2 = [(Fosc)/(4 x Fpwm x TMR2 Prescale Value)] - 1

Note that PWM frequency (Fpwm) is defined as 1/[PWM period]. The unscaled output of TMR2 is available primarily to the CCP modules, where it is used as a time base for operations in PWM mode. The Timer2 postscalers are not used in the determination of the PWM frequency.

The PWM duty cycle is specified by writing to the CCPRxL register and to the CCPxCON<5:4> bits. Up to 10-bit resolution is available. The CCPRxL contains the eight most significant bits, and the CCPxCON (bits 5 and 4) contains the two least significant bits. This 10-bit value is represented by CCPRxL:CCPxCON (bits 5 and 4). The following equation is used to calculate the PWM duty cycle in time:

PWM Duty Cycle = (CCPRXL:CCPXCON<5:4>) x Tosc x (TMR2 Prescale Value).

As mentioned before, the duty cycle is defined as the percentage of the time the pulse is high in a clock period. Note that the upper eight bits in the CCPRxL is the decimal part of the duty cycle while bits 5 and 4 of the CCPxCON register contain the fractional part of the duty cycle. For example, consider 25% duty cycle. Since duty cycle is a fraction of the PR2 register value, decimal value for the duty cycle with a PR2 value of 30 is 7.5 (0.25 x 30). Hence, the 8-bit binary number 00000111_2 must be loaded into CCPRxL, and 10_2 (0.5_{10}) must be loaded for DCxB1 and DCxB0 bits in the CCPxCON register (Figure 9.26).

As mentioned before, there are two CCP modules in the PIC18F4321 namely, CCP1 the PWM waveform on the CCP1 pin with the specified duty cycle in CCPR1L:CCP1CON register as follows:

CCPR1L is copied to CCPR1H. Two bits of CCP1CON<5:4> are latched internally to form the 10-bit duty cycle. Also, 8-bit TMR2 is concatenated with two-bit internal latch to create 10-bit duty cycle. After TMR2 is started from 0, the CCP1 goes to HIGH to indicate the start of a cycle. The CCPR1H and two-bit latch values are compared with TMR2 and two-bit latch values for a match. As soon as the match occurs, the CCP1 pin goes to LOW. At this point, the waveform will be HIGH for the duration specified by the duty cycle. TMR2 keeps incrementing. As soon as the contents of TMR2 and PR2 match, the CCP1 pin is driven to HIGH, and TMR2 is cleared to 0. This completes a cycle, and another cycle is then started. The same process continues for subsequent cycles.

The following procedure should be followed when configuring the CCP module for PWM operation:

1. The PR2 register should be initialized with the PWM period.
2. Load the PWM duty cycle by writing to the CCPRxL register for higher eight bits, and bits 5, 4 of CCPxCON (Figure 9.26) for lower two bits.
3. Make the CCPx pin an output by clearing the appropriate TRIS bit.
4. Set the TMR2 prescale value, then enable TMR2 by writing to T2CON.
5. Initialize TMR2 register to 0.
6. Set up the CCPx module for PWM operation and turn TMR2 ON.

Example 9.11 uses a C-program to generate a PWM waveform on the CCP1 pin with a specified duty cycle.

As mentioned before, typical applications of the PWM mode include DC motor control. The speed of a DC motor is directly proportional to the driving voltage. The speed of a motor increases as the voltage is increased. In earlier days, voltage regulator circuits were used to control the speed of a DC motor. But voltage regulators dissipate lots of power. Hence, the PIC18F in the PWM mode is used to control the speed of a DC motor. In this scheme, power dissipation is significantly reduced by turning the driving voltage to the motor ON and OFF. The speed of the motor is a direct function of the ON time divided by the OFF time.

Microcontrollers such as the PIC18F4321 are not capable of outputting the required large current and voltage to control a typical DC motor. Hence, a driver such as the CNY17F Optocoupler is needed to amplify the current and voltage provided by the PIC18F's output, and provide appropriate levels for the DC motor. One of the many useful applications for using a PWM signal is its ability to control a mechanical device, such as a motor.

Note that the motor will run faster or slower based on the duty cycle of the PWM signal. The motor runs faster as the duty cycle of the PWM signal at the CCPx pin is increased. To illustrate this concept, two different duty cycles are used in Example 9.11.

PWM waveform can be generated by polling the TMR2IF in the PIR1 register (Figure 9.7).

The following C-segment polls the TMR2IF flag bit in the PIR1 register by clearing TMR2IF to 0, turning the TMR2 On, and then waiting in a loop for the TMR2IF bit to become HIGH:

```
PIR1bits.TMR2IF=0; // Clear Timer2 interrupt
T2CONbits.TMR2ON=1; // Turn Timer2 ON
while(PIR1bits.TMR2IF==0); // Wait until TMR2IF = 1
```
Example 9.11 illustrates this concept.

Example 9.11

It is desired to change the speed of a DC motor by dynamically changing its pulse width using a potentiometer connected at bit 0 of PORTB (Figure 9.27). Note that the PWM duty cycle is controlled by the potentiometer. Write a C language program that will input the potentiometer voltage via the PIC18F4321's on-chip A/D converter as eight bits, generate an 800-Hz PWM waveform on the CCP1 pin, and will then change the speed of the motor as the potentiometer voltage is varied. Use 4 MHz crystal and a TMR2 prescaler value of 16. Ignore the fractional part of the duty cycle.
Write the C-program using
(a) Polled ADC
(b) Interrupt-driven ADC

Solution

PR2 = [(Fosc)/(4 x Fpwm x TMR2 Prescale Value)] - 1
= [(4 MHz)/(4 x 800x 16)] - 1
= 77.125
Hence, PR2 will be 77 approximately.

After converting the potentiometer voltage into 8-bit binary, the ADRESH will contain the converted data. Note that the 8-bit ADRESH register can accommodate a maximum value of 255. The contents of ADRESH can be moved to CCPR1L to represent the decimal portion of the duty cycle. In order for the duty cycle to be in the range of 0 to 77, the contents of ADRESH must be divided by which is 3 (255/77 = 3 approximately). This will ensure that the decimal portion of the duty cycle is between 0 and 77. The higher the voltage across the potentiometer, the higher will be the value of ADRESH/3. This will generate a PWM waveform with a higher duty cycle, and thus the motor will run faster.

(a)

```c
// Polled ADC
#include <p18f4321.h>
#pram config WDT=OFF
#pragma config LVP=OFF
#pragma config BOR=OFF
#pragma config OSC=INTIO2 // using internal oscillator
void main()
{
  OSCCON=0xEC; // 4 MHz
  TRISC=0; // CCP1 output
  T2CON=0; // Configure Timer2 with prescale 16
  PIR1bits.TMR2IF=0; // Clear Timer2 interrupt
  ADCON0=0x31; // AN12
  ADCON1=0; // Enable analog pins
  ADCON2=0x29; // Fosc/8
  CCP1CON=0x0C; // PWM enable
  PR2=77; // Load PR2 with 77
  TMR2=0x00; // Clear Timer2 to 0
  while(1)
  {
```

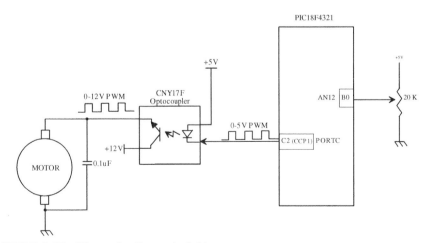

FIGURE 9.27   Figure for Example 9.11.

```
        CCPR1L=ADRESH/3; // For 33% duty cycle
        ADCON0bits.GO=1; // Start the ADC
        PIR1bits.TMR2IF=0; // Clear Timer2 interrupt
        T2CONbits.TMR2ON=1; // Turn Timer2 ON
        while(PIR1bits.TMR2IF==0); // Wait until TMR2IF = 1
    }
}


(b)


//Interrupt-driven ADC
#include <p18f4321.h>
#pragma config WDT=OFF
#pragma config LVP=OFF
#pragma config BOR =OFF
#pragma config OSC = INTIO2 // using internal oscillator
void ADINT_ISR(void);    // Prototype

#pragma interrupt check_int
void check_int(void)
{ ADINT_ISR();
}
#pragma code ADC_INT=0x08 // At interrupt, code jumps here
void ADC_Int (void)
{ _asm
GOTO check_int
_endasm
}
#pragma code // End of code
void main()
{     OSCCON=0xEC; // 4 Mhz
      PIE1bits.ADIE = 1;  // enable ADC interrupt
      PIR1bits.ADIF = 0;  // clear ADC interrupt flag
      INTCONbits.PEIE = 1; // enable peripheral interrupt
      INTCONbits.GIE = 1; // enable global interrupt
      TRISC=0;         // CCP1 output
      T2CON=0x02;      // Configure Timer2 with prescale 16
      PIR1bits.TMR2IF=0;  // Clear Timer2 interrupt flag
      ADCON0=0x31;     // AN12
      ADCON1=0x00;     // Enable analog pins
      ADCON2=0x29;     // Fosc/8
      CCP1CON=0x0C;    // PWM enable
      PR2= 77;
      TMR2 = 0;
      ADCON0bits.GO = 1;
   while(1);        // Wait for interrupt
}
void ADINT_ISR (void)
```

```
{
     PIR1bits.ADIF = 0;
     CCPR1L=ADRESH/3;
     PIR1bits.TMR2IF=0; //Clear Timer 2
   //interrupt
     T2CONbits.TMR2ON=1;
 while(PIR1bits.TMR2IF==0);
 ADCON0bits.GO = 1;
}
```

Questions and Problems

9.1   Find the contents of T0CON register to program Timer0 in 8-bit mode with 1:16 prescaler using the external clock, and incrementing on negative edge.

9.2   Write a C program to initialize Timer0 as an 8-bit timer to provide a time delay with a count of 100. Assume 4 MHz internal clock with a prescaler value of 1:16.

9.3   Write a C language program to generate a square wave with a period of 4 ms on bit 0 of PORTC using a 4 MHz crystal. Use Timer0.

9.4   Write a C language program to generate a square wave with a period of 4 ms on bit 7 of PORTD using a 4 MHz crystal. Use Timer1.

9.5   Write a C language program to turn an LED ON connected at bit 0 of PORTC with a PR2 value of 200. Assume a 4 MHz crystal and TMR2 prescaler and postscaler values of 1:1.

9.6   Write a C language program to generate a square wave on pin 3 of PORTC with a 4 ms period using Timer3 in 16-bit mode with no prescaler value. Use a 4 MHz crystal.

9.7   Assume PIC18F4321 with Fosc = 1 MHz. Consider the following C program:

```c
#include <p18f4321.h>
void delay(void);
void main(void){
  TRISBbits.TRISB=0;
  while (1){
      PORTBbits.RB4^=1;
      delay( );
  }
}
void delay( ){
  T0CON=0x01;
  TMR0H=0xC2;
  TMR0L=0xF7;
  T0CONbits.TMR0ON=1;
  while(INTCONbits.TM0IF == 0);
  T0CONBits.TMR0ON = 0;
  INTCONbits.TMR0IF = 0;
}
```

(a) What type of signal is generated on the RB4 pin? What is its frequency?

(b) What is the frequency of the signal that appears on the RB4 pin?

(c) Repeat part (b) if T0CON is initialized with 0x42, and TMR0H = 0xC2 is deleted from the program.

9.8   Assume the PIC18F4321-DMC 16249 interface of Figure 9.15. Write a C program to display the phrase "PIC18F" on the LCD as soon as the four input switches connected to Port C are all HIGH.

9.9   Design a PIC18F4321 based digital clock. The clock will display time in hours, minutes, and seconds. Write a C program to accomplish this.

9.10   Assume that two PIC18F4321s are interfaced in the SPI mode. A switch is connected to bit 0 of PORTD of the master PIC18F4321 and an LED is connected to bit 5 of PORTB of the slave PIC18F4321. Write C language programs to input the switch via the master, and output it to the LED of the slave PIC18F4321. If the switch is open , the LED will be turned ON while the LED will be turned OFF if the switch is closed.

9.11   Assume PIC18F4321. Write a C language program that will measure the period of a periodic pulse train on the CCP1 pin using the capture mode. The 16-bit result will be performed in terms of the number of internal (Fosc/4) clock cycles, and will be available in the TMR1H:TMR1L register pair. Use 1:1 prescale value for Timer1. Store the 16-bit result in CCPR1H:CCPR1L.

9.12   Assume PIC18F4321. Write a C language program that will generate a square wave on the CCP1 pin using the Compare mode. The square wave will have a period of 20 ms with a 50% duty cycle. Use Timer1 internal clock (Fosc/4 from XTAL) with 1:2 prescale value.

9.13   Write a C language program to generate a 16 KHz PWM with a 75% duty cycle on the RC2/CCP1/P1A pin of the PIC18F4321. Assume 10 MHz crystal. Also assume no prescale and no postscale for Timer2.

9.14   Figure P9.14 shows a simplified diagram interfacing the PIC18F4321 to a DC motor via the CNY17F Optocoupler. The purpose of this problem is to control the speed of a DC motor by inputting two switches connected at bit 0 and bit 1 of PORTD. The motor will run faster or slower based on the switch values (00 or 11), but will not provide any measure of the exact RPM of the motor. When both switches are closed (00), a PWM signal at the CCP1 pin of the PIC18F4321 with 50% duty cycle will be generated. When both switches are open (11), a PWM signal at the CCP1 pin of the PIC18F4321 with 75% duty cycle will be generated. Otherwise, the motor will stop, and the program will wait in a loop. If switches are closed (00), the motor will run using the 4 KHz PWM pulse of Example 9.10 with 50% duty cycle. If both switches are open (11), the motor will run at a faster speed with a duty cycle of 75%. The program will first perform initializations, and wait in a loop until the switches are 00 or 11. Write a C language program to accomplish this.

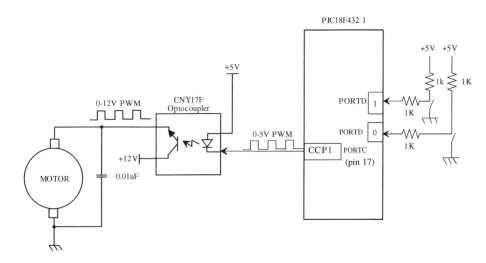

FIGURE P9.14.

APPENDIX

ANSWERS TO SELECTED PROBLEMS

Chapter 2

2.1(b) $1101.101_2 = 13.625_{10}$

2.2(b) $343_{10} = 101010111_2$

2.3(a) $1843_{10} = 3463_8$

2.4(b) $3072_{10} = C00_{16}$

2.6(c) $-48_{10} = 1101\ 0000_2$

2.11(c) $61440_{10} = 1001\ 0100\ 0111\ 0111\ 0011_2$

2.16(b) $0011\ 1110_2$

2.19(b)   0; no overflow

2.19(d)   overflow

2.22(a) $0001\ 0000\ 0010_2 = 102$ in BCD

Chapter 3

3.1 $36_{16} \oplus 2A_{16} = 1C_{16}$

3.3   1's Complement of $A7_{16}$

3.6(d)
$$\overline{(A + \overline{A}B)} = \overline{A}(\overline{\overline{A}B})$$
$$= \overline{A}\ (A + \overline{B})$$
$$= \overline{A}\ \overline{B}$$

3.6(f) $\overline{B}\ \overline{C} + ABC + \overline{A}\ \overline{C} = \overline{C}(\overline{B} + \overline{A}) + ABC$
$$= \overline{C}(\overline{AB}) + (AB)C$$
$$= \overline{C \oplus (AB)}$$

3.7(c)   BC

3.9(a) $\overline{F} = \prod M(0, 1, 5, 7, 10, 14, 15)$

3.11(c) $F = \overline{Z}$

3.12(b) $F = BC + \overline{A}B$

3.13(d) $F = W \oplus Y$

3.13(e) $F = Z$

3.16(a) $f = A + \overline{B}C + B\overline{C}$

3.16(c) $f = \overline{B}$

Fundamentals of Digital Logic and Microcontrollers, Sixth Edition. M. Rafiquzzaman.
© 2014 John Wiley & Sons, Inc. Published 2014 by John Wiley & Sons, Inc.

3.17 $F = \overline{A}\,\overline{C} + \overline{C}\,\overline{D}$

3.18(b) $F = (\overline{A + C}) + (B + \overline{C})$

Chapter 4

4.1   F = 0

4.3(c) $F = A\overline{C} + BC$

4.6 $f = \overline{A \oplus C}$

4.9 $f_3 = \overline{A}\,\overline{B}\,\overline{C},\ f_1 = C$
         $f_2 = B \oplus C\ f_0 = \overline{D}$

4.13   Add the 4-bit unsigned number to itself using full-adders.

4.16 $Z = 1$
         $Y = 0$
         $X = m5$
         $W = m9$

4.22   For 4-Bit signed number, A
         $A + 1111_2 = A - 1$, decrement by 1.
         $A + 0001_2 = A + 1$, increment by 1.
         Manipulate C_{in} to accomplish the above.

4.31 (b) Behavioral Decoder:

```
module Decoder_behavioral( dataIn, enable, decOut );

input [2:0] dataIn;
input enable;
output reg [7:0]decOut;
wire enablenot = ~enable;

always @ (dataIn[0] or dataIn[1] or dataIn[2])
begin
if (enablenot == 1)
begin
case(dataIn)
    3'b000: decOut = 8'b00000001;
    3'b001: decOut = 8'b00000010;
    3'b010: decOut = 8'b00000100;
    3'b011: decOut = 8'b00001000;
    3'b100: decOut = 8'b00010000;
    3'b101: decOut = 8'b00100000;
    3'b110: decOut = 8'b01000000;
    3'b111: decOut = 8'b10000000;
default: decOut = 8'b00000000;
endcase
end
else
decOut = 0;
end
endmodule
```

4.33 Dataflow Multiplexer:

```
module Dataflow Multiplexer( dataIn ,select ,dataOut );

input [3:0] dataIn;
input [1:0] select;
output dataOut;

assign dataOut =  ( select == 2'b00 ) ? dataIn[0]  :
( select == 2'b01 ) ? dataIn[1]  :
( select == 2'b10 ) ? dataIn[2]  :
dataIn[3];
endmodule
```

4.36 Priority Encoder:

```
module PriorityEncoder( dataIn, dataOut );
input [3:0] dataIn;
output [1:0] dataOut;

wire d1 = dataIn[1];
wire d2 = dataIn[2];
wire d3 = dataIn[3];
wire d2not;
not ( d2not, dataIn[2] );

assign dataOut[0] = ( d1 | d3 ) & ( d2not | d3 );
assign dataOut[1] = d2 | d3;

endmodule
```

Chapter 5

5.5   A = 1, B = 0
5.7   A = 1, B = 1
5.9

Figure for solution 5.9

5.13(a)   Tie JK inputs to HIGH ; Clock is the T input.
5.15 $B_+ = A$, *output* $y = \overline{B}$
5.17(b) $Jx = z,\ kx = y$
        $Jy = 1,\ ky = x + z$
        $Jz = \overline{xy},\ kz = x$

5.19 $D_A = (A \oplus x) + \bar{B}x$ Where x is the input
          $D_B = \bar{x}\overline{(A \oplus B)} + A\bar{B}x$

5.20(c) $Tx = \bar{y}$
          $Ty = 1$
5.23 $T3 = Q_3Q_0 + Q_2Q_1Q_0$, $T_2 = Q_1\,Q_0$, $T_1 = \bar{Q_3}\,Q_0$, $T_0 = 1$
5.24(a) $J_A = B$, $K_A = BC$, $J_B = C$, $K_B = C$, $J_C = 1$, $K_C = A + B$
          self correcting

5.35 (a) Positive edge-triggered JK

```
module JK_FF(
input clk,
input J,
input K,
output reg Q,
output nQ
  );

wire w[1:0];
wire nK, D;

not(nK, K);

and(w[0], J, nQ);
and(w[1], nK, Q);

or(D, w[0], w[1]);

always@(posedgeclk)
begin
Q<= D;
end

not(nQ, Q);


endmodule
```

5.35 (c) T Flip Flop

```
module T_FF(
Input T,
```

```
Output Q,
Output nQ
   );

JK_FFJK1(T, 1'b1, 1'b1, Q, nQ);


endmodule


//JK Flip Flop

module JK_FF(
input clk,
input J,
input K,
output reg Q,
output Q
   );


always@(posedgeclk)
begin
case({J,K})
        0: Q<= Q;

        1: Q<= 0;

        2: Q<= 1;

        3: Q<= nQ;
endcase
end

assignnQ = ~Q;

endmodule
```

5.35 (i) JK Counter of Problem 5.24 (a)

```
modulei_a(
input clk,
input reset,
output [2:0] Q
   );

wireJa, Jb, Jc, Ka, Kb, Kc;
```

```
assignJa = Q[1];
assignJb = Q[0];
assignJc = 1'b1;
assignKb = Q[0];

and(Ka, Q[1], Q[0]);
or(Kc, Q[2], Q[1]);

jk_FFjkA(clk, reset, Ja, Ka, Q[2]);
jk_FFjkB(clk, reset, Jb, Kb, Q[1]);
jk_FFjkC(clk, reset, Jc, Kc, Q[0]);

endmodule


//JK Flip Flop

module jk_FF(
input clk,
input reset,
input J,
input K,
output reg Q
  );

always@(posedgeclk)
begin
if(reset)
Q<= 0;
else
begin
case({J, K})
        0: Q<= Q;
        1: Q<= 0;
        2: Q<= 1;
        3: Q<= ~Q;
endcase
end
end

endmodule
```

Chapter 6

6.1   Use four MUX's. Manipulate inputs of the MUX's to obtain the desired outputs. Use tristate buffers at the ouputs of the MUX's.

$$x_i$$

6.4(a) $s_{15} \underset{3\Delta}{\leftarrow} c_{15} \underset{2\Delta}{\leftarrow} c_{12} \underset{2\Delta}{\leftarrow} g_i p_i \underset{2\Delta}{\leftarrow} G_i P_i \underset{\Delta}{\leftarrow} y_i$$;   worst case add-time: 10Δ

$$c_0$$

6.6   Refer to figure below:

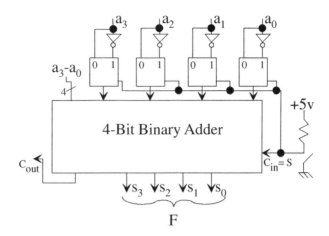

6.13(a) $P_0 = \overline{Z}T_3$, $P_1 = T_5$
         $L = P_0 + P_1$, $d_2 = P_1$, $d_1 = P_0$, $d_0 = P_1$
         $C_0 = C_1 = T_0$, $C_2 = T_1$, $C_3 = C_4 = C_6 = T_2$, $C_5 = T_4$

6.22(a)

	C_4	C_3	C_2	C_1	C_0	
Solution 1	1	0	1	0	0	; A ← A minus A
Solution 2	1	1	1	0	0	; A ← A ex-or A

6.29(b)   6 x 64 decoder

6.30   Maximum Directly Addressable Memory = 16 Megabytes;
        14 unused address pins Available.

6.32   Memory Chip #1 EC00H - EDFFH
        Memory Chip #2 F200H - F3FFH

6.33(a)   ROM Map: 0000H - 07FFH
        RAM Map: 2000H - 27FFH

Chapter 7

7.1   A single chip microcomputer contains CPU, memory, and I/O on a single
        chip. A typical microcontroller contains the CPU, memory, I/O, timers, A/D
        converter—all on a single chip.

7.2   The ALU is 8-bit. PIC18F4321.

7.4   22 bits. $3FFFFF_{16}$.

7.5   2KB.

7.13   No.

Chapter 8

8.1   one microsecond.
8.4(a)   TRISC = 0xFF;
8.4(b)   TRISD = 0;
8.5

```
#include <P18F4321.h>
void main (void)
{
TRISCbits.TRISC1=1;                //Configure PORTC bit 1 as input
TRISDbits.TRISD1=1;                //Configure PORTD bit 1 as input
TRISCbits.TRISC0= 0;               //Configure PORTC bit 0 as output
TRISDbits.TRISD0= 0;               //Configure PORTD bit 0 as output
PORTCbits.RC0= PORTCbits.RC1>>1;   //Align and output data to PORTC bit 0
PORTDbits.RD0= PORTDbits.RD1>>1;   //Align and output data to PORTD bit 0
while(1);                          //Finished
}
```

8.6

```
#include <P18F4321.h>
void fail (void);
void main ()
{
TRISDbits.TRISD0=0;            //Bit0 of PORTD output
TRISDbits.TRISD1=0;            //Bit1 of PORTD output
TRISDbits.TRISD2=1;            //Bit2 of PORTD input
TRISDbits.TRISD3=0;            //Bit3 of PORTD output
TRISDbits.TRISD4=0;            //Bit4 of PORTD output
   PORTDbits.RD0=0;           //Test case 00
   PORTDbits.RD1=0;
   if(PORTDbits.RD2==0)       //If output is 0 fail
     fail();
   PORTDbits.RD0=1;           //Test case 01
   PORTDbits.RD1=0;
     if(PORTDbits.RD2==0)     //If output is 0 fail
     fail();
   PORTDbits.RD0=0;           //Test case 10
   PORTDbits.RD1=1;
     if(PORTDbits.RD2==0)     //If output is 0 fail
     fail();
   PORTDbits.RD0=1;           //Test case 11s
   PORTDbits.RD1=1;
     if(PORTDbits.RD2==1)     //If output is 1 fail
     fail();

   while(1)                   //NAND chip has passed test
```

```
    {
        PORTDbits.RD3=1;            //Turn off LED on bit 3
        PORTDbits.RD4=0;            //Turn on LED on bit 4
    }
}
void fail(void)
{
    while(1)                        //NAND chip has failed test
        {
            PORTDbits.RD3=0; //Turn on LED on bit 3
            PORTDbits.RD4=1; //Turn off LED on bit 4
        }
}
```

Chapter 9

9.3

```
#include <p18f4321.h>
void main(void)
{
    TRISCbits.TRISC0=0;             //Configure bit 0 of PortC as output
    T0CON=0x08;                     //Timer0 is 16-bit no prescaler
    TMR0H=0xF8;                     //Value placed TMR0H
    TMR0L=0x30;                     //Value placed in TMR0L
    INTCONbits.TMR0IF=0;            //Clear TMR0 interrupt flag
    T0CONbits.TMR0ON=1;             //Turn on TMR0
    while(1)
    {
    while(INTCONbits.TMR0IF==0);
    T0CONbits.TMR0ON=0;             //Turn off TMR0
    PORTCbits.RC0=~PORTCbits.RC0; //Change output of square wave
    TMR0H=0xF8;                     //Value placed in TMR0H
    TMR0L=0x30;                     //Value placed in TMR0L
    INTCONbits.TMR0IF=0;            //Clear TMR0 interrupt flag
    T0CONbits.TMR0ON=1;             //Turn on TMR0
    }
}
```

9.7 (a) A square wave with 50% duty cycle or a symmetrical square wave will be
         generated on the pin RB4.

9.11

```
#include <P18F4321.h>
void main()
{
```

```
  TRISC=1;                         // PORTC is input
  CCP1CON=0x05;                    // Capture mode, event on rising edge
  T1CON=0xC8;                      // Internal clock, no prescale
  TMR1L=0;                         // Clear TMR1L register
  TMR1H=0;                         // Clear TMR1H register
  PIR1bits.CCP1IF=0;               // Clear CCP1 interrupt flag
  while(PIR1bits.CCP1IF==0);       // Wait for first rising edge
  T1CONbits.TMR1ON=1;              // Turn on TMR1
  PIR1bits.CCP1IF=0;               // Clear CCP1 interrupt flag
  while(PIR1bits.CCP1IF==0);       // Wait for second rising edge
  T1CONbits.TMR1ON=0;              // Turn off TMR1
  while(1);                        // Period is found in registers CCPR1L and CCPR1H
}
9.13

#include <p18f4321.h>
void main (void)
{
  TRISC=0x00;                      //PORTC is output
  T2CON=0x00;                      //Configure TMR2 with no precscale and no postscale
  TMR2=0x00;
  PIR1bits.TMR2IF=0;               //Clear TMR2 interrupt flag
  CCP1CON=0x0C;                    //Use PWM generator
  PR2=156;                         //Set period of PWM

  while(1)
  {
   CCPR1L=117;                     //Value for 75% duty cycle
   PIR1bits.TMR2IF=0;              //Clear TMR2 interrupt flag
   T2CONbits.TMR2ON=1;             //Turn on TMR2

   while(PIR1bits.TMR2IF==0);      //Wait until cycle is over
  }
}
```

APPENDIX

B

GLOSSARY

Accumulator: Register used for storing one of the operands (data) before arithmetic operations such as addition; available with the PIC18F.

Address: A unique identification number (or locator) for source or destination of data. An address specifies the register or memory location of an operand involved in the instruction.

Address Register: A register used to store the address (memory location) of data.

Address Space: The number of storage location in a microcontrollers's memory that can be directly addressed by the CPU. The addressing range is determined by the number of address bits provided with the CPU.

American Standard Code for Information Interchange (ASCII): An 8-bit code commonly used with microcontrollers for representing alphanumeric codes.

Analog-to-Digital (A/D) Converter: Transforms an analog voltage into its digital equivalent.

AND gate: The output is 1, if all inputs are 1; otherwise the output is 0.

Arithmetic and Logic Unit (ALU): A digital circuit which performs arithmetic and logic operations on two n-bit numbers.

ASIC: Application Specific IC. Chips designed for a specific, limited application. Normally reduces the total manufacturing cost of a product by reducing chip count.

Assembler: A program that translates an assembly language program into a machine language program.

Assembly Language: A type of microcontroller programming language that uses a semi-English-language statement.

Asynchronous Operation: The execution of a sequence of steps such that each step is initiated upon completion of the previous step.

Asynchronous Sequential Circuit: Completion of one operation starts the next operation in sequence. Time delay devices (logic gates) are used as memory.

Asynchronous Serial Data Transmission: The transmitting device does not need to be synchronized with the receiving device.

Barrel Shifter: A specially configured shift register that is normally included in 32-bit microprocessors for cycle rotation. That is, the barrel shifter shifts data in one direction.

Baud Rate: Rate of data transmission in bits per second.

Behavioral Modeling: Using hardware description languages such as Verilog and VHDL, a system can be described in terms of what it does and how it behaves rather than in terms of its components and their interconnections.

Big Endian: This convention is used to store a 16-bit number such as 16-bit data in two bytes of memory locations as follows: the low memory address stores the high byte while the high memory address stores the low byte. The Motorola/Freescale HC11 8-bit microcontroller follows the big endian format.

Binary-Coded Decimal (BCD): The representation of 10 decimal digits, 0 through 9, by their corresponding 4-bit binary number.

Bit: An abbreviation for a binary digit. A unit of information equal to one of two possible states (one or zero, on or off, true or false).

Buffer: A temporary memory storage device deigned to compensate for the different data rates between a transmitting device and a receiving device (for example, between a CPU and a peripheral). Current amplifiers are also referred to as buffers.

Bus: A collection of wires that interconnects computer modules. The typical micro-computer interface includes separate buses for address, data, control, and power functions.

Bus Arbitration: Bus operation protocols (rules) that guarantee conflict-free access to a bus. Arbitration is the process of selecting one respondent from a collection of several candidates that concurrently request service.

Bus Cycle: The period of time in which a microprocessor carries out read or write operations.

Central Processing Unit (CPU): The brains of a computer containing the ALU, register section, and control unit. CPU in a single chip is called *microprocessor*.

Chip: An Integrated Circuit (IC) package containing digital circuits.

CISC: Complex Instruction Set Computer. The Control unit is designed using microprogramming. Contains a large instruction set. Difficult to pipeline compared to RISC.

Clock: Timing signals providing synchronization among the various components in a microcomputer system. Analogous to heart beats of a human.

CMOS: Complementary MOS. Dissipates low power, offers high density and speed compared to TTL.

Combinational Circuit: Output is provided upon application of inputs; contains no memory.

Compiler: A program which translates the source code written in a high-level programming language into machine language that is understandable to the processor.

Control Unit: Part of the CPU; its purpose is to translate or decode instructions read (fetched) from the main memory into the Instruction Register.

CPLD: Complex PLD. This chip contains several basic PLDs along with all interconnections.

Data: Basic elements of information represented in binary form (that is, digits consisting of bits) that can be processed or produced by a microcomputer. Data represents any group of operands made up of numbers, letters, or symbols denoting any condition, value, or state. Typical microcomputer operand sizes include: a word, which typically contains 2 bytes or 16-bits; a long word, which contains 4 bytes or 32 bits; a quad word, which contains 8 bytes or 64 bits.

Dataflow Modeling: Behavioral modeling with concurrent statements.

Data Register: A register used to temporarily hold operational data being sent to and from a peripheral device.

Debugger: A program that executes and debugs the object program generated by the assembler or compiler. The debugger provides a single stepping, breakpoints, and program tracing.

Decoder: A chip, when enabled, selects one of 2^n output lines based on n inputs.

Demultiplexer: Performs reverse operation of a multiplexer.

Digital to Analog (D/A) Converter: Converts binary number to analog signal.

Diode: Two terminal electronic switch.

Directly Addressable Memory: The memory address space in which the CPU can directly execute programs. The maximum directly addressable memory is determined by the number of the CPU's address pins.

DRAM: See Dynamic RAM.

Dynamic RAM: Stores data as charges in capacitors and therefore, must be refreshed since capacitors can hold charges for a few milliseconds. Hence, requires refresh circuitry.

EAROM (Electrically Alterable Read-Only Memory): Same as EEPROM or E^{2}PROM. Can be programmed one line at a time without removing the memory from its sockets. This memory is also called *read-mostly memory* since it has much slower write times than read times.

EEPROM or E^2 PROM: Same as EAROM (see EAROM).

Encoder: Performs reverse operation of a decoder. Contains a maximum of 2^n inputs and n outputs.

EPROM (Erasable Programmable Read-Only Memory): Can be programmed and erased all programs in an EPROM chip using ultraviolet light. The chip must be removed from the microcomputer system for programming.

Equivalence: See Exclusive-NOR.

Exception Processing: Includes the microprocessor's processing states associated with interrupts, trap instructions, tracing, and other exceptional conditions, whether they are initiated internally or externally.

Exclusive-OR: The output is 0, if inputs are same; otherwise; the output is 1.

Exclusive-NOR: The output is 1, if inputs are same; otherwise, the output is 0.

Extended Binary-Coded Decimal Interchange Code (EBCDIC): An 8-bit code commonly used with microprocessors for representing alphanumeric codes. Normally used by IBM.

Firmware: Microprogram is sometimes referred to as firmware to distinguish it from hardwired control (purely hardware method).

Flag(s): An indicator, often a single bit, to indicate some conditions such as trace, carry, zero, and overflow.

Flash Memory: Utilizes a combination of EPROM and EEPROM technologies. Used in cellular phones and digital cameras.

Flip-Flop: One-bit memory.

FPGA: Field Programmable Gate Arrays. This chip contains memory, mux's and flip-flops along with all interconnections.

Full-Adder: Adds three bits generating a sum bit and a carry bit.

Gate: Digital circuits which perform logic operations.

Half-Adder: Adds two bits generating a sum bit and a carry bit.

Handshaking: Data transfer via exchange of control signals between the microcontroller and an external device.

Hardware: The physical electronic circuits that make up the microcontroller.

Hardwired Control: Used for designing the control unit using all hardware.

Harvard CPU Architecture: The CPU uses separate instruction and data memory units along with separate buses for instructions and data.

HCMOS: High speed CMOS. Provides high density and consumes low power.

Hexadecimal Number System: Base-16 number system.

High-Level Language: A type of programming language that uses a more understandable human-oriented language such as C.

HMOS: High-density MOS reduces the channel length of the NMOS transistor and provides increased density and speed in VLSI circuits.

Instruction: Causes the microprocessor to carry out an operation on data. A program contains instructions and data.

Instruction Cycle: The sequence of operations that a microcontroller has to carry out while executing an instruction.

Instruction Register (IR): A register storing instructions; typically 8 bits long for an 8-bit CPU.

Instruction Set: Lists all the instructions that the microcontroller can execute.

Internal Interrupt: Activated internally by exceptional conditions such as overflow and division by zero.

Interrupt I/O: An external device can force the microcontroller system to stop executing the current program temporarily so that it can execute another program known as the *interrupt service routine*.

Interrupts: A temporary break in a sequence of a program, initiated externally or internally, causing control to jump to a routine, which performs some action while the program is stopped.

I/O (Input/Output): Describes that portion of a microcontroller that exchanges data between the microcontroller and an external device.

I/O Port: A register that contains control logic and data storage used to connect a microcontroller to external peripherals.

Inverting Buffer: Performs NOT operation. Current amplifier.

Karnaugh Map: Simplifies Boolean expression by a mapping mechanism.

Keybounce: When a mechanical switch opens or closes, it bounces (vibrates) for a small period of time (about 10-20 ms) before settling down.

Large-Scale Integration (LSI): An LSI chip contains 100 to 1000 gates.

LED: Light Emitting Diode. Typically, a current of 10 mA to 20 mA flows at 1.7V to 2.4V drop across it.

Little Endian: This convention is used to store a 16-bit number such as 16-bit data in two bytes of memory locations as follows: the low memory address stores the low byte while the high memory address stores the high byte. The PIC18F microcontroller follows the little endian format.

Machine Code: A binary code (composed of 1's and 0's) that a computer understands.

Machine Language: A type of microcontroller programming language that uses binary or hexadecimal numbers.

Macroinstruction: Commonly known as an *instruction*; initiates execution of a complete microprogram. Example includes assembly language instructions.

Mask: A pattern of bits used to specify (or mask) which bit parts of another bit pattern are to be operated on and which bits are to be ignored or "masked" out. Uses logical AND operation.

Mask ROM: Programmed by a masking operation performed on the chip during the manufacturing process; its contents cannot be changed by user.

Maskable Interrupt: Can be enabled or disabled by executing typically the interrupt instructions.

Memory: Any storage device which can accept, retain, and read back data.

Memory Access Time: Average time taken to read a unit of information from the memory.

Memory Address Register (MAR): Stores the address of the data.

Memory Cycle Time: Average time lapse between two successive read operations.

Memory Map: A representation of the physical locations within a microcontroller's addressable main memory.

Memory-Mapped I/O: I/O ports are mapped as memory locations, with every connected device treated as if it were a memory location with a specific address. Manipulation of

I/O data occurs in "interface registers" (as opposed to memory locations); hence there are no input (read) or output (write) instructions used in memory-mapped I/O.

Microcode: A set of instructions called *microinstructions* usually stored in a ROM in the control unit of a microprocessor to translate instructions of a higher-level programming language such as assembly language programming.

Microcomputer: Consists of a microprocessor, a memory unit, and an input/output unit.

Microcontroller: Typically includes a microcomputer, timer, A/D (Analog to Digital) and D/A (Digital to Analog) converters in the same chip.

Microinstruction: Most microprocessors have an internal memory called *control memory*. This memory is used to store a number of codes called *microinstructions*. These microinstructions are combined to design the instruction set of the microprocessor.

Microprocessor: The Central Processing Unit (CPU) of a microcomputer.

Microprogramming: The microprocessor can use microprogramming to design the instruction set. Each instruction in the Instruction register initiates execution of a microprogram stored typically in ROM inside the control unit to perform the required operation.

Multiplexer: A hardware device which selects one of n input lines and produces it on the output.

NAND: The output is 0, if all inputs are 1; otherwise, the ouput is 1.

Nibble: A 4-bit word.

Non-inverting Buffer: Input is same as output. Current amplifier.

Nonmaskable Interrupt: Occurrence of this type of interrupt cannot be ignored by microcontroller and even though interrupt capability of the microcontroller is disabled. Its effect cannot be disabled by instruction.

NOR: The output is 1, if all inputs are 0's; otherwise, the output is 0.

NOT gate: If the input is 1, the output is 0, and vice versa.

Object Code: The binary (machine) code into which a source program is translated by a compiler, assembler, or interpreter.

Octal Number System: Base 8-number system.

Ones Complement: Obtained by changing 1's to ' 0's, and 0's to 1's of a binary number.

Op Code (Operation Code): Part of an instruction defining the operation to be performed.

Operand: A datum or information item involved in an operation from which the result is obtained as a consequence of defined addressing modes. Various operand types contain information, such as source address, destination address, or immediate data.

Operating System: Consists of a number of program modules to provide resource management. Typical resources include CPU, disks, and printers.

OR Gate: The output is 0, if all inputs are 0; otherwise, the output is 1.

Parallel Operation: Any operation carried out simultaneously with a related operation.

Parallel Transmission: Each bit of binary data is transmitted over a separate wire.

Parity: The number of 1's in a word is odd for odd parity and even for even parity.

Peripheral: An I/O device capable of being operated under the control of a CPU through communication channels. Examples include disk drives, keyboards, printers, and modems.

Personal Computer: Low-cost, affordable microcomputer normally used by an individual for word processing and Internet applications.

Physical Address Space: Address space is defined by the address pins of the CPU.

Pipeline: A technique that allows a microcontroller processing operation to be broken down into several steps (dictated by the number of pipeline levels or stages) so that the individual step outputs can be handled by the microcontroller in parallel. Often used to fetch the processor's next instruction while executing the current instruction, which considerably speeds up the overall operation of the microcontroller. Overlaps instruction fetch with execution.

Pointer: A storage location (usually a register within a CPU) that contains the address of (or points to) a required item of data or subroutine.

Polled Interrupt: A software approach for determining the source of interrupt in a multiple interrupt system.

Port: A register through which the microcontrollers communicate with peripheral devices.

Primary or Main Memory: Storage that is considered as part of the microcontroller. The microcontroller can directly execute all instructions in the main memory. The maximum size of the main memory is defined by the number of address bits of the microcontroller.

Processor Memory: A set of microcontroller registers for holding temporary results when a computation is in progress.

Program: A self-contained sequence of computer software instructions (source code) that, when converted into machine code, directs the computer to perform specific operations for the purpose of accomplishing some processing task. Contains instructions and data.

Program Counter (PC): A register that normally contains the address of the next instruction to be executed in a program.

Programmable Array Logic (PAL): Contains programmable AND gates and fixed OR gates. Similar to a ROM in concept except that it does not provide full decoding of the input lines. PAL's can be used with 32-bit microprocessors for performing the memory decode function.

Programmable Logic Array (PLA): Contains programmable AND and Programmable OR gates.

Programmable Logic Device (PLD): Contains AND gates and OR gates.

Programmed I/O: The microcontroller executes a program to perform all data transfers between the microcontroller system and external devices.

PROM (Programmable Read-Only Memory): Can be programmed by the user by using proper equipment. Once programmed, its contents cannot be altered.

Protocol: A list of data transmission rules or procedures that encompass the timing, control, formatting, and data representations by which two devices are to communicate. Also known as hardware *handshaking*, which is used to permit asynchronous communication.

Random Access Memory (RAM): A read/write memory. RAMs (static or dynamic) are volatile in nature (in other words, information is lost when power is removed).

Read-Only-Memory (ROM): A memory in which any addressable operand can be read from, but not written to, after initial programming. ROM storage is nonvolatile (information is not lost after removal of power).

Reduced Instruction Set Computer (RISC): A simple instruction set is included. The RISC architecture maximizes speed by reducing clock cycles per instruction. The control unit is designed using hardwired control. Easier to implement pipelining.

Register: A high-speed memory, inside the CPU, usually constructed from flip-flops that are directly accessible to the microcontroller. It can contain either data or a specific location in memory that stores word(s) used during arithmetic, logic, and transfer operations.

RISC: See Reduced Instruction Set Computer.

SDRAM: Synchronous DRAM. This chip contains several DRAMs internally. The control signals and address inputs are sampled by the SDRAM by a common clock.

Secondary Memory Storage: An auxiliary data storage device that supplements the main (primary) memory of a microcomputer. It is used to hold programs and data that would otherwise exceed the capacity of the main memory. Although it has a much slower access time, secondary storage is less expensive. Example includes hard disks.

Sequential Circuit: Combinational circuit with memory.

Serial Transmission: Only one line is used to transmit the complete binary data bit by bit.

Seven-Segment LED: Contains an LED in each of the seven segments. Can display numbers.

Single-Chip Microcomputer: Microcomputer (CPU, memory, and input/output) on a chip.

Single-chip Microprocessor: Microcomputer CPU (microprocessor) on a chip.

Software: Computer programs.

Source Code: The assembly language program written by a programmer using assembly language instructions. This code must be translated to the object (machine) code by the assembler before it can be executed by the microcontroller.

SRAM: See Static RAM.

Standard I/O: Utilizes a control signal on the CPU typically called the $M/\overline{IO}$, in order to distinguish between input/output and memory; IN and OUT instructions are used for input/output operations.

Static RAM: Also known as *SRAM*. Stores data in flip-flops; does not need to be refreshed. Information is lost upon power failure unless backed up by battery.

Status Register: A register which contains information concerning the flags in a processor.

Structural Modeling: Using hardware description languages such as Verilog and VHDL, a schematic or a logic diagram can be described.

Synchronous Operation: Operations that occur at intervals directly related to a clock period.

Synchronous Sequential Circuit: The present outputs depend on the present inputs and the previous states stored in flip-flops.

Synchronous Serial Data Transmission: Data is transmitted or received based on a clock signal.

Transistor: Electronic switch; performs NOT; current amplifier.

Tristate Buffer: Has three output states: logic 0, 1, and a high-impedance state. This chip is typically enabled by a control signal to provide logic 0 or 1 outputs. This type of buffer can also be disabled by the control signal to place it in a high-impedance state.

Two's Complement: The two's complement of a binary number is obtained by replacing each 0 with a 1 and each 1 with a 0 and adding one to the resulting number.

Unsigned Number: An unsigned binary number has no arithmetic sign and, therefore, is always positive. Typical examples are your age or a memory address, which are always positive numbers. An 8-bit unsigned binary integer represents all numbers from 00_{16} through FF_{16} (0 through 255 in decimal).

Verilog: Not an acronym. Hardware design language developed by Gateway Design Automation in 1984 and later acquired by Cadence Design Systems. Verilog syntax is based mostly on C and some Pascal. Used for programming CPLD and FPGA chips.

Very Large Scale Integration (VLSI): a VLSI chip contains more than 1000 gates. More commonly, a VLSI chip is identified by the number of transistors rather than the gate count.

VHDL: Stands for VHSIC (Very High Speed Integrated Circuit) Hardware Description Language. Developed by US Department of Defense. Syntax is based on Ada. can be used to program CPLD and FPGA chips.

von Neumann (Princeton) CPU Architecture: Uses a single memory unit and the same bus for accessing both instructions and data.

Word: The bit size of a microcontroller refers to the number of bits that can be processed simultaneously by the basic arithmetic and logic circuits of the microcontroller. A number of bits taken as a group in this manner is called a *word*.

APPENDIX C: TUTORIAL FOR COMPILING AND DEBUGGING A C-PROGRAM USING THE MPLAB

The C-programs in this book are compiled using Microchip's C18 compiler. Appendix C provides a tutorial showing step-by-step procedure to download the C18 compiler from Microchip's web site (www.microchip.com), and compile a simple C-program using the MPLAB C18 compiler v3.47. Microchip introduces newer versions of the C18 compiler frequently. One may download the C18 v3.47 from the archive or a newer version from Microchip website.

Compiling a C- language program using MPLAB

First download MPLAB v3.47 from Microchip website as follows:
1. Go to www.microchip.com.
2. Type in C18 v3.47 in the "Search Microchip" and hit return.
3. Click on MPLAB C for PIC18 v3.47 in lite mode.

After installing and downloading the program, you will see the following icon on your desktop:

Double click (right) on the MPLAB icon and wait until you see the following screen:

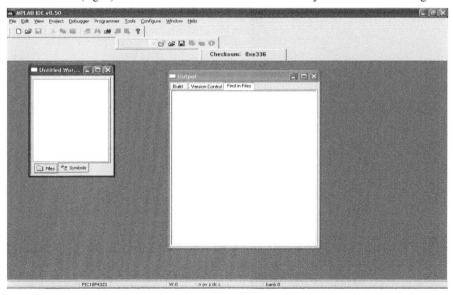

Next, click on 'Project' and then 'Project Wizard', the following screen will appear:

Click Next, the following screen shot will be displayed:

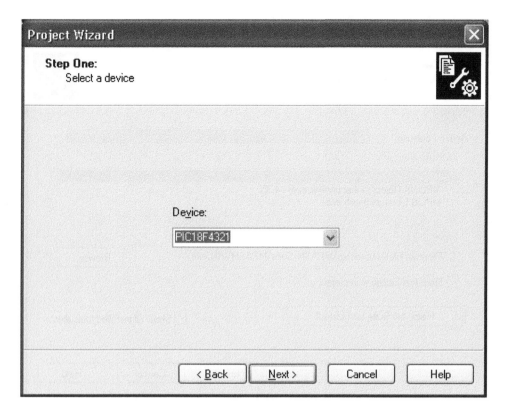

Select the device PIC18F4321, hit Next, and wait, the following will be displayed:

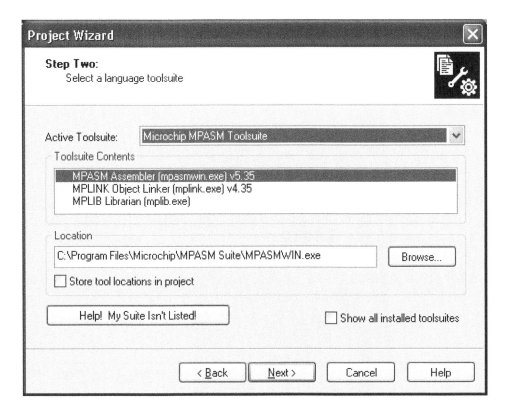

In the 'Active Toolsuite', select 'Microchip C18 Toolsuite', and click Next, the following will be displayed:

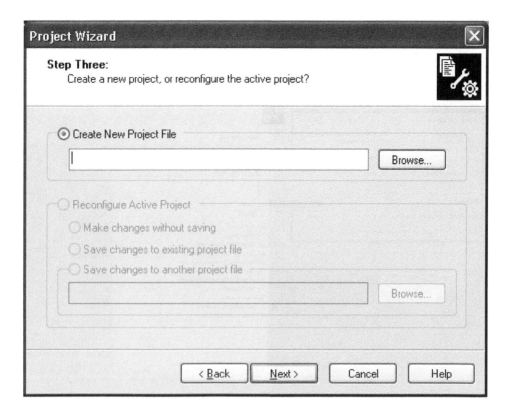

Select a location where all project contents will be placed. For this example, the folder will be placed on the desktop (arbitrarily chosen). Go to the desktop directory, make a new folder, and name the folder. In order to do this, Click on 'Browse', select desktop:

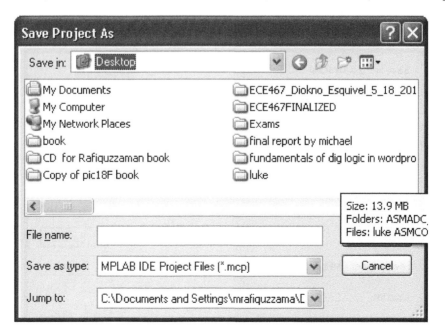

Next, create a new folder by clicking on the icon (second yellow icon from right on top row) or by right clicking on the mouse on the above window, and then go to New to see the following screen:

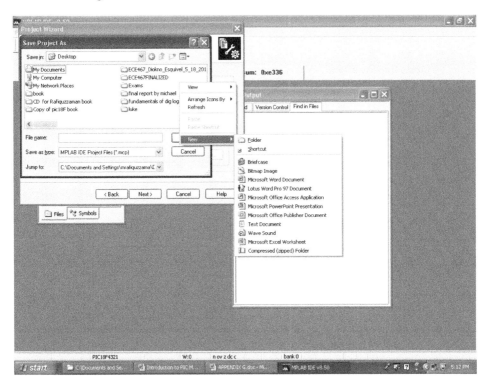

Click on Folder to see the following:

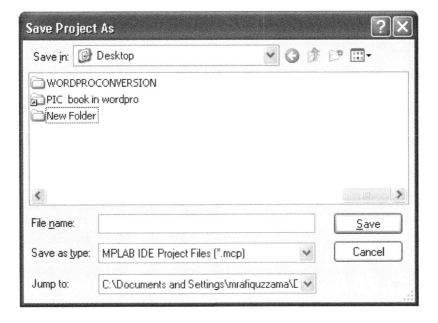

Click on folder, name it 'plus' (arbitrarily chosen name) and see the following:

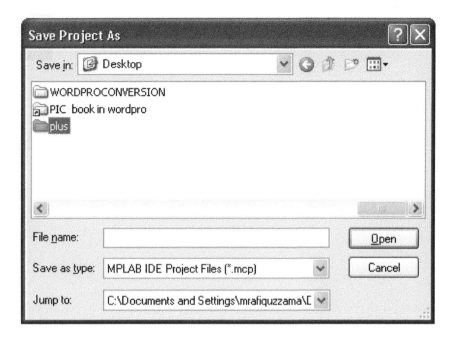

File name 'addition' is arbitrarily chosen. Type in the File name to see the following:

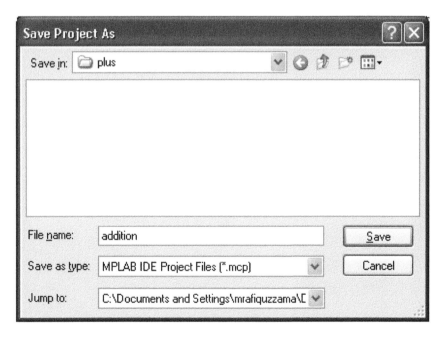

Next, click on Save, the following screen will appear:

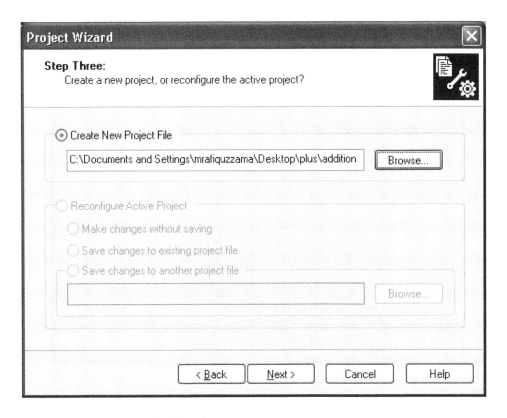

Click on Next, and see the following:

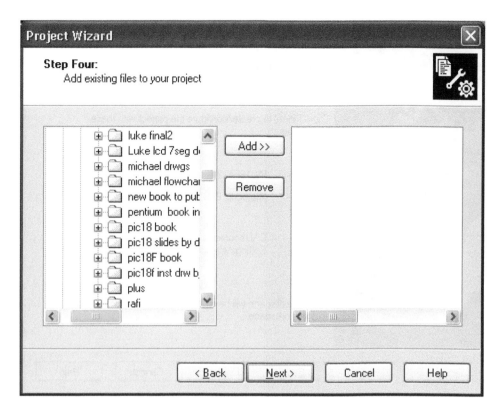

Click Next to see the following:

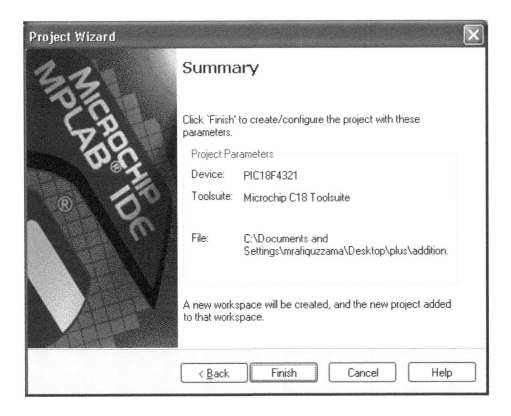

Click on Finish, and see the following:

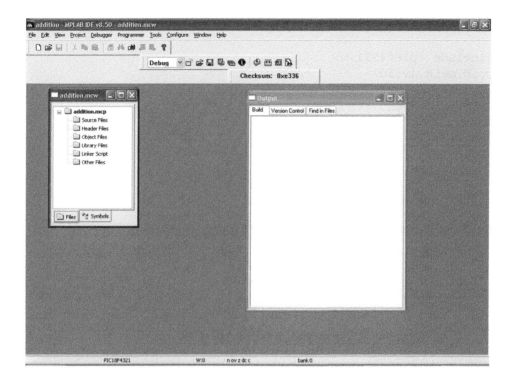

Click on File, and then New to see the following:

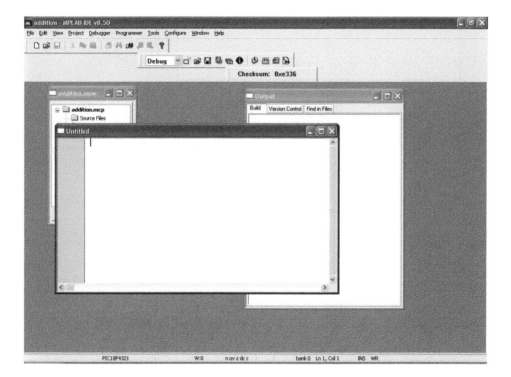

Type in the program you want to compile. The following addition program is entered:

```
#include <p18f4321.h>
void main (void)
{int a=5;
int b=1;
int c;
c=a+b;
while(1);
}
```

After entering the program, see the following:

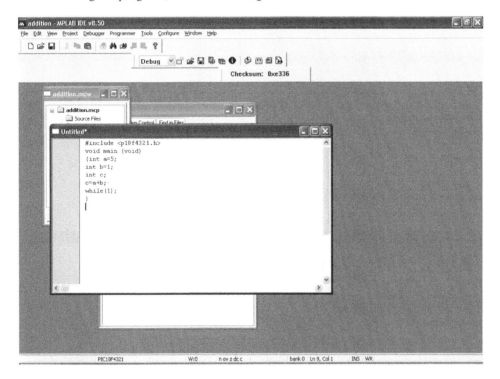

Next, click on File, and then Save as to see the following screen shot:

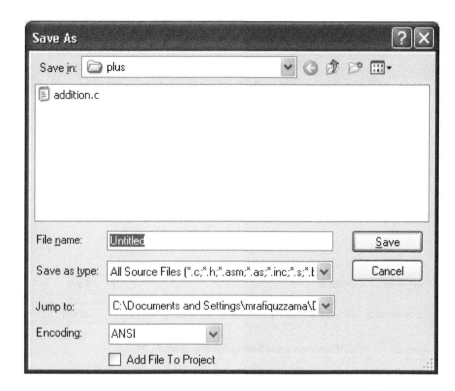

Make sure you scroll up to desktop, and then click on plus (the folder which was created before), and see the following:

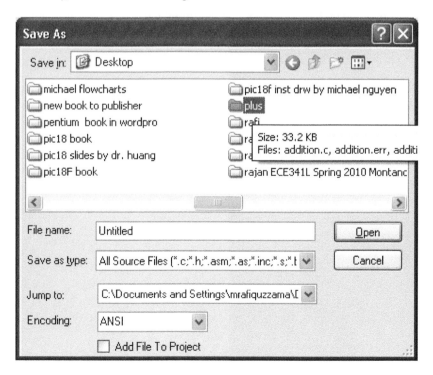

Next, double click (left) on plus to see the following:

Delete Untitled, enter the same file name 'addition' with .c extension as File name. Click on save, and see the following screen shot (notice the display changes color):

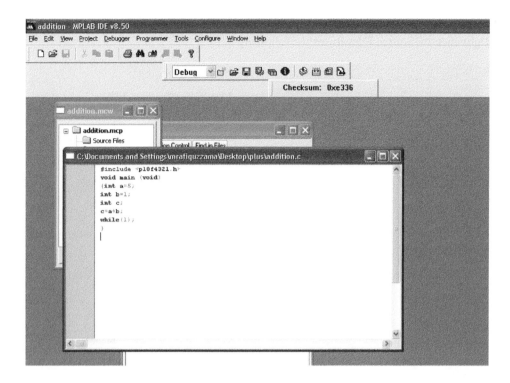

Next highlight by clicking on the top (blue) section of addition.mcw, and see the following:

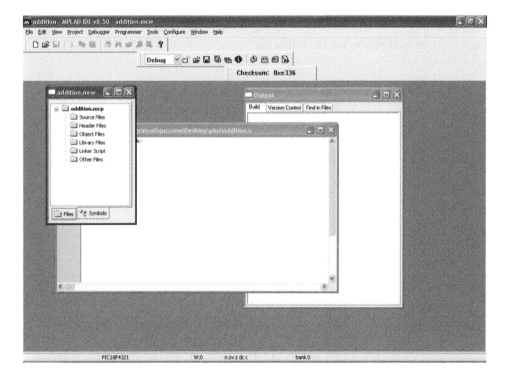

Right click on Source Files to see the following:

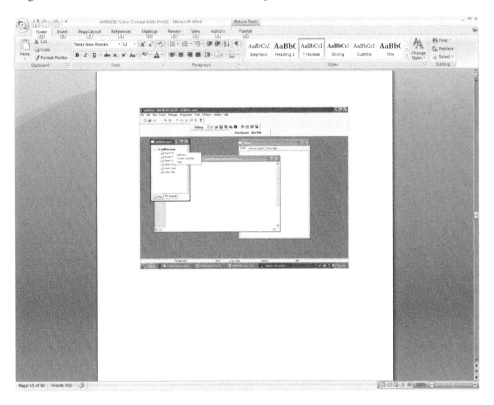

Click (left) on Add files to see the following:

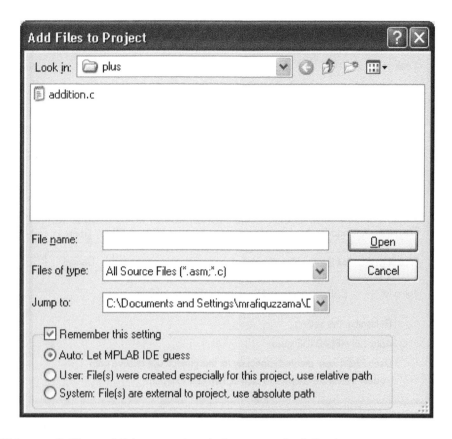

Click once (left) on addition.c on the window to see the following:

Click Open to see the following:

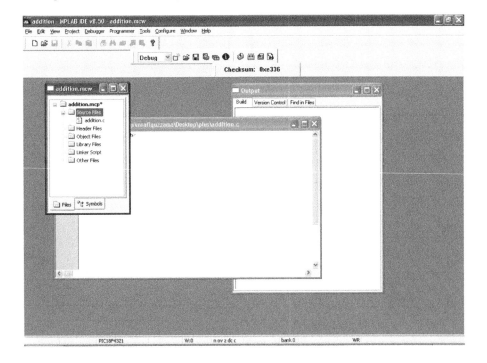

Next, do the following:
- Click on project, Click on Build options and then project
- Scroll down on output directory to Linker.Script Search Path, Select new
- Click on ... (three dots on the extreme right)
- Select C:\MCC18, bin, and then LKR, Click OK
- Click on project, Click on Build options and then project
- Scroll down on output directory to Library Search Path, Select new
- Click on ... (three dots on the extreme right)
- Select C:\MCC18, and then lib, Click OK
- Click on project, Click on Build options and then project
- Scroll down on output directory to Include Search Path, Select new
- Click on ... (three dots on the extreme right)
- Select C:\MCC18, and then h, Click OK

Note that addition.c is listed under Source Files. Next, click on Project and then build all (or only the 'Build All' icon, third icon on top right of the Debug toolbar), and see the following:

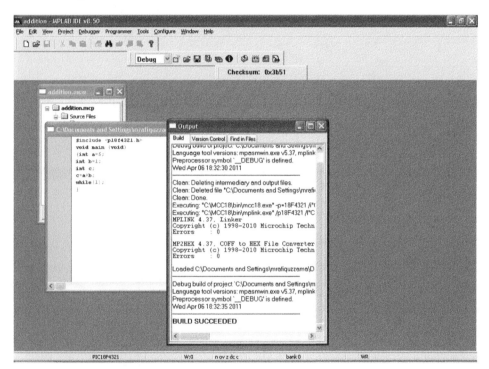

This means that the compiling the C program is successful. Next the result will be verified using the debugger.

Click on Debugger, Select Tool, and then MPLAB SIM to see the following display:

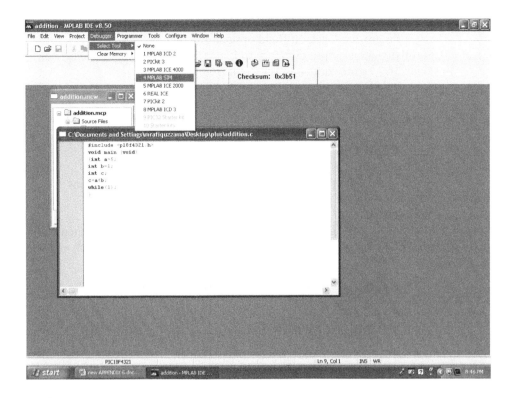

Click on MPLAB SIM to see the following:

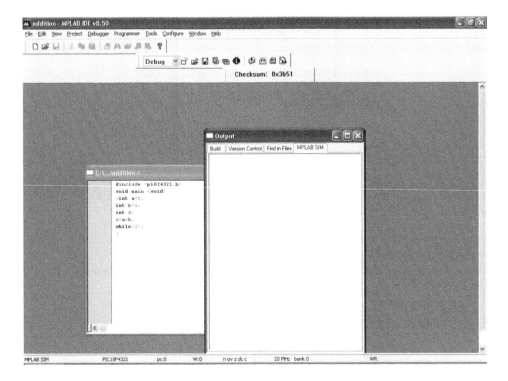

Click on View, toolbars, and Debug to see the following display with Debug toolbar:

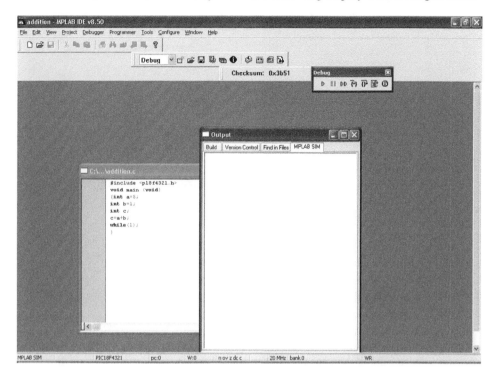

In the above, locate the Debug Toolbar. If, for some reason, Debug toolbar is missing, go to view, select Toolbars, click on Debug.

Next, click on View, and then watch to see the following:

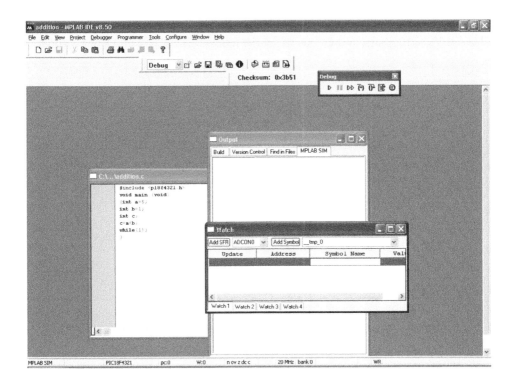

On the Watch list, you can now include locations a, b, c to monitor their contents. For example, to add 'a', simply select 'a' by scroll down using the arrow on the Add Symbol window, and then click on Add Symbol to see the following display:

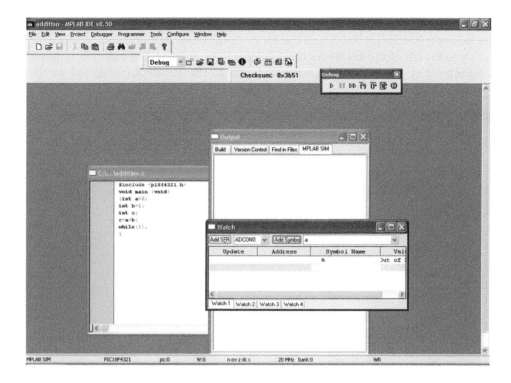

See that 'a' is displayed on the watch window. Similarly, display 'b' and 'c', and see the following screen shot:

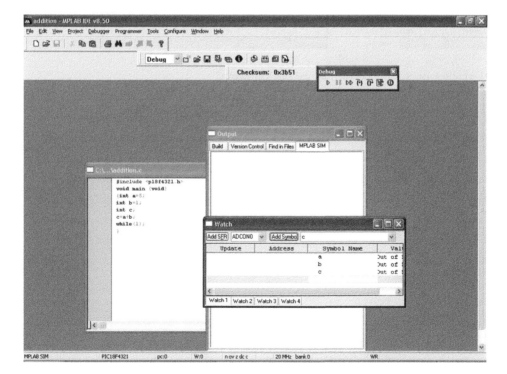

Next, insert breakpoints. Three breakpoints will be inserted for this program. One for int a = 5, one for int b = 2, and one for c = a+b. To insert a breakpoint, move the cursor to the left of the line where breakpoint is to be inserted. For example, to insert a breakpoint at int a = 5, move cursor to the left of the line, click (right) and see the following display:

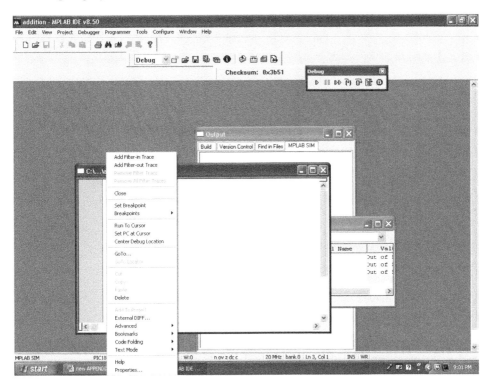

Next, click on Set Breakpoint to see the following:

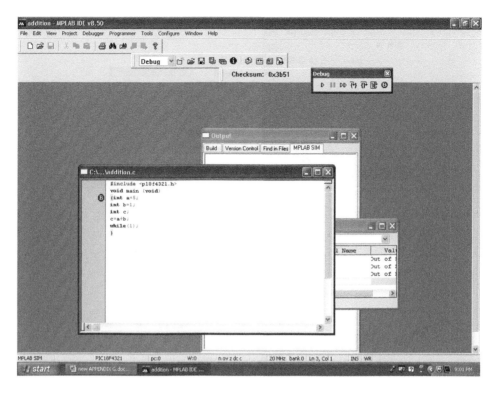

B in red on the left side of the line would indicate that the breakpoint is inserted. Similarly, insert the breakpoints for' 'b and 'c', and obtain the following display:

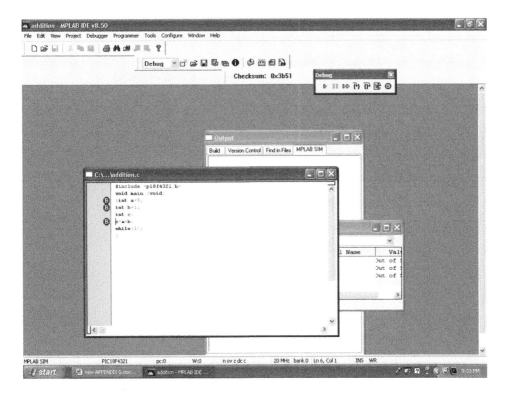

Next go to the Debug menu and Watch menu to see the contents of a, b, and c as each line is executed.

First go to Debug menu, and left click on reset (first symbol on right), and then click on the single arrow called the 'Run' arrow (left most arrow on the Debug menu), the code int a = 5; will be executed next. Click on single arrow again, the code is executed, and the following will be displayed:

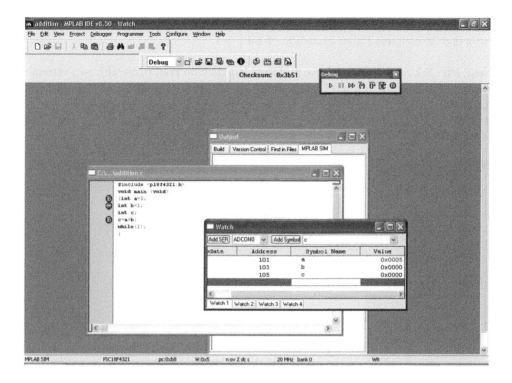

Note that 'a' contains 5. Next, left click on the single arrow, the following will be displayed:

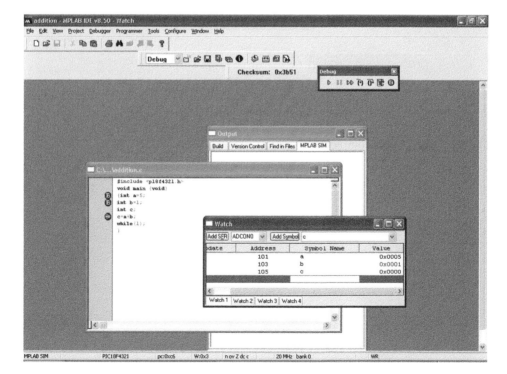

Note that 'b' contains 1 after execution of int b = 1;

Next, left click on the single arrow, and then left click on Halt (icon with two vertical lines, second from left on the Debug menu) to see the final result after execution of the line
```
c = a +b ;
```
as follows:

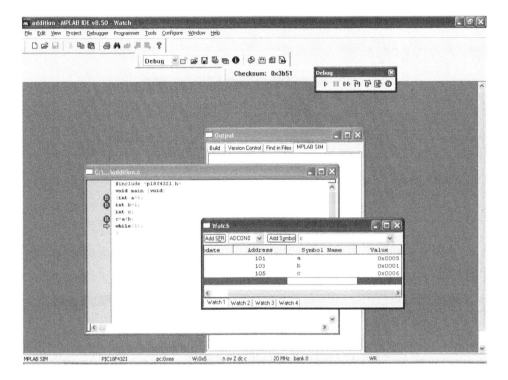

In the above, see that 'c' contains 6 (final answer).

The debugging is now complete.

APPENDIX D: INTERFACING THE PIC18F4321 TO A PERSONAL COMPUTER OR A LAPTOP USING PICkit™ 3

Appendix D contains the procedure for how to download a compiled C-program (compiled using the MPLAB C18 compiler of Appendix C) from a personal computer or a laptop into the PIC18F4321's memory. Microchip's PICkit3 is used to accomplish this. The PIC18F4321 is contained on the breadboard. Note that the PIC18F4321-based hardware on the breadboard is designed and built by the designer for a specific application.

After successful downloading of the desired program, the PIC18F-based breadboard can be disconnected from the Personal computer or Laptop. The PIC18F4321 chip on the breadboard can then execute the programs downloaded into its memory upon activating the RESET pushbutton. Thus, the desired task can be accomplished.

Finally, for C programs to work with PIC18F4321 I/O, certain configuration commands are necessary. The following configuration commands should be inserted after #include <p18f4321.h>:

```
#pragma config OSC = INTIO2 // Select internal oscillator
#pragma config WDT = OFF // Watch Dog Timer OFF
#pragma config LVP = OFF // Low Voltage Programming OFF
#pragma config BOR = OFF // Brown On Reset OFF
```

Appendix D contains the following:

D.1   INITIAL HARDWARE SETUP FOR THE PIC18F4321
D.2   CONNECTING THE PERSONAL COMPUTER (PC) OR THE LAPTOP TO THE PIC18F4321 VIA PICkit™ 3
D.3   PROGRAMMING THE PIC18F4321 FROM A PERSONAL COMPUTER OR THE LAPTOP USING PICkit™3

D.1   INITIAL HARDWARE SETUP FOR THE PIC18F4321

Figure D.1 shows the initial set up for the PIC18F4321 microcontroller. Pin #1 of the PIC18F4321 is the RESET input for the microcontroller. The $\overline{\text{MCLR}}$ (pin #1) must be connected to the reset circuit shown in the figure. There are two pairs of pins on the PIC18F4321 that must be connected to power and ground. For example, pins 11 and 32 must be connected directly to +5 V while pins 12 and 31 are connected directly to ground. Be sure not to connect any capacitors to these pins, connect them directly to either ground or +5 V. With any of the PIC18 family microcontrollers containing an "F" in the name, such as the PIC18F4321, the operating Vdd range is between 4.2 and 5.5 V. Figure D.1 also shows the proper connections for the header that will connect to the programmer.

Fundamentals of Digital Logic and Microcontrollers, Sixth Edition. M. Rafiquzzaman.
© 2014 John Wiley & Sons, Inc. Published 2014 by John Wiley & Sons, Inc.

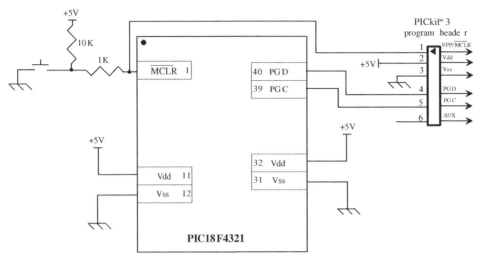

FIGURE D.1   Initial set up for the PIC18F4321.

Note that the programmer has six pins but the sixth pin (Aux pin) makes no connections. After the PIC18F4321 is properly connected, the appropriate software must be installed.

There are two programs that must be installed in order to interface with the PIC18F. The first program is called *MPLAB* and the latest version can be downloaded at www.microchip.com/MPLAB. The second is called *MPLAB C18* and is the C compiler for the PIC18F which can be found at www.microchip.com/c18. Note that at this site there is a link for academic use of the C18 compiler; be sure to click on the link and download the student C18 compiler. After the software has been installed, the PIC18F is now ready to be implemented.

D.2   CONNECTING THE PERSONAL COMPUTER (PC) OR THE LAPTOP TO THE PIC18F4321 VIA PICkit3

First, the PIC18F4321 initial setup circuit on the breadboard should be implemented. Next, a personal computer or a Laptop should be connected to the PIC18F4321 using the PICkit™ 3. Figure D.2 shows a simplified block diagram of the implementation.

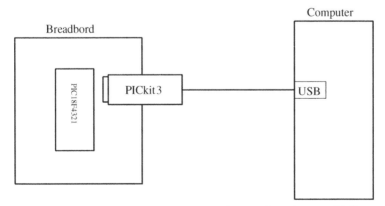

FIGURE D.2   PIC18F4321 computer interface using the PICkit™ 3.

Figure D.3 shows a pictorial view of the implementation. The following picture shows how the initial setup along with the reset circuit for the above block diagram would look like after building it on a breadboard:

Once the circuit is built, the PICkit™ 3 can be connected to the USB port of the computer as shown in Figure D.4. Next, the header part of the PICkit™ 3 can be connected to the header pins on the breadboard as shown in Figure D.5.

The other necessary I/O devices such as switches, LEDs, LCDs, and seven-segment displays can now be connected to the PIC18F4321 on the breadboard to perform some meaningful experiments. After implementing the desired hardware, the PIC18F4321 can then be programmed using the MPLAB software.

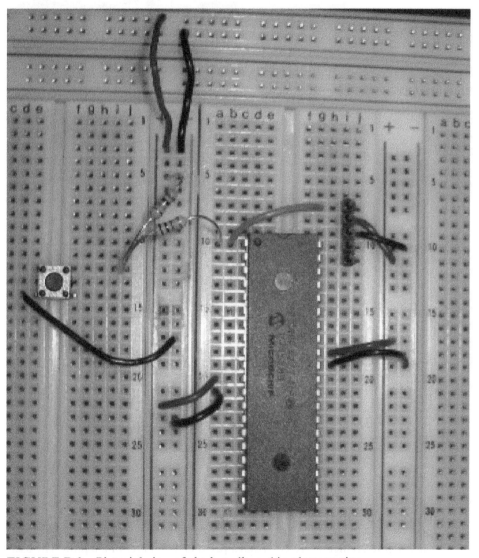

FIGURE D.3   Pictorial view of the breadboard implementation.

FIGURE D.4   Pictorial view of connecting the PICkit™ 3 to the USB port.

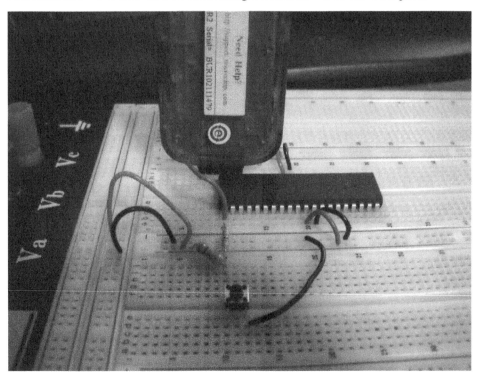

FIGURE D.5   Connecting the PICkit™ 3 to the breadboard.

D.3   PROGRAMMING THE PIC18F4321 FROM A PERSONAL COMPUTER OR A LAPTOP USING THE PICkit3

In order to configure PICkit3 from a personal computer or a laptop, The user needs to click on the 'Programmer', and then select PICkit3 as follows:

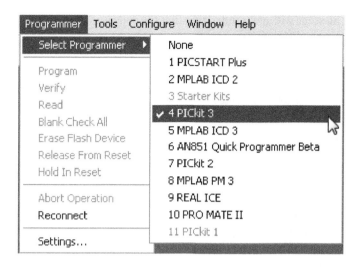

The following screen shot with the warning sign will appear. Just make sure that the PIC18F4321 microcontroller is connected to the proper voltage, and then click "OK" as follows:

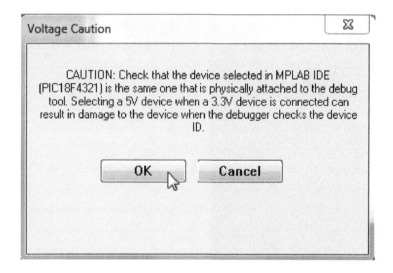

The PICkit3 is now connected.
Several options will appear at the top menu to program the PIC18F4321. The screen shot is provided below:

After successfully assembling or compiling a program, click the "program" option and MPLAB will download the program into the microcontroller.

The following message will apear indicating that the code was successfully programmed and verified onto the PIC18F:

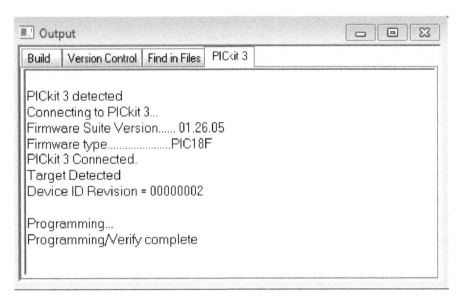

This will complete downloading the programs from the computer into the PIC18F4321 microcontroller.

BIBLIOGRAPHY

Arnold, M. G., *Verilog Digital Computer Design,* Prentice Hall, Upper Saddle River, NJ, 1999.

Breeding, K., *Digital Design Fundamentals*, 2nd ed., Prentice Hall, Upper Saddle River, NJ, 1992.

Brown, S. and Vranesic, Z., *Fundamentals of Digital Logic with Verilog Design,* McGraw-Hill, New York, 3rd ed., 2014.

Ciletti, M. D., *Advanced Digital Design with the Verilog HDL*, Prentice-Hall, 2003.

Ciletti, M. D., *Modeling, Synthesis, and Prototyping with the Verilog HDL,* Prentice Hall, Upper Saddle River, NJ, 1999.

Daconta, M., *Java for C/C++ Programmers*, Wiley, Hoboken, NJ, 1996.

Fletcher, C., Verilog: *always@ Blocks*, UC Berkley, Version 0-2008.9-4, September 5, 2008.

Floyd,T., *Digital Fundamentals*, Prentice-Hall, 8th ed., 2003.

Hamacher, V. C., Vranesic, Z. G., and Zaky, S. G., *Computer Organization*, McGraw-Hill, New York, 1978; 2nd ed., 1984; 3rd ed., 1990.

Hayes, J., *Introduction to Digital Logic Design*, Addison-Wesley, Reading, MA, 1993.

Hwang, K., and Briggs, F. A., *Computer Architecture and Parallel Processing,* McGraw-Hill, New York, 1984.

Katz, R., *Contemporary Logic Design*, Benjamin/Cummings, San Francisco, 1994.

Lee, S., *Design of Computers and other Complex Digital Devices,* Prentice Hall, Upper Saddle River, NJ, 2000.

Majidi, M. A., Mckinlay, R. D., and Causey, D., *PIC Microcontroller and Embedded Systems using assembly and C for PIC18*, Prentice Hall, 2008.

Mano, M., *Computer Engineering*, Prentice Hall, Upper Saddle River, NJ, 1988.

Mano, M., *Computer System Architecture*, Prentice Hall, Upper Saddle River, NJ, 1983.

Mano, M., *Digital Design*, 2nd ed., Prentice Hall, Upper Saddle River, NJ, 1991; 3rd ed., 2002.

Mano, M., and Kime, C., *Logic and Computer Design Fundamentals,* 2nd ed. updated, Prentice Hall, Upper Saddle River, NJ, 2001.

Microchip Technology, Inc. "*PIC18F4321 Family Data Sheet*", 2009.

National Semiconductor, *CMOS Logic Data Book*, National Semiconductor, Santa Clara, CA, 1988.

Fundamentals of Digital Logic and Microcontrollers, Sixth Edition. M. Rafiquzzaman.
© 2014 John Wiley & Sons, Inc. Published 2014 by John Wiley & Sons, Inc.

National Semiconductor, *Fast® Advanced Schottky TTL Logic Data Book*, National Semiconductor, Santa Clara, CA, 1990.

National Semiconductor, *LS/S/TTL Logic Data Book*, National Semiconductor, Santa Clara, CA, 1989.

National Semiconductor, *Programmable Logic Devices Data Book and Design Guide*, National Semiconductor, Santa Clara, CA, 1989.

Nelson, V. P., Nagle, H. T., Irwin, J. D., and Carroll, B. D., *Digital Logic Circuit Analysis and Design,* Prentice Hall, Upper Saddle River, NJ, 1995.

Rafiquzzaman M., *Fundamentals of Digital Logic and Microcomputer Design*, 5th Edition, Wiley, 2005.

Rafiquzzaman M., "*Microcontroller Theory and Apllications with the PIC18F*", Wiley, 2011.

Rafiquzzaman M. and Chandra, R., "*Modern computer Architecture*", West/PWS, 1988.

Rafiquzzaman M., "*Microprocessors and Micocomputer-based System Design*", 1st ed., CRC Press, 1990.

Rafiquzzaman M., "*Microprocessors and Micocomputer-based System Design*", 2nd ed., CRC Press, 1995.

Texas Instruments, *Linear Circuits Data Book*, Texas Instruments, Dallas, TX, 1990.

Texas Instruments, *The TTL Data Book*, Vol. 1, Texas Instruments, Dallas, TX, 1984.

Texas Instruments, *The TTL Data Book for Design Engineers*, 2nd ed., Texas Instruments, Dallas, TX, 1976.

Tocci, R. J., and Widmer, N. S., *Digital Systems,* 7th ed., Prentice Hall, Upper Saddle River, NJ, 1998.

Vahid, F, *Digital Design with RTL Design, VHDL, and Verilog*, Wiley, 2nd ed., 2001.

Wakerly, J., *Digital Design Principles and Practices*, 3rd ed. updated, Prentice Hall, Upper Saddle River, NJ, 2001.

Wikipedia.org/wiki/Unicode, "*Unicode*", 2014.

CREDITS

The material cited here is used by permission of the sources listed below.

Copyright of Microchip Technology, Inc. 2014, Used by Permission.

From PIC18F4321 Family Data Sheet, DS39689F:

FIGURE OF THE PIC18F4321 pin diagram on Page 3.

FIGURE 1-1 PIC18F4221/4321 (40/44-PIN) BLOCK DIAGRAM on Page 11.

TABLE 1-3: PIC18F4221/4321 PINOUT I/O DESCRIPTIONS, Pages 16-21.

REGISTER 2-2: OSCCON-OSCILLATOR CONTROL REGISTER, Page 31.

REGISTER 9-1: INTCON-Interrupt Control Register, Page 93.

REGISTER 9-2: INTCON2-Interrupt Control Register2, Page 94.

REGISTER 9-3: INTCON3-Interrupt Control Register3, Page 95.

REGISTER 4-1: RCON-Reset Control Register, Page 102.

REGISTER 9-4: PIR1-Peripheral Interrupt Request (Flag) Reg 1, Page 96.

REGISTER 9-5: PIR2-Peripheral Interrupt Request (Flag) Reg 2, Page 97.

REGISTER 9-6: PIE1-Peripheral Interrupt Request (Flag) Reg 1, Page 98.

REGISTER 9-7: PIE2-Peripheral Interrupt Request (Flag) Reg 2, Page 99.

REGISTER 9-8: IPR1-Peripheral Interrupt Request (Flag) Reg 2, Page 100.

REGISTER 9-9: IPR2-Peripheral Interrupt Request (Flag) Reg 2, Page 101.

Fundamentals of Digital Logic and Microcontrollers, Sixth Edition. M. Rafiquzzaman.
© 2014 John Wiley & Sons, Inc. Published 2014 by John Wiley & Sons, Inc.

INDEX

Q

R

S

T

U

Printed and bound by CPI Group (UK) Ltd, Croydon, CR0 4YY

27/10/2024

14580277-0005